Palgrave Studies in Economic History

Series Editor
Kent Deng
London School of Economics
London, UK

"This rich collection of essays opens up a new perspective on the globalisation of knowledge in history. It highlights networks of scientific and technical experts, which contributed to global developments from the time of the Second Industrial Revolution to the end of the Cold War era. The essays cover a breath-taking scope of topics and have been written by leading experts in their fields. The editors are to be congratulated for bringing together this highly informative volume."

—Jürgen Renn, *Max Planck Institute for the History of Science in Berlin, Germany*

"This is an important book, which fruitfully engineers a complex set of productive conversations. Methodologically, it approaches global history by bringing economic history together with the history of science and technology. Content wise, it emphasises the tension-filled and heterogeneous character of globalisation by focusing on a geographically extensive set of technologically mediated projects carried out by engineers and other experts. True to life, not all of these projects succeeded, but their stories work together to reveal the dynamics that continue to inform global developments today."

—Lissa Roberts, *University of Twente, the Netherlands*

"This book situates the history of science and technology firmly in the history of modern globalisation. This wide-ranging collection of essays—written by leading scholars from ten countries—explores the myriad ways in which networks of technical experts have served as agents of globalisation. They shed new light on the "backstage" of technical knowledge and practice which, in many cases, made globalisation possible."

—Stuart McCook, *University of Guelph, Canada*

Palgrave Studies in Economic History is designed to illuminate and enrich our understanding of economies and economic phenomena of the past. The series covers a vast range of topics including financial history, labour history, development economics, commercialisation, urbanisation, industrialisation, modernisation, globalisation, and changes in world economic orders.

More information about this series at
http://www.palgrave.com/gp/series/14632

David Pretel • Lino Camprubí
Editors

Technology and Globalisation

Networks of Experts in World History

Editors
David Pretel
College of Mexico
Mexico City, Mexico

Lino Camprubí
Max Planck Institute for the History of Science
Berlin, Germany

Palgrave Studies in Economic History
ISBN 978-3-319-75449-9   ISBN 978-3-319-75450-5 (eBook)
https://doi.org/10.1007/978-3-319-75450-5

Library of Congress Control Number: 2018939244

© The Editor(s) (if applicable) and The Author(s) 2018, Corrected Publication 2018
This work is subject to copyright. All rights are solely and exclusively licensed by the Publisher, whether the whole or part of the material is concerned, specifically the rights of translation, reprinting, reuse of illustrations, recitation, broadcasting, reproduction on microfilms or in any other physical way, and transmission or information storage and retrieval, electronic adaptation, computer software, or by similar or dissimilar methodology now known or hereafter developed.
The use of general descriptive names, registered names, trademarks, service marks, etc. in this publication does not imply, even in the absence of a specific statement, that such names are exempt from the relevant protective laws and regulations and therefore free for general use.
The publisher, the authors and the editors are safe to assume that the advice and information in this book are believed to be true and accurate at the date of publication. Neither the publisher nor the authors or the editors give a warranty, express or implied, with respect to the material contained herein or for any errors or omissions that may have been made. The publisher remains neutral with regard to jurisdictional claims in published maps and institutional affiliations.

Cover illustration: Brain light / Alamy Stock Photo

This Palgrave Macmillan imprint is published by the registered company Springer Nature Switzerland AG
The registered company address is: Gewerbestrasse 11, 6330 Cham, Switzerland

The original version of this book was revised: the titles of Chapter 4 & Chapter 12 were incorrectly published. A correction to this book can be found at DOI https://doi.org/10.1007/978-3-319-75450-5_15

Acknowledgements

We would like to acknowledge financial support from the Spanish 'Ministerio de Economia y Competitividad': Projects HAR2016-75010-R (La desigualdad económica en la España contemporánea) and HAR2014-57776-P (La física en la construcción de Europa).

Contents

List of Editors and Contributors

Editors

David Pretel Centro de Estudios Históricos (CEH), College of Mexico, Mexico City, Mexico

Lino Camprubí Max Planck Institute for the History of Science, Berlin, Germany

Contributors

Carles Brasó Broggi Universitat Oberta de Catalunya, Barcelona, Spain

Maria Paula Diogo CIUHCT, Faculty of Science and Technology, NOVA University of Lisbon, Lisbon, Portugal

Maurits W. Ertsen Water Resources Management, Delft University of Technology, Delft, Netherlands

Leida Fernandez-Prieto Instituto de Historia, CSIC-Madrid, Madrid, Spain

Ian Inkster School of Oriental and African Studies (SOAS), University of London, London, UK

Darina Martykánová Department of Contemporary History, Universidad Autónoma de Madrid, Madrid, Spain

Bruno J. Navarro CIUHCT, Faculty of Science and Technology, NOVA University of Lisbon, Lisbon, Portugal

M. d. Mar Rubio-Varas INARBE, Universidad Pública de Navarra (UPNA), Pamplona, Spain

Gloria Sanz Lafuente Economics Department, Universidad Pública de Navarra (UPNA), Pamplona, Spain

Tiago Saraiva Department of History, Drexel University, Philadelphia, PA, USA

Dagmar Schäfer Max Planck Institute for the History of Science, Berlin, Germany

Amy E. Slaton Department of History, Drexel University, Philadelphia, PA, USA

Joseba De la Torre Economics Department, Universidad Pública de Navarra (UPNA), Pamplona, Spain

Hector Vera Universidad Nacional Autónoma de México, México City, Mexico

María Cecilia Zuleta Centre for Historical Studies, El Colegio de México, Mexico City, Mexico

List of Figures

List of Tables

1

Technological Encounters: Locating Experts in the History of Globalisation

David Pretel and Lino Camprubí

This collection of essays was conceived at a conference in Cambridge in April 2016. Since then, the editors, authors and publisher have worked on the book from a distance. We, the editors, coordinated it from our computers in Mexico City, Berlin and Barcelona. We used WhatsApp to discuss the most urgent questions and Dropbox to organise the workflow. The authors, representing ten different nationalities, wrote from cities across Asia, Europe and the Americas. The publishing house editor was in London. We attended conferences in Chicago, Florence and Madrid to discuss chapters and commissioned new contributors at workshops in Lisbon, Xalapa and Saint Petersburg. Authors, working on topics spanning four continents, sent their abstracts, drafts and eventually their final corrected chapters via

D. Pretel
Centro de Estudios Históricos (CEH), College of Mexico,
Mexico City, Mexico
e-mail: dpretel@colmex.mx

L. Camprubí (✉)
Universidad de Sevilla, Sevilla, Spain

Max Planck Institute for the History of Science, Berlin, Germany
e-mail: lcamprubi@mpiwg-berlin.mpg.de

© The Author(s) 2018

D. Pretel, L. Camprubí (eds.), *Technology and Globalisation*, Palgrave Studies in Economic History, https://doi.org/10.1007/978-3-319-75450-5_1

email. They kept moving, for personal as well as professional reasons. They communicated from Havana, Shanghai, Philadelphia, Buenos Aires and Taipei. We hope readers and libraries from Delhi to Paris, from Buenos Aires to Luanda will be requesting copies of this book.

The book itself is proof of the globalised world we live in. Transnational lives and international connectivity are, however, only globalisation's most visible faces. Today's world is more than a virtual multicultural cosmopolis or an immaterial information society. It is a planet wired by transoceanic fibre-optic cables, digital storage facilities and satellites. It is a world made of global commodity networks, continental infrastructures and transnational migrant workers. It is not, however, a borderless world of unconstrained and uninterrupted flows and free movements. The technology and materials that support globalisation have paradoxical consequences. The same knowledge and technologies that promote globalisation often create new barriers, gaps or borders. Moreover, the complex technological scaffolding that supports modern globalisation and de-globalisation are built, designed and installed in specific parts of the planet. And what of the social actors who put technologies in place? This book aims to locate them and bring them back into the history of the making of the global economy.

Globalisation(s) and the World Economy

Historian Jeremy Adelman pointed out in a 2017 article that the rising wave of anti-globalisation and nationalist politics might be threatening global history.[1] This field, which has made it possible to overcome the excessive provincialisation of national histories and has brought historians back to the centre of current debates, was itself restored by the practices and ideologies of late twentieth-century globalisation. It might be too easily concluded that the current challenges posed to globalisation by protectionism, economic crisis, inequality and climate change will bring that young field to a premature dismissal. We believe instead that it is more rewarding to recognise that while the current contestation of globalisation exposes the contradictions of the global history paradigm, it also represents an opportunity to develop more nuanced, complex and plural world histories. The global view does not imply looking at the entire planet as an homogeneous historical entity, but instead rewrites local and regional histories with an emphasis on global connections. It accepts that globalising processes are neither homogeneous nor incompatible with state and imperial interests. This approach allows us to talk about globalisations in the plural and to explore the role of individual actors in various globalising enterprises.[2]

The distinctive feature of the new global approach to history has been the study of connections between societies, particularly the transnational movements of people, ideas, capital and things. This perspective has brought to light the areas of encounter and conflict among societies, showing the interplay between the local and the global. Less emphasis has been placed on how capitalism functions at local and global levels. In this respect, the intersection between two fields, global history and the history of capitalism, is of particular interest for this volume. We would like to suggest that the history of the capitalist world economy can provide a context to local, national and regional narratives. Perhaps the most critical task facing the social sciences is to identify the underlying local and global conditions that have determined the metamorphosis of the capitalist world economy over the past five centuries.

While the global turn is a relatively recent trend, globalisation itself is not new. As Jerry Bentley observes, long-distance cross-cultural interactions, such as maritime trade, were already commonplace in ancient civilisations throughout Eurasia and northern Africa.[3] It is precisely this conception of globalisation as a process independent of the rise of the West that leads Hopkins to speak of an 'archaic-globalisation' (until 1600) and a 'proto-globalisation' (1600–1800), predating 'modern-globalisation' (1800–1950) and 'post-colonial globalisation' (1950s onwards).[4] The integration of the Atlantic world economy from the sixteenth century onwards extended material interactions to an ever-increasing planetary scale.[5] This volume, however, focuses on the last two centuries. The industrial era transformed the scale and scope of international flows of money, artefacts and people. From that moment, local and national histories were increasingly affected by supranational dynamics. National industrialisation processes facilitated specific technological and economic globalising structures.[6] Not only national histories, but also the histories of empires transformed from the nineteenth century onwards.[7]

The acceleration of globalisation during the last third of the nineteenth century was again strongly linked to the technological transformations of the so-called Second Industrial Revolution, as well as to the expansion of multinational corporations, which were the cornerstones of a progressive globalism. We would not want to incur the technological determinism locked in the phrase 'the annihilation of space and time', with which Karl Marx and others characterised global industrial modernity.[8] But the role of technology in modern globalisation should not be underestimated. Through the mobilisation of experts and the production of capital-intensive technologies for mass production, the new industrial paradigm of the age of 'machinofacture' took global capitalism to hitherto unimaginable extremes.[9]

Speaking about globalisation and capitalism in the plural does not make this volume immune to the problems that face much of global history from its restoration. A global approach has frequently meant the study of European and North American roles in world history, neglecting national histories, tales of resistance and alternative types of connections in vast parts of the world. Post-independence Latin America, for example, is barely incorporated into the global histories of the nineteenth and twentieth centuries, and when it does appear it is often in a marginal and peripheral position.[10] Moreover, academics from universities in Europe or the United States remain the dominant narrators of global history, with some remarkable exceptions.[11] Our volume certainly aims to incorporate these and other often neglected views, but its focus on networks of technical experts and professionals, which were considered essentially modern and Western, makes it difficult to escape from the narrative of the economic success and industrial leadership of some nations since the nineteenth century. Many of the chapters, however, explore failed plans and unintended consequences, while others give voice to opposing views, appropriations and confrontations. The volume covers Africa, Latin America and the Caribbean, and Asia, in an effort to go beyond Europe and the United States.

Global history has faced other related problems, some shared with comparative history and historical sociology. Studying broad areas, with disparate cultures and languages, creates new difficulties. The global shift has made it necessary for historians to rethink some of their theoretical perspectives and to introduce new analytical categories, such as networks, borders and cosmopolitanism, to name just a few. This expansion of topics, in turn, has led to an expansion of sources, requiring contact with specialists from various disciplines in the excessively compartmentalised realm of academia. One major challenge has to do with scale: combining case-specific detailed studies with broad or long-term approaches and global analysis with local histories. The need for a balance between abstraction and particularism seems clear, and it frequently involves the conciliation of macro-historical narratives with postmodern and postcolonial critiques.[12] Creating a local history of the global has also been one of the main problems of epistemology. We believe that cross-fertilisation of methods between economic history and the history of science and technology, as well as a sustained attention to precise networks and actors, can provide answers to some of the main challenges faced by historians of modern economic globalisation.[13]

Expert Networks, Global Technologies

When focusing on world economic integration, multinational companies, international trade and multilateral institutions come to mind. Most historical studies have therefore focused on the most visible figures in modern globalisation, such as traders, regulators, diplomats and commercial brokers. Engineers and scientists, as well as other social actors with expertise in technology and science, are often neglected. Take for example the pioneering book *History and Globalisation* by Kevin O'Rourke and Jeffrey Williamson.[14] Only once in the entire text do the authors mention engineering, and then without any explicit reference to the role of technical experts in the making of the Atlantic economy during the nineteenth century.

If Antoine Picon and David Cannadine argue for 'engineering history', we would like to make the case for 'engineering' both global history and the history of political economy.[15] In a very general sense, technology has provided societies with the material structures necessary for world integration. But calling attention to the globalising role of technologies does not exclude a range of other factors and does not suppose that technologies always favour connectivity. First, although national histories are usually insufficient in accounting for the production, circulation and use of technical artefacts, a techno-globalist approach to the history of technology would be equally misguided.[16] Nation-states continue to be the leading investors in science and technology, as well as in infrastructures and war machines.[17] Second, industrial research and technological developments often promote de-globalisation; for example through physical barriers such as concrete walls, barbed wire or satellite-demarcated territories.[18]

The difficulty, then, is in finding multiple ways of discussing the place of technology in histories of globalisation that integrate heterogeneous and eclectic methodologies and concepts. Economic historians, sometimes drawing from economic theory, have been mainly concerned with the technological determinants of the 'Great Divergence' and the so-called First and Second Industrial Revolutions.[19] Especially interesting for our purposes has been the examination of the shifting patterns of technological progress through the analysis of specific institutions, sites of innovation and aggregated data over the long term.[20] By contrast, historians of science and technology have, as has much of cultural history, provided detailed case studies and 'thick' descriptions that preserve historical specificity. Productive as it is, this approach has the twofold danger of obscuring long-term secular trends as well as global connections. For the last decade, there have been several proposals to go beyond case-studies without losing the richness of local studies.[21] For our purposes, the most promising are studies of systems, networks and circulation.

Thomas Hughes proposes historians and sociologists of technology turn to 'technological systems', an approach that integrates material artefacts, institutional settings and actors—particularly 'system builders' such as engineers and scientists.[22] His study of electrical networks in the United States and Europe has been an inspiration for accounts of railways and worldwide systems of data gathering and climate modelling—see particularly Paul Edwards's notion of 'infrastructural globalism'.[23] In their turn, Bruno Latour and Michael Callon also insist that networks include material and cultural aspects as well as a variety of actors.[24]

While systems are associated with the structures built by engineers and scientists, the looser concept of network allows for more informal and fluid interactions. For our study, this flexibility is particularly interesting when discussing the creation of networks of experts, which become agents of globalisation and de-globalisation across the world. While some professional networks are explicitly created for a particular purpose, many emerge through long spans of time in contexts that range from imperial endeavours to transnational corporations.[25] For instance, British engineers formed worldwide formal and informal networks through several waves of diasporas, imperial politics, professional associations and shared infrastructural projects.[26]

For the purpose of this volume, what is important is not simply the network conception of expertise, but how networks of experts were reconfiguring the global economy. We can think of networks as spaces for socialisation and transnational lives that have reconfigured expert practices. Networks of professional and bureaucratic elites favour the exchange of knowledge and espouse transnational professional ideals beyond national traditions. Experts often communicate among themselves in a particular jargon or a lingua franca that separates them from users of the vernacular.[27] They create and participate in international organisations and institutions that spread their ideas and techniques and reinforce their image of detached objectivity. International reputation and social connections, in turn, have been powerful tools for technical experts to gain political and professional legitimacy back at the nation-state.[28] And, back in the international arena, the authority, prestige and connections enjoyed by experts have invested them with power to shape international relations, trade and resource management. As many of the chapters in this volume show, experts have often acted as what we would like to call technical diplomats. They have been agents of empire, of international trusts, of modernisation and of global flows. Engineers, in particular, have been able to draw on internationalised values of objectivity, technical neutrality and quantitative rigour to gain leverage for their political and professional decisions in local, national, regional and international contexts.

The professionalisation of engineers since the nineteenth century onwards was usually pegged to national politics, technological traditions and models of industrialisation, and as such national engineering styles manifest themselves in distinct training requirements, legal attributions and social status.[29] But in most cases and beyond these national identities, engineers have largely built their authority on symbolic capital topping their substantive knowledge.[30] As such, the study of transnational expert networks poses interesting questions about scale and agency that can be illuminated through an attention to local and particular globalising endeavours. For this, it is especially promising to explore how experts moved across different institutional and geographical settings establishing the nodes of resources and knowledge that constitute transnational networks.

Travelling and moving are hence of particular interest to this volume. Again, material connections are as significant as personal ones. The encounter of distinct technological systems also raises many questions for historians. The terminology used by the authors in their chapters to convey movements and encounters is as eclectic as their disciplinary backgrounds. The concept of technology transfer, still preferred by economic historians, concentrates on the patterns of technological dependence and the blockages to technological assimilation. From this perspective, the notion of transfer helps bring attention to the political economy of technology and issues such as technological capacities, gaps, sovereignty and learning.

On the other hand, authors who want to emphasise the flows of technological knowledge and artefacts recur to a metaphor taken first from physiology and electricity and then from political economy: circuits and circulation.[31] Technological circuits have historically overlapped with formal and informal social networks, from commercial to migratory ones. Circulation always entails transformation.[32] Users, from workers to consumers, make material adaptations and cultural appropriations of objects designed elsewhere. The process of technology transfer also implies the relocation of technologies previously established in a given place. Technologies originally from elsewhere combine in original ways with local technologies, often forming hybrids and creole artefacts. Adaptation, appropriation and use are as central to the history of global technologies as invention, innovation and diffusion of new technologies.[33]

Historians of science have long been interested in reinterpreting universal truths as partly the product of the circulation of material particulars that ensure experimental reproducibility.[34] This approach is particularly useful for questions related to the making of global political economies, which depend on the circulation of standardised materials, numbers and formulas.[35] Producing and regulating standards is a powerful tool for those seeking to increase the scale and scope of their economic enterprises.[36] On a transnational scale, standards become a matter not only of science, technology and political economy, but also of military strategy, geopolitics and diplomacy.[37]

Technology and Globalisation: Four Connections

We would like to highlight four transnational processes that are central in investigating globalisation since the nineteenth century: international trade; empires, imperialism and economic hegemony; hidden integration in peace and in war; and the globalisation of models of political economy. In all these transnational connections, technology and expertise have played central roles.

First, international trade is one of the most visible features of globalisation since the mid-nineteenth century, with trade in technology providing opportunities for interactions among technical experts and encounters between technological cultures. The reorganisation of international economic relations and world markets has often led to the persistent movements of machinery and technical experts. A good example is the rail construction boom that occurred during the second half of the nineteenth century, linked to a global expansion of trade and global commodity networks, which entailed a constant transnational movement of engineers, and other skilled workers, into the global South.[38] Similarly, although today's financial markets have become a symbol of virtual world economies, they rest on expert networks and material connections constantly expanded and maintained.[39]

International technology transfer intensified in the world of the Second Industrial Revolution. While profound economic and technological disparities put constraints on the successful adoption of foreign technologies, industrial inequalities simultaneously encouraged the movement of technologies among nations. While clashes among different technological cultures and traditions of know-how hindered the assimilation of technological knowledge, such clashes also created opportunities for late industrial development. In this context, as observed by Ian Inkster and Joel Mokyr from a range of global historical material, institutional change could promote over the long term the successful transfer of technological systems and the overcoming of patterns of dependency. However, institutional changes could be inhibited or blocked by a variety of factors, ranging from political instability to colonial rule.[40]

The interplay between two different technological systems, resulting from, for example, the importation of foreign technology, reveals the crucial role of engineers in adapting technologies to local conditions.[41] For imported technology to function successfully in a new economic environment, it often needs to be accompanied by the transfer of know-how, either through expert migration or the assimilation of specialised skills and knowledge.[42] The successful adoption of foreign technologies relies on the presence of technical experts who could operate and adapt new machinery to specific industries and institutional environments. Of course, the creation of domestic expertise is costly and requires investment in education and training for long periods. Often high levels of machinery importation by a given country coincided with low levels of assimilation of technological capabilities. For example, Edward Beatty observes that the Mexican technology gap during the period 1870–1910 is explained mostly by Mexico's dependence on foreign skilled personnel and the lack of domestic engineers.[43]

Second, globalisation has also resulted from empires and imperialism. Slavery and the cotton industry during the Industrial Revolution or growing exports by imperial Japan are just two examples.[44] The role of technology and technical experts in the expansion of empires has aroused considerable interest, with scholars emphasising the need to place the history of technology at the centre of the historical narrative of imperialism. Technology has not only been a means of controlling nature but also an instrument of imperial domination and control. For instance, Daniel Headrick shows how steam navigation, firearms and tropical medicine were cornerstones of the new imperialism.[45] For him, technology and technical experts are tools rather than active agents of empire. Headrick thus follows an instrumentalist approach, to use Bennett and Hodge's expression.[46] Michael Adas injects another element into the debate when he reveals the ideals and technocratic ambitions of colonial engineers, showing how technical experts in imperial contexts were missionaries of modern civilisation and technological progress.[47]

Alternatively, scholarship has portray technical experts as active agents for the consolidation of economic and political imperial power. This view favours an approach to imperial history that focuses on interconnections in the imperial space beyond the metropolis–colony distinction. From this perspective, what matters are areas of encounter, mutually reinforced relations, imperial practices and the movements of experts throughout the imperial space. For example, Ben Marsden and Crosbie Smith highlight the role of engineers in the construction of imperial economies. To these authors, 'engineers are empire-builders: active agents of political and economic empire'.[48] The administration of the colonies enabled resource exploitation and the development of an imperial economy. Even if colonial experts were at times viewed as second-class professionals, colonial projects opened up lucrative opportunities to technical experts as well as new possibilities to develop their professional careers and status.[49] Colonies also provided them with real-live laboratories for technical experimentation and trials.[50] As a result, colonialism is at the root of much formal and informal modern expertise. The relevance of colonial expertise is well captured in the 1962 jewel of orientalist cinema *Lawrence of Arabia*. In it, Lawrence warns his Bedouin allies after throwing the Ottomans out of Damascus: 'take English engineers and you take English government' —and recall also that the actions of British and Arab demolition teams were directed mainly against Ottoman railways.[51]

The history of globalisation is also intimately linked to the notion of economic hegemony.[52] While not necessarily through land-grabbing or territorial control, other forms of imperialism, from informal empire to soft power, are no less important to the history of the world economy. Technical experts have also been critical players in designing and implementing less formal types of economic imperialism. Economic imperialism emerged on a global scale at the end of the nineteenth century as a consequence of profound changes in the capitalist mode of production, which was now nominated by large corporations and finance capital.[53] Technical experts accompanied foreign investment and moved to new sites of production, engaging with local expertise for the establishment and maintenance of infrastructures and industries. A good example is the European and American machinists and engineers who worked in plantations, economic enclaves and company towns in Asia, Africa and Latin America.[54]

Foreign trade and investment are central means to exert what diplomatic historians call soft power.[55] But the role of technology and experts in that domain has been even wider. The spread of American-inspired institutions for scientific research and education in Europe or India during the Cold War, for instance, points at ways in which technology has aided in the cultural and institutional wars which are crucial to economic globalisation.[56] This also seems to have been true for the continuous movements of American engineers abroad between the 1870s and the Great Depression, when many of them served as agents of Americanisation in other countries' technical schools, business cultures and labour relations.

Third, we are interested in the creation, sustenance, transformations and meanings of global material infrastructures, from steel bridges to water supply systems and to railways. Economic globalisation is surely the result of institutions, flows of capital and international relations.[57] But it is also based on an often invisible technological layer of infrastructure, transport networks, telecommunications and technical standards. Europe has been a particularly well-researched example. Little more than a decade ago, Thomas Misa and Johan Schot set out to reconstruct the history of European integration through 'hidden technological integration'.[58] Since then, six volumes have appeared under the general title *Making Europe* which offer a rich account of the kinds of webs that lie below the economic and political landmarks that constitute more familiar approaches to the history of Europe. In the third of these volumes, *Europe's Infrastructure Transition*, Per Högselius, Arne Kajser and Erik van der Vleuten provide convincing accounts of various systems connecting parts of (mostly northern) Europe amongst themselves and to global infrastructural webs from Australia to Canada.[59] It does not seem hard to imagine the extrapolation of a similar approach to other continents and regions.[60]

Sometimes, technological integration runs counter to more visible economic or political barriers. For instance, iron gas pipelines traversed the Iron Curtain and set the stage for European energy dependency.[61] On other occasions, however, integration occurred alongside political barriers. For instance, European countries are creating a common border to the south through the integration of national surveillance systems. Furthermore, the violence of war is also a driver for technical integration, which often survives into times of peace. Examples include food canning in the nineteenth century, heavy industries during war efforts in the twentieth century and NATO surveillance systems during the Cold War.[62] The twenty-five-year-long war for the Western Sahara between Morocco and the Polisario Front, which resulted in what is today considered the last African colony, was a way of ensuring the increasing integration of the world market for fertilisers.[63]

While we discuss the effects of technical integration on economic globalisation, the chapters in this volume highlight integrations that go beyond the economic sphere into the political, the cultural and the environmental. Environmental global integration happens in ways directly related to the history of technology. The first is the standardisation of landscapes, for instance through the techniques, know-how and techno-scientific objects spread during the Green Revolution.[64] The second is by providing the means for the coming into being of the global environment as an object of analysis.[65] The economic significance of the global environment is evident today in discussions about world climate change. During the Cold War, the new notion of an interdependent planetary environment informed decisions about allocation of resources and discussions on development and the limits to growth.[66]

Fourth, and finally, technocratic experts have joined state bureaucrats and economists in globalising political economy models, from laissez-faire to autarky and to current neoliberalism. Engineers with a strong scientific education became a political elite in the nineteenth century in France and elsewhere.[67] They consciously disseminated new ideas on political economy, which then accompanied new infrastructural networks or industrialising engines. This pattern continued into the twentieth century, and it should come as no surprise that economic nationalism during the 1930s was a rather international movement.[68] Similarly, scientists and engineers exported centrally planned economies to socialist regimes in the Soviet era.[69]

Decolonisation processes in Africa, Asia and Latin America led to new economic geographies. According to Timothy Mitchell, it was this transition which led to the emergence of no less an entity than 'the national economy' as an abstract magnitude of monetary circulation accounted in terms of gross domestic product (GDP) and commodity exchanges rather than resource availability.[70] Continuities between colonial and post-colonial periods are a topic of particular interest to this volume. As historians have shown, post-Second World War economic international institutions and international developmental programmes were often means of articulating post-colonial hierarchies through technology transfers built on world asymmetries.[71] As such, after political independence technocrats of all sorts became a key link between the former metropolis, or the new global powers such as the United States, and the newly independent countries. From these new positions of authority, experts, from engineers to economists, were strategic in economic planning and government intervention.[72] They were active agents in the making of the post-Second World War world economic order and in its de-regulation from the 1980s with the triumph of neoliberalism. As Wendy Larner and Nina Laurie demonstrate, a series of travelling technocrats such as water and telecommunications engineers, have been instrumental since then in the privatisation of public services on a global scale.[73]

Networks of Experts in World History

This book examines the role of experts and expertise in the shifting dynamics of globalisation since the mid-nineteenth century. It shows how techno-scientific experts have acted as agents of globalisation, providing many of the material and institutional means for world integration. By placing engineers and scientists, among other experts, at the centre of its historical narrative, this volume provides original insights into the relationship between technology and globalisation. Focusing on the study of international connections, it illustrates how expert practices have shaped the political economies of interacting countries, entire regions and the world economy.

As the body of literature cited throughout this chapter makes clear, there is a growing interest in the international history of expertise. While the available studies in the field of history of science and technology are intellectually strong and have served as an inspiration for this volume, they generally focus on a specific process, region or period rather than on the general issue of the role of techno-scientific experts in the making of the global economy. Moreover, few of them are explicitly connected to a second body of literature, the one addressing the history of economic globalism. Just as these books are best considered in the context of economic history, important studies of economic history would benefit from a sustained attention to the knowledge structures and the active agents behind the processes they describe. We hope to contribute to bringing experts and expertise into economic histories of globalisation and de-globalisation.

To highlight the global dimension of technology, the present volume brings together a range of approaches and topics across diverse geographical regions, transcending nationally bounded historical narratives. The contributors to this volume focus as well on diverse historical periods from the mid-nineteenth century to the Cold War. Each chapter deals with a particular topic that places expert networks at the centre of the history of globalisation. Their studies combine the micro-histories of experts with the study of their impacts on local, national and global processes. Through different case studies, authors trace local, national, regional and worldwide connections, providing an alternative map of the modern world. Admittedly, this map is no *mappa mundi*: it is profoundly incomplete. But it does put forward new frames to think about globalisation, technology and experts in world history.

The interdisciplinary contributors to this volume concentrate on central themes including intellectual property rights, technology transfer, tropical science, energy production, large technological projects, technical standards and colonial infrastructures. Most of the chapters in this volume explore the links that have developed since the late nineteenth century between the globalisation of the engineering profession and the worldwide rise of statist political economic projects and technocratic institutions. Many of the chapters also consider methodological, theoretical and conceptual issues that speak to current theories of globalisation.

Dagmar Schaffer's opening chapter asks what it means to be an engineer today and how that shapes our histories of engineering. Its *longue durée* reflection provides a clear picture of material and cultural continuities and ruptures of the ethos of 'building bridges' in China, at once related to state building and to globalising understood as building a civilised world. More generally, it provides a conceptual entry point to the questions examined by the rest of the chapters as well as a background for understanding the shifting notions of technical expertise throughout history.

Networks of technology and experts yielded both connections and confrontations among far-flung places in the world. Ian Inkster examines the global connectivity that played out between urban England and rural Taiwan from the 1860s to the First World War. He argues that chemical advancements and the development of the celluloid industry in Birmingham directly impacted colonial dynamics and the politics of warfare in Taiwan. This is explained by European and American chemical industries' need for camphor obtained in Taiwan. Inkster's chapter reminds us of the technological imperatives, linkages and impacts of the Second Industrial Revolution. Technological breakthroughs and the resort to patents were a pervasive source of marginalisation and conflict. However, this is not only a story of destruction but at the same time of unintended consequences and resistance.

Global circuits of production and trade have provided career opportunities for technical experts, for instance in commodity production and in the establishment and maintenance of infrastructures. As Darina Martykanova shows, during the late nineteenth and early twentieth centuries an elite group of engineers enjoyed truly transnational professional lives. Through a biographic study of the graduates of the École Centrale des Arts et Manufactures, she reconstructs the work, projects and varied activities of certain engineers born in the Ottoman empire, Spain and Latin America. Several factors shaped the engineers' global trajectories, ranging from geopolitical questions to increasing transnational investments, and from friendship to ethno-religious affinity.

At least since the mid-nineteenth century, European empires embraced a 'civilising mission' embodied by and realised through technology. Maria Paula Diogo and Bruno Navarro show how engineers are not just tools of empire, but active participants in the shaping of colonial rule and of global infrastructures. Through a study of failed plans for railway construction in Angola and Mozambique for more than seventy years, colonial Portugal provides a new entry point into the role of technical experts in the European scramble for Africa. Intercolonial cooperation and competition appears as a major player shaping what the authors call techno-diplomacy.

Intellectual and industrial rights regulate production, distribution and access to new technologies. As David Pretel shows, patent experts became globalising agents by creating and coordinating national institutions and legislation from the late nineteenth century onwards. In his chapter, attorneys, brokers, technical consultants, lawyers and investors emerge as the creators of international networks of diffusion, dependency and control. By tracking patent agents in a variety of global contexts, Pretel provides new tools for understanding modern multinational companies and the role of research laboratories in securing their expansion across the world. The real game was not just inventing but featuring as the patentee. Economic globalisation occurred in the frame of these transnational legal and technical networks.

Technical experts were not only an expanding and highly regarded professional group in the United States during the early twentieth century but were essential in the rise of American power abroad. Leida Fernández-Prieto concentrates on the networks of American scientists, botanists and agronomists working at Harvard Botanic Station for Tropical Research and Sugar Cane Investigation in Cienfuegos, Cuba. This chapter makes it clear that most American experts abroad, except perhaps for military engineers, were not mere marionettes following American policy. The flow of American scientists and engineers resulted not from their willingness to serve as a tool of foreign political intervention but from their desire to build careers abroad. American technical experts played key roles in the globalisation of American institutions and capital, for example through their collaboration with local experts in agricultural research and commodity production in the tropics.

Nobody symbolised American technological progress and promise better than the engineer. Hector Vera shows that American engineers were central in the regulation of standards and institutions in the American technological system. He looks at a specific group of mechanical engineers and how they mobilised opinion against an apparently more rational and superior solution to the question of how to standardise weights and measures: the metric system. Economic and commercial rationales underpinned their contestation of a system that would eventually became a truly global language of measurement. The American rejection of this system would have enduring consequences until today.

Tiago Saraiva and Amy Slaton concentrate on the circulation of knowledge from local into transnational settings. Their case in point is statistics for agriculture and their actors are well known to historians of economics and of science alike. The chapter's major innovation is the simultaneous exploration of scientific identity (class, gender, service) and disciplinary history. The pivotal ground is the role that experts crafted for themselves as champions of new ideas and practices of political economy and of (limited) democracy. From the New Deal to the Food and Agriculture Organization of the United Nations international programmes, the cold work of statistics was invested with politics, notions of citizenship and strategies for the world economy.

Business is among the realms most affected by global connectivity. Carles Brasó examines the business strategy of a transnational engineering company with a headquarters in Shanghai and its collaboration with Chinese textile industrialists during wartime (1937–1945). His chapter lies at the interface of several disciplines, offering multiple avenues for interdisciplinary discussion, in particular in the history of technology and economic and business history. It reveals how engineering networks have been key in transnational machinery trade and the globalisation of technological companies. Difficult times, including war and periods of restricted trade, can be very productive, generating business opportunities and transforming companies' strategies.

Maurits Ertsen's biographical approach takes us through the meanders of colonial, international and post-colonial Dutch engineering and reminds us of the historical significance of education and hierarchy as well as of individual motivations and choices. Attention to technicalities shows technology transfer as necessarily two-directional, as technological devices adapted to new environments, where new local cultures were coproduced. Knowledge and expertise themselves appear as mutable mobiles in their movement from colonial to post-colonial settings. Dutch irrigation engineering became global through developmental programmes, and that globality subsisted through very specific interactions of local settings.

Joseba de la Torre, María del Mar Rubio Vargas and Gloria Sanz Lafuente put nuclear science and technology at the centre of American hegemony in Cold War Europe. It starts with a surprising fact: in the complex world market for nuclear fuels and reactors of the early 1970s, Spain was the major market for American nuclear materials, a position well above what its GDP would have indicated. To understand this anomaly the authors turn to the diplomatic functions of American, German and Spanish nuclear scientists and industrial engineers. The combination of business history, strategic and diplomatic studies and the history of science and technology illuminates this economic conundrum.

Wartime is also a dynamic time for institutional transformations. María Cecilia Zuleta shows how oil rationing during the Second World War set the stage for the establishment and development of the South American Petroleum Institute, a regional association consisting of a diverse community of scientists, technicians and engineers as well as state and multinational oil companies. In line with recent historiography, she recognises that engineers' mutual associations favoured transnational exchanges of knowledge and information in Latin America. More innovatively, Zuleta's chapter makes clear that engineers' associations have been far more central in geopolitics than previous studies have recognised. That said, South American engineers failed to consolidate their diplomatic achievements during the post-war period owing to their opposing interest and ideologies.

Finally, Ian Inkster's epilogue provides a conclusion for the volume, bringing us back to the question that opened this book: what forces and agents have created the globalised capitalist modernity we live in?

Notes

1. Adelman, Jeremy: 'What is Global History now?', *AEON* (2 March 2017).
2. Chakrabarty, Dipesh: *Provincializing Europe. Postcolonial Thought and Historical Difference,* NJ: Princeton University Press, 2000; Hunt, Lynn: *Writing History in the Global Era,* New York: W. W. Norton, 2014;

Hopkins, A. G.: 'The History of Globalization and the Globalization of History', in Hopkins, A. G. (ed.): *Globalization in World History*, London: Pimlico, 2002, pp.11–46.

3. Bentley, Jerry H.: 'Cross-Cultural Interaction and Periodization in World History', *The American Historical Review*, Vol. 101, No. 3 (1996), pp. 749–770.
4. Hopkins, A. G.: 'Introduction: Globalization – An Agenda for Historians', in Hopkins, A. G. (ed.): *Globalization in World History*, London: Pimlico, 2002, pp. 1–10.
5. Gordon, Peter and Morales, Juan José: *The Silver Way. China, Spanish America and the Birth of Globalization, 1565–1815,* Hong-Kong: Penguin, 2017.
6. Bordo, M. et al. (eds.): *Globalization in Historical Perspective*, Chicago: University of Chicago Press, 2003; Hugill, Peter: *Global Communications since 1844: Geopolitics and Technology,* Baltimore: Johns Hopkins University Press, 1999; Wenzlhuemer, Roland: *Connecting the Nineteenth-Century World: The Telegraph and Globalization*, Cambridge: Cambridge University Press, 2013.
7. See, for example, Leonard, Adrian and Pretel, David: *The Caribbean and the Atlantic World Economy*, Palgrave Macmillan, 2015.
8. Stein, Jeremy: 'Reflections on Time, Time-Space Compression and Technology in the Nineteenth Century', in May, Jon and Thrift, Nigel (eds.): *Timespace: Geographies of Temporality*, London and New York: Routledge, 2001, pp. 106–119.
9. Inkster, Ian: 'Machinofacture and Technical Change: The Patent Evidence', in *The Golden Age: Essays in English Social and Economic History,* Inkster, Ian et al. (ed.), Aldershot: Ashgate, 2000, pp. 121–142.
10. Brown, Mathew: 'The global history of Latin America', *Journal of Global History,* Vol. 10, No. 3, (2015), pp. 365–386.
11. See for instance work by the economic historian Carlos Marichal, based in Mexico, and by science historian Irina Podgorny, based in Argentina. Marichal, C.: *Bankruptcy of Empire: Mexican Silver and the Wars Between Spain, Britain and France,* 1760–1810, Cambridge: Cambridge University Press, 2010; Podgorny, I.: 'Fossil dealers, the practices in comparative anatomy and British diplomacy in Latin America, 1820–1840', *British Journal for the History of Science,* 46, 4 (2013).
12. O'Brien, P.: 'Historiographical Traditions and Modern Imperatives for the Restoration of Global History', *Journal of Global History*, Vol. 1, No.1 (2006), pp. 3–39.

13. For a reflection on the long history of interactions between these two fields, see Schabas, Margaret: 'Coming Together: History of Economics as History of Science', *History of Political Economy*, Vol. 34, Annual Supplement (2002), pp. 208–225.
14. O'Rourke, Kevin and Williamson, Jeffrey: *Globalization in History: The Evolution of a Nineteenth-Century Atlantic Economy*, Cambridge, MA: Harvard University Press, 1999.
15. Picon, A.: 'Engineers and Engineering History: Problems and Perspectives', *History and Technology*, Vol. 20, No. 4 (2004), pp.421–36; Cannadine, David: 'Engineering History, or the History of Engineering? Rewriting the technological past', *Transactions of the Newcomen Society*, 74 (2004), pp. 163–80.
16. Edgerton, D.: 'The Contradictions of Techno-Nationalism and Techno-Globalism: A Historical Perspective', *New Global Studies*, Vol.1, No.1 (2007), pp. 1–32.
17. Edgerton, D.: 'The Political Economy of Science – prospects and retrospects' in Tyfield, David, et al. (eds.): *The Routledge Handbook of the Political Economy of Science,* New York: Routledge, 2017.
18. Netz, Reviel: *The Shaping of Deduction,* Cambridge: Cambridge University Press, 1999; Dijstelbloem, Huub and Meijer, Albert: *Migration and the New Technological Borders of Europe*, London: Palgrave Macmillan, 2011; Rankin, William: *After the Map. Cartography, Navigation, and the Transformation of Territory in the Twentieth Century*, Chicago: Chicago University Press, 2016.
19. See, for example, Landes, David, S.: *The Unbound Prometheus. Technological Change and Industrial Development in Western Europe from 1750 to the Present,* Cambridge: Cambridge University Press, 1969; Archibugi, Daniele and Michie, Jonathan (eds.): *Technology, Globalisation and Economic Performance*, Cambridge: Cambridge University Press, 1997; Pomeranz, K.: *The Great Divergence: China, Europe, and the Making of the Modern World Economy*, Princeton: Princeton University Press, 2001.
20. The history of patenting is a good example of this tendency. See for example Streb, J.: 'The Cliometric Study of Innovations', in Diebolt, C. and Haupert, M. (eds.): *Handbook of Cliometrics*, Berlin: Springer, 2016; Sáiz, P. and Pretel, D.: 'Why Did Multinationals Patent in Spain? Several Historical Inquiries' in Donzé, Pierre-Yves and Nishimura, Shigehiro (eds.): *Organizing Global Technology Flow: Institutions, Actors, and Processes*, New York: Routledge, 2013, pp. 39–59.

21. Armitage, David and Guldi, Jo: *The History Manifesto*, Cambridge: Cambridge University Press, 2014.
22. Hughes, Thomas P.: 'The Seamless Web: Technology, Science, Etcetera, Etcetera', *Social Studies of Science*, Vol. 16, No. 2 (1986), pp. 281–292.
23. Edwards, Paul: 'Meteorology as Infrastructural Globalism', *Osiris,* 21 (2006), pp. 229–250.
24. Latour, Bruno: *Reassembling the Social: An Introduction to Actor-Network-Theory*, Oxford: Oxford University Press, 2007, pp. 9–13; Callon, Michael: 'Techno-economic Networks and Irreversibility', in Law, John (ed.): *A Sociology of Monsters: Essays on Power, Technology and Domination*, London: Routledge, 1991.
25. Ghoshal, S. and Bartlett, C.A.: 'The multinational corporation as an interorganizational network', *Academy of Management Review*, 15 (1990), pp. 603–625.
26. Buchanan, R.A.: 'The Diaspora of British Engineering', *Technology and Culture*, Vol. 27, No. 3 (1986), pp. 501–524; Magee, G. B. and Thompson, A. S.: *Empire and Globalisation: Networks of People, Goods and Capital in the British World, c.1850–1914*, Cambridge: Cambridge University Press, 2010; Bennett, B. and Hodge, J. (eds.): *Science and Empire: Knowledge and Networks of Science across the British Empire, 1800–1970*, Basingstoke: Palgrave-Macmillan, 2011; Diogo, Maria Paula: 'Engineering', in Iriye, A. and Saunier, P.: *The Palgrave Dictionary of Transnational History: From the mid-19th century to the present day*, Basingstoke and New York: Palgrave Macmillan, 2009, pp. 331–333.
27. Gordin, Michael: 'Introduction: Hegemonic Languages and Science', *Isis*, 108, 3 (2017), pp. 606–611.
28. For the application of the concept of legitimacy to the study of professions see Abbot, Andrew: *The System of Professions: An essay on the division of expert labour*, Chicago: University of Chicago Press, 1988.
29. On engineers' identities and national styles see: Cardoso de Matos, A., et al. (eds.): *The Professional Identity of Engineers: Historical and Contemporary Issues*, Lisbon: Colibri, 2009.
30. Collins, H. and Evans, R.: *Rethinking Expertise*, Chicago and London: University of Chicago Press, 2007; Porter, T.: *Trust in Numbers. The Pursuit of Objectivity in Science and Public Life*, Princeton: Princeton University Press, 1996; Shapin, Steven: *Never Pure: Historical Studies of Science as if It was Produced by People with Bodies, Situated in Time, Space, Culture, and Society, and Struggling for Credibility and Authority*, Baltimore: The Johns Hopkins University Press, 2010; Kellner, Hansfried

and Heuberger, Frank W. (eds.): *Hidden Technocrats: The New Class and New Capitalism,* New Brunswick and London: Transactions Publishers, 1991.

31. Wise, M. Norton: 'What Can Local Circulation Explain? The Case of Helmholtz's Frog-Drawing-Machine in Berlin', *HoST*, 1 (2007).
32. Leonard, A. and Pretel, D. (eds.): *The Caribbean and the Atlantic World Economy: Circuits of Trade, Money and Knowledge*, Basingstoke: Palgrave Macmillan, 2015.
33. Edgerton, David: 'Creole Technologies and Global Histories: Rethinking how Things Travel in Space and Time', *HoST*, 1 (2007).
34. O'Connell, Joseph: 'Metrology: The Creation of Universality by the Circulation of Particulars', *Social Studies of Science*, Vol. 23, No. 1 (1993), pp. 129–173.
35. Porter, *Trusty in Numbers,* op. cit.; Wise, M. Norton (ed.): *The Values of Precision,* Princeton: Princeton University Press, 1995.
36. Chandler, Alfred D.: *Scale and Scope: The Dynamics of Industrial Capitalism,* Cambridge, MA: Harvard University Press, 1990.
37. Ogle, Vanessa: *The Global Transformation of Time, 1870–1950*, Cambridge, MA: Harvard University Press, 2015.
38. Buchanan, 'The Diaspora of British Engineering', op. cit.
39. MacKenzie, Donald: 'Mechanizing the Merc: The Chicago Mercantile Exchange and the Rise of High-Frequency Trading', *Technology and Culture,* Vol. 56, No. 3 (2015), pp. 646–675.
40. Inkster, I.: *Technology and Industrialisation: Historical Case Studies and International Perspectives*, London: Ashgate, London, 1998; Mokyr, J.: *The Lever of Riches. Technological Creativity and Economic Progress*, Oxford: Oxford University Press, 1990.
41. See for example Basset, Ross: *The Technological Indian*, Cambridge, MA: Harvard University Press, 2016.
42. See Shaffer, S., et al.: *The Mindful Hand: Inquiry and Invention from the Late Renaissance to Early Industrialisation*, Chicago: University of Chicago Press, 2007; Jeremy, David J. (ed.): *International Technology Transfer: Europe, Japan and the USA, 1700–1914*, Aldershot: Elgar, 1991; Bruland, Kristine: British Technology and European Industrialization, Cambridge: Cambridge University Press, 1989; Pretel, D. and Nadia Fernández de Pinedo, N.: 'Circuits of Knowledge: Foreign Technology and Transnational Expertise in Nineteenth-Century Cuba', in Leonard, A. and Pretel, D. (eds.), *The Caribbean and the Atlantic World Economy: Circuits of Trade, Money and Knowledge, 1650–1914*, Basingstoke: Palgrave-Macmillan, 2015, pp. 263–289.

43. Beatty, E.: *Technology and the Search for Progress in Modern Mexico*, Oakland: University of California Press, 2015.
44. Beckert, Sven: *Empire of Cotton: A Global History*, New York: Alfred A. Knopf, 2014; Moore, A.: *Constructing East Asia Technology, Ideology, and Empire in Japan's Wartime Era*, 1931–1945, Stanford: Stanford University Press, 2013.
45. Headrick, D. R.: *Tools of Empire: Technology and European Imperialism in the Nineteenth Century*, Oxford: Oxford University Press, 1981.
46. Hodge, J. M.: 'Science and Empire: An Overview of the Historical Scholarship', in Bennett, B. M. and Hodge, J. M. (eds.): *Science and Empire: Knowledge and Networks of Science across the British Empire, 1800–1970*, Basingstoke and New York: Palgrave Macmillan, 2011, pp. 8–9.
47. Adas, M.: *Dominace by Design: Technological Imperatives and America's Civilizing Mission*, Cambridge, MA: The Belknap Press of Harvard University Press, 2006.
48. Marsden, B. and Smith: C.: *Engineering Empires: A Cultural History of Technology in Nineteenth-Century Britain*, Basingstoke: Palgrave Macmillan, 2005; See also Andersen, C.: *Engineers and Africa, 1875–1914*, London and Brookfield: Pickering & Chatto, 2011.
49. Arnold, David: 'Europe, Technology, and Colonialism in the 20th Century', *History and Technology*, Vol. 21, No.1 (2005), p.89.
50. Rood, Daniel: *The Reinvention of Atlantic Slavery*, Oxford: Oxford University Press, 2017; Magee and Thomson, *Empire and Globalisation*, op. cit., pp.137–143.
51. Lean, David: *Lawrence of Arabia* (1962 film); On T. E. Lawrence and orientalism, see Said, Edward W.: *Orientalism. Western Conceptions of the Oriental World*, New York: Vintage Books, 1978.
52. McKeown, Adam: *Melancholy Order: Asian Migration and the Globalization of Borders*, New York: Columbia University Press, 2008; Conrad, Sebastian: *Globalization and the Nation in Imperial Germany*, Cambridge: Cambridge University Press, 2010.
53. Saccarelli, E. and Varadarajan, L.: *Imperialism Past and Present*, Oxford: Oxford University Press, 2015.
54. See for example Curry-Machado, Jonathan: *Cuban Sugar Industry: Transnational Networks and Engineering Migrants in Mid-Nineteenth Century Cuba*, New York: Palgrave Macmillan, 2011.
55. Hayden, Craig: *The Rhetoric of Soft Power. Public Diplomacy in Global Contexts*, Lanham: Lexington Books, 2012.

56. Krige, John: *American Hegemony,* Cambridge, MA: The MIT Press, 2006.
57. Armstrong, Philip et al.: *Capitalism since 1945*, Oxford: Blackwell, 1991.
58. Misa, T. J. and Schot, J.: 'Inventing Europe: Technology and the Hidden Integration of Europe', *History and Technology*, Vol. 21, No. 1 (2005), pp. 1–19.
59. Högselius, P. et al.: *Europe's Infrastructure Transition,* London: Palgrave Macmillan, 2015.
60. This is in part the goal of the ERC-funded Project 'A Global History of Technology': https://www.geschichte.tu-darmstadt.de/index.php?id=3586&L=2
61. Högselius, P.: *Red Gas: Russia and the Origins of European Energy Dependence*, London: Palgrave Macmillan, 2013.
62. Roberts, P. and Turchetti, S.: *The Surveillance Imperative. Geophysics in the Cold War and Beyond,* London: Palgrave Macmillan, 2014.
63. Camprubí, Lino: 'Resource Geopolitics: Cold War Technologies, Global Fertilizer, and the Fate of Western Sahara', *Technology and Culture,* Vol. 57, No. 3 (2015), pp. 676–703.
64. Kloppenburg, Jack R.: *First the Seed: The Political Economy of Plant Biotechnology, 1492–2000*, Cambridge: Cambridge University Press, 1988; Cullather, Nick: *The Hungry World: America's Cold War Battle against Poverty in Asia*, Cambridge, MA: Harvard University Press, 2011.
65. Edwards, Paul: *A Vast Machine: Computer models, climate data, and the politics of global warming*. Cambridge, MA: The MIT Press, 2010; Mitchell, T.: *Carbon Democracy: Political Power in the Age of Oil*, London: Verso, 2011.
66. Hamblin, Jacob D.: *Arming Mother Nature: The Birth of Catastrophic Environmentalism,* Oxford: Oxford University Press, 2013; Macekura, Stephen J.: *Of Limits and Growth. The Rise of Global Sustainable Development in the Twentieth Century,* Cambridge: Cambridge University Press, 2015.
67. Picon, Antoine: *L'invention de l'ingéneur moderne: l'Ecole des Ponts et Chaussées 1747–1851*, Paris: Press de l'Ecole Nationale des Ponts et Chaussées, 1992; Alder, Ken: *Engineering the Revolution*, Princeton: Princeton University Press, 1997.
68. Camprubí, Lino: *Engineers and the Making of the Francoist Regime,* Cambridge, MA: The MIT Press, 2014; Saraiva, Tiago: *Fascist Pigs. Technoscientific Organisms and the History of Fascism*, Cambridge, MA: The MIT Press, 2016.

69. Andreas, Joel: *Rise of the Red Engineers: The Cultural Revolution and he Origins of China's New Class,* Stanford: Stanford University Press, 2009; Kennedy, Paul: *Engineers of Victory: The Problem Solvers who Turned the Tide in the Second World War,* New York: Allen Lane, 2015.
70. Mitchell, T.: *Rule of Experts: Egypt, Techno-Politics, Modernity,* Berkeley and Los Angeles: University of California Press, 2002.
71. Hecht, Gabrielle: *Being Nuclear: Africans and the Global Uranium Trade,* Cambridge, MA: The MIT Press, 2012*;* Andersen, Casper: 'Internationalism and Engineering in UNESCO during the End Game of Empire, (1943–1968)', *Technology and Culture,* Vol. 58, No. 3 (2017), pp. 650–677.
72. Hodge, J.: *Triumph of the expert: Agrarian doctrines of development and the legacies of British Colonialism,* Athens: Ohio University Press, 2007.
73. Larner, Wendy and Laurie, Nina: 'Travelling technocrats, embodied knowledges: Globalising privatisation in telecoms and water', *Geoforum,* Vol. 41, No. 2 (2010), pp. 218–226.

2

The Historical Roots of Modern Bridges: China's Engineers as Global Actors

Dagmar Schäfer

'Engineering' in China

Like many other concepts, the idea of an engineering profession entered China from Europe in the nineteenth century.[1] However, 'master of work processes' (*gongcheng shi* 工程師), as the term engineer was translated, was in no way a novel role, as the railway engineer Cheng Qingguo and Tang Youcheng, then Head of the Geophysics Research Group at the Institute of Remote Sensoring Application, Chinese Academy of Science, emphasise in their 1984 study on bridges.[2] They explain that China's past had included numerous architectural and hydraulic masterminds who knew how to mobilise masses of workers, materials and land, and were technically adept. They assert that every type of modern bridge construction can be 'found in embryonic form in the ancient Chinese bridges, whose designers made such great progress and achieved such distinctive features in structure and construction so long ago'.[3] In their rendering of the past, artefacts verify technical expertise, while textual

D. Schäfer (✉)
Max Planck Institute for the History of Science, Berlin, Germany
e-mail: dschaefer@mpiwg-berlin.mpg.de

© The Author(s) 2018

D. Pretel, L. Camprubí (eds.), *Technology and Globalisation*, Palgrave Studies in Economic History, https://doi.org/10.1007/978-3-319-75450-5_2

records illustrate these individuals' learnedness and altruism, as well as the political importance of their efforts and manifold deeds.

The historical continuity of both an engineer's technical prowess and his/her political responsibility were important aspects of China's late nineteenth-century quest for modernity. Both issues continued throughout the twentieth-century nation-building and identity debates, although they were increasingly coupled with a Chinese concern about global political recognition from 'Western' countries such as the USA and Europe as a modern, advanced and politically important actor. During the late Qing and Republican eras, we find scholars trained in the West, such as the geologist Ding Wenjiang 丁文江 (1888–1936), or practitioners such as the architect Zhu Qiqian 朱启钤 (1872–1964), who searched through China's history for exemplars with scientific and technical skills.[4] Others, such as the US-trained railway engineer Zhou Houkun周厚坤 (1889–?), attempted to make engineering responsible for preserving culture. In the 1910s, Zhou urged engineers to invent a typewriter for 'this wonderful language of ours',[5] instead of requesting a change in the language itself or assuming that engineers had reached the limits of their abilities. Republican and Communist politicians purposefully anchored modern ideals of science and technology into state practice, continuing the ritualisation of mythological emperors such as the flood tamer, the Great Yu 大禹, rebuilding his temples and memorials in Shaoxing between 1930 and 1939. Both Communist and Guomindang members politically promoted technocratic and expertocratic forms of leadership while, at the same time, historians such as Gu Jiegang 顧頡剛 (1893–1980) and the writer-politician Guo Moruo 郭沫若 (1892–1978) debunked the historical constructivism and social purpose of such mythological approaches.[6]

The Past in the Present

If Chinese actors saw historical continuity in technical prowess and an engineer's leading political role from the premodern to the modern age, how then, did this era explain the early modern-to-modern change? Many modern historical accounts distinguish these periods by referring to the growing scale of operations or the upsurge in mechanisation.[7] Cheng and

Tang pinpoint the increasingly globalised knowledge culture of the contemporary engineering trade. They assert that engineers of the modern age were able to cope with previously insurmountable difficulties such as the construction of a permanent bridge over the Yangtze River near its delta, because they could draw on 'not only the intelligence of the Chinese people but also the experience of other countries'.[8] Unlike classic scholars of the Song, Yuan, Ming or Qing dynasties (i.e. between the ninth and eighteenth centuries),[9] engineers after the nineteenth century thought beyond the nation-state and, by sharing their knowledge and expertise, were 'propagating friendship and association between peoples throughout the world'.[10]

In their 1984 article, Cheng and Tang articulated an aspiration that the new millennium would see expansion—in more than rhetorical terms—to countries in the Association of Southeastern Asian Nations (ASEAN) and diverse communities on the continents of Africa and Latin America.[11] Trained in technologies, architecture, economics and sociology, Chinese engineers in the twenty-first century envisage and realise the biggest, most costly and most technically challenging projects, such as a water pipeline from China's south to north, new railway lines with wide-spanning bridges, sports arenas or expansive highways across African deserts—or even entire continents. When other countries talk politics, China's political elite often remain silent in global discourses—and instead let engineering projects do the job. Another factor is that since the 1950s engineers increasingly moved from the back seat to the forefront of political, economic and social decision-making.

Of course, as Cheng and Tang anticipated, the methods and means of engineering have changed substantially. Bridges are higher, span wider and are lighter than their historical precedents. What concerns me in this chapter, though, is Cheng and Tang's claim that premodern engineers were more bound to political territories and regimes (that is, empires) and did not share their knowledge as easily as today. Is the political boundedness of knowledge, its identification within empires and nations, an important characteristic explaining/identifying the premodern/modern divide?

Although, prior to the eighteenth century, Chinese 'masters of work processes' did not see the world defined in terms of a globe, they—like their modern successors—believed themselves to be operating on a scale relevant to 'all under heaven' (*tian xia* 天下). In this political framework, we can indeed see that they defined their skills and knowledge in terms of how their efforts to regulate natural forces were important for ordering (*zhi* 治, i.e. governing) the world, thus making it a habitable and 'cultured' (*wen* 文) place. As such, engineering projects to control water, land and society were mainly pursued by the cultured inhabitants of China's various dynasties, whereas the barbarians outside China were believed to be unaware of such means. Cheng and Tang indeed assert that historically Chinese bridge-building expertise was exported to places such as Edo-Japan, whereas they do not name any cases of foreign influx into China before the nineteenth century.

Cultures and Bridges

In fact, Cheng and Tang are right to assume that in the Chinese historical context, sharing knowledge meant civilising the world. The Chinese literati considered this shared knowledge to include more than just the scientific and technical proficiency and skills admired by twentieth-century Chinese engineers. The civilising influence of engineering know-how lay in its benefits for both the state and its common people, being dependent on the moral application of knowledge, defined in terms of commensurability.

The numerous historical accounts of hydraulic and construction projects are one example of this. These projects often achieved an order of magnitude that required attention from emperors, and thus became memorised in dynastic historiography. Substantial investment in resources was only justifiable because, once accomplished, such projects would benefit the community long into the future, far exceeding any single individual's vision, or even a whole generation's desire.

Rhetorically, scholars approached the state's responsibility for a *longue durée* view and issues of scale by adopting the mythologised account of ancient sage-kings. As the assigned Minister of Work (*gongbu* 工部) in the court of (the equally mythological) emperor Shun 舜, the figure of the Great Yu exemplified the imperial role in water management projects. Yu is recounted to have successfully channelled water into the major Chinese rivers. In this way, as Mark Lewis suggests, Yu accomplished 'the structuring of the world through a process of [spatial] division'.[12] This enabled people to travel throughout the country on rivers and roads and determined which lands were suitable for agricultural cultivation. Following Yu's lead, court elites and scholar-officials took on the same tasks in the dynastic state. By the time of the Song Dynasty (964–1279), the emperor's legitimacy depended, to a large extent, on his ability to keep the capital Kaifeng free from floods and, after the dynasty withdrew to the south, to tame the water there to cultivate new land. Over the longue-durée, concepts as broad as Wittfogel's problematic hydraulic states as well as modern environmental history perspectives are employed.

In this legitimising role, the Great Yu appears, with short interruptions, in Chinese historiography at least since the Northern Song. While the task remained largely the same, the intervention strategies varied considerably, with emperors such as Huizong attempting to get to the root of the problem and redirect the waters of the Yangtze from its source.[13] Most imperial rulers settled on dealing with the outcomes or following a rhetoric of imperial tasks. However, Hongwu 洪武 (1328–1398, reigned from 1368), the founding emperor of the Ming, reinstalled the rituals of the Great Yu in the seventh year of his reign, 1374.[14] In addition, when he was a young prince, the later Yongle 永樂 emperor (1360–1424, reigned from 1402) was continuously reminded of the hardworking (*qinlao* 勤勞) exemplar, the Great Yu.[15]

Water, Politics and the Engineer Hero Yu

While emperors may have controlled these projects, they relied on water management experts to actually implement them. 'Hydraulic engineers' populate China's dynastic and private histories. In addition to their technical ability, they had to be able to negotiate the suitable means and planning strategies to ensure the successful completion of such projects. They were granted moral judgement, yet they were frequently subjected to criticism and were blamed for the causes and results of any mishap.

A case that illustrates engineers' ambiguous moral role is that of the Song Dynasty civil servant and relative of the contemporary prime minister, Tang Zhongyou 唐仲友 (1136–1188). On visiting Tiantai district,

Tang designed a bridge that could withstand the difficult tides at that particular juncture. Even though Tang is shown in this context as a quintessential manager-engineer with exceptional technical abilities, the story's purpose is to depict Tang as a quintessential immoral villain. Tang initiated a huge, financially demanding and technically complex construction project to build a bridge comprising twenty-five connecting sections, with harbour wings for fifty boats, and embankments against the tides 115 *li* (5832 metres). towards the south—because he had felt forced to share a ferry with some drunken, ill-behaved passengers who had lost their moral standards.[16] His contemporary colleague and the originator of Song-orthodox Neo-Confucian philosophy Zhu Xi 朱熹 (1130–1200) accused Tang of having used inappropriate means—a technical solution—for a trivial social problem. He alleged that Tang had strained the wealth of the commoners and the imperial treasure trove, instead of simply educating his fellow travellers. Or, expressed in modern terms, Zhu Xi accused Tang of choosing technical over social engineering.

This incident showcases a disagreement over cost–demand efficiency and the adequacy of resources, as much as over the technical scale and scope of human planning. In a ritual tract that was published at around the same time, Zhu Xi stated that he favoured a method of minor interventions. In this work, he described how the social order of the state depended substantially on the proper placing of an ancestral shrine in each individual's home.[17] Proper action meant understanding that major outcomes could be achieved by taking care of rudimentary concerns.

Li Cho-ying's comparative research on hydraulic engineering during the Song and Ming period shows that debates about appropriateness also concerned the specific level at which decisions should be made. Whereas Song rulers centralised structures and took preventive and affirmative action, Ming rulers increasingly withdrew from such projects and left these matters in the hands of local officials and gentry.[18]

In a more recent study of hydraulics, Li draws particular attention to the third Ming emperor, Yongle, who abandoned hydraulic management and replaced it with tax exemptions and aid relief in around 1404. Li's research elucidates that the reasons for such political shifts, and the effect they had on the actors who were in charge of social, political and technical decisions, were complex. Resource management—such as the availability of wood or labour—as well as straightforward financial considerations may have played an important role in Yongle's rash withdrawal from a redesign of water management practices in the Lower Yangtze region, in the same year that he moved the capital from Nanjing to Beijing. This spatial relocation of political power meant a need to renovate and rebuild the Yuan palace structures in Beijing, as well as carry out infrastructure projects such as constructing canals and roads. Furthermore, between 1403 and 1420 Yongle sponsored projects such the expansion of the Daoist complex of the Wudang shan pathway, which contained over sixty bridges, to open up routes through a sacred landscape for pilgrims.[19] Various temples and resting sites were included, spread across an area of 140 *li* (67200 metres). Yongle may not have envisioned the long-term financial and social implications of such large-scale projects in 1403, when he promoted the civil service examinee Xia Yuanji 夏原吉 (1366–1430) into ever-higher civil servant positions (up to the role of Minister of Finance), thus enabling XiaYuanji to develop a technical solution for dredging the waterways of the Lower Yangtze River. But he certainly had to deal with the various financial and social repercussions of these projects.

But clearly, as Li has also shown, the definition of an 'appropriate' mode and scope of intervention depended on its actual aims, which did not always involve an urgent environmental issue, an agricultural or even a social purpose. Yongle's redesign of the Lower Yangtze did not fill any immediate need such as extensive flooding, drought or famine. Rather, Yongle magnified small incidents 'to the degree that the entire region had to engage in water management...'. Hence, he initiated a huge project 'to legitimise his reign, without saying so explicitly'.[20] He politicised the flow of water, so as to justify an empire-wide—and in this sense for his world 'global'—technical intervention.

Two technical innovations are attributed to the historical engineer Xia Yuanji: a 'pedal pump rescue' and his specific 'polder dyke construction'. The nature of Xia's innovations were, however, systemic rather than technical, because Xia made sure that his 'machines' enabled local officials to control floods themselves—thus alleviating the central state of this central responsibility. However, he also made sure that the government could still reach local homes, by implementing a tax for state management of the pumps.[21]

Historians of science have dedicated special attention to China's rich and resourceful history of engineers. But only lately have historical studies such as Li's started to look beyond innovative technologies and connected the dots between social, economic, environmental, scientific and technological change. Thus also the long-term implications that such projects had on knowledge standards and ideals as well as for hydraulic practices more generally. In fact the imperial context authorised technical solutions for succeeding generations. Scholar-officials in the late Ming and Qing emulated Xia's systemic choices over the next two centuries. Similarly consistent was Pan Jixun's 潘季馴 (1521–1595, *jinshi* 1550) later solution to channel the waters more narrowly into a torrential stream, in order to 'use water to attack water and regulate the river with the river' (借水攻水,以河治河).[22] Engineers in the 1950s consulted Pan's work carefully before attempting to speed up the water flow, so that it would flush the silt alongside the embanked riverbed.[23] Thus, certain ideas prevailed across the premodern and modern divide, serving as inspiration for new technical solutions or, in some cases, a reconstitution of the old ones.

Tensions, Compressions and Torsion: Political Arches

As central as ideas of nation-building are in contemporary China's engineering cultures, they also often address ideological continuities of a larger, less territorially bounded identity discourse, addressing the multiple concerns brought about by large-scale engineering interventions such as the remodelling of rivers, mountains or urbanised space. And here we can find another continuity. Like emperors, early Republican scientists, historians and politicians managed the public image of such projects and the public's concerns about them by employing state rituals and propaganda. While Song emperors financed dykes and canals, they commissioned paintings of rituals around the Great Yu and water mills as Liu Heping shows.[24] Sun Yatsen 孫中山 (1856–1925) worshipped the mythological flood and river conqueror, the Great Yu in 1919 in a state ritual, as did Zhou Enlai 周恩來 (1898–1976), who, in 1939, suggested that China's political elites had not adequately studied 'Yu's lessons about flood control. They know restraint, but nothing about effective guidance and hence create a despotic tyranny…'.[25] The Guomindang regime continued legitimising rituals, whereas the Communists broke with such feudal practices.

As Mizuni, di Moia and Moore have recently argued, the limits of a nation-state-based historiography of economic development and engineering traditions lie 'in its neglect of the continuity between colonial Asia from Cold War Asia, and empires from post-WWII international development'.[26] In fact, a closer look clearly reveals such continuities within China's national engineering debates too. As recent research has shown, Communist politicians and state actors between the Great Leap Forward (1958–1962) and the Four Modernisations (since 1972) never entirely cut ties with the past in their joint aims to foster progress within nation- and identity-building. Following a policy of employing 'red and expert' (i.e. both politically conscious and professionally competent) civil servants and party members in the 1960s, the state still promoted the publication of classical tracts on agriculture for utilitarian means, such as *Nongshu*農書 (Book of agriculture) and Song Yingxing's *Tiangong kaiwu* 天工開物 (The works of heaven and the inception of things). Officials held that such classic literature would be a suitable guide to innovation after the Soviets had cut their technical and financial aid during the Great Leap Forward. Such examples of ancient wisdom also legitimised historical empiricism—that is, learning from the past—and enabled politicians to invoke former technical visions as important and now feasible project ideas in the changing political context. Along these lines, the idea of channelling the abundant waters of the south up to the drylands of the north (*xibu da kaifa* 西部大開發), developed in the 1950s by Ministry of Water personnel, was put into effect in 2000. In between proposal and implementation, a new ruling class of engineers such as Jiang Zemin 江澤民 (b. 1926) emerged and ascended to the highest positions of political power. Jiang, who was state president from 1993 until 2003, received a technical education at one of China's elite universities (Shanghai) before the Cultural Revolution of 1964 (before the term 'red engineer' fell out of favour). He was an advocate of the Three Gorges Dam, and he actively reinvigorated the cult of the Great Yu in Shaoxing in 1995, supporting the state financing of temples and memorials.[27]

In contrast to Jiang Zemin's rise, engineers such as Cheng and Tang anchored their social responsibility in a Henri de Saint-Simon-type ideal of economic planning based on scientific principles, which mobilised the Communist ideals of common production means and public property within a capitalist market economy in a seemingly paradoxical, yet very efficient, combination.[28]

Conclusion

Engineering traditions in Asia did not develop out of a void. Bridges existed or were built where ferry passages had previously crossed rivers, and irrigation is constantly updated but still continues flowing in its old beds and grids even today. Overlooking the desired and unwanted continuities of Chinese engineering history and historiography beyond the modern age means dismissing, all too easily, another cultural means that helped to shape identity and promote an engineering modernity within changing political climates and new economic ideals. The political instrumentalisation of past state mythologies within a revived modern Neo-Confucianism philosophy since the 1990s shows that such references had important social and moral implications as well as technical consequences. When Jiang Zemin attended a ritual and left a plate in his own handwriting on the Great Yu's tomb (Da Yu Ling 大禹陵), he was deliberately promoting technical solutions for both social and environmental challenges. His successors have taken similar viewpoints since the 1990s, and increasingly on a global scale.

The state remains a major actor in both the small- and large-scale endeavours of China's engineers today. The people of Shaoxing city celebrate the Great Yu by offering a small sacrificial ritual each year, a public sacrifice every five years and a grand sacrifice every ten years (with the exception of 2003). This event has even turned into a modest tourist attraction, and with the economic benefit, and globalisation debates, the public increasingly embraces the political messaging as a sign of culture and identity. In 2007 the ritual achieved national status not unlike previously in imperial times. Jointly promulgated by the city, regional and state organisations, the stakeholders revived an imperial format (*dili* 禘禮) of thirteen sacrificial steps. At the same time, international representatives, including several groups from Taipei, Japan, Korea, Poland, India, Iran and France, attended the celebration. From a Chinese perspective, therefore, it is not so important that China takes a front row seat in worldwide politics as long as Chinese engineers can confidently hold major agency in the very political nature of an engineered, globalising world.

Notes

1. Chen Yue 陳悅 and Sun Lie 孙烈: 'Gongcheng yu gongchengshi ciyuan kaolüe 工程與工程師詞源考略', *Gongcheng Yanjiu* 5/1 (2013), pp. 53–7. Most of these concepts arrived in China through railways and other infrastructure and thus were mediated through Japan. Andre Schmid. *Korea between Empires 1895–1919* (Columbia University Press 2002)
2. Cheng Qingguo and Tang Youcheng: 'The traditions of bridge technique and modern bridge engineers of China', *European Journal of Engineering Education,* 9/1 (1984), p. 13–19.
3. Cheng and Tang, 'Bridge engineers of China', p. 13.
4. The best biography in English is still Charlotte Furth: *Ding Wen-chiang: Science and China's New Culture,* Cambridge: Harvard University Press, 1970. Zhu Qiqian (1932/2004) provides historical biographical data on craftsmen, engineers and other experts, in Yang Yongsheng 楊永生 (ed.): *Zhejianglu* 哲匠錄, Beijing: Zhongguo jianzhu gongye chubanshe, 2004.
5. Thomas Mullany: *The Chinese typewriter: A history,* Boston: The MIT Press, 2017, p. 138.
6. Robin McNeal: 'Constructing myth in modern China', *The Journal of Asian Studies,* 71/3, (2012), pp. 679–704, p. 687.
7. See A.A. Hamm, Brian W. Beetz and Rudi Volti: *Engineering in time. The systematics of engineering history and its contemporary context,* London: Imperial College Press, 2004 – for instance, their survey of the period

1800–1940 entitled 'Expansive Engineering'. B. Marsden and C. Smith emphasise this growing industrialisation and mechanisation in: *Engineering empires: A cultural history of technology in nineteenth century Britain,* London: Palgrave Macmillan, 2004. Michael Adas discusses the notion of a Western expansive hegemonic regime in: *Machines as the measure of men. Science, technology, and ideologies of Western dominance,* Ithaca: Cornell University Press, 1989.

8. Cheng and Tang, 'Bridge engineers of China', p. 16.
9. These dynasties are traditionally referred to as 'premodern' because of a defined set of characteristics shared with the European periodisation 'early modern'.
10. Cheng and Tang, 'Bridge engineers of China', pp. 15–16.
11. Ashley Kim Stewart and Li Xing: 'Beyond debating the differences: China's aid and trade in Africa', in Li Xing and Abdulkadir Osman Farah (eds.): *China-Africa relations in an era of great transformations,* Farnham: Ashgate, 2013, pp. 23–48. See also Greg Brazinsky, Winning the Third World. Sino American Rivalry during the Cold War (University of North Carolina Press 2017)
12. Mark E. Lewis: *The flood myths of Early China,* Albany: State University of New York Press, 2006, p. 30.
13. Ling Zhang: *The river, the plain and the state. An environmental drama in Northern Song China, 1048–1128,* Cambridge, MA: Cambridge University Press, 2016. See also Ruth Mostern: *Dividing the realm in order to govern: the spatial organization of the Song state (960–1279),* Cambridge, MA: Harvard University Asia Center, 2011.
14. Xia Yuanji 夏元吉, Yang Shiqi 楊士奇 et al.: *Ming Taizu shilu* 明太祖實錄, Taipei: Zhong yanjiuyuan shiyu suo jingyinben, 1604/1962, p. 1501. Sun Yuantai 孙远太: 'Dayu jidian yi da yu wenhua de zhuanbo 大禹祭典與大禹文化的傳播', *Qianyan* 263/9 (2010), pp. 181–184.
15. Xia Yuanji, *Ming Taizu shilu,* p. 1209. For the role of Yu in the Ming era, see Xu Jin 徐進: 'Mingdai Da Yu jiyi ji qi wenhua yiwen 明代大禹記憶及其文化意蘊', *Yindu Xuekan,* 32 (2016), p. 35.
16. The stone stele Xinjian Zhongjing qiao beiji 新建中津橋碑記 is included in Lin Biaomin 林表民 (ed.) (*c.* 1450): *Chicheng ji* 赤城集. Wenyuange siku quanshu-edition, Chap. 12.
17. Francesca Bray: 'Technics and Civilisation in Late Imperial China: An Essay on the Cultural History of Technology', *OSIRIS,* Vol. 13 (Special Issue: *Beyond Joseph Needham: Science, Technology, and Medicine in East and Southeast Asia),* 1998, pp. 11–33.

18. Li Cho-ying: 'Contending Strategies, Collaboration among Local Specialists and Officials, and Hydrological Reform in the Late-Fifteenth-Century Lower Yangzi Delta', *East Asian Science, Technology and Society: An International Journal,* 4:2, 2010, pp. 229–253.
19. Song Jing 宋晶: Mingdai Wudang Shan Qiaoliang chutan 明代武當山橋樑初探 in *Journal of Hubei University,* 33:5, 2006, pp. 587–590.
20. Li Cho-ying: '"As a sage-king re-emerges, all water returns to its proper path": Xia Yuanji's water management and the legitimisation of the Yongle reign' in Francesca Bray and Lim Jongtae (eds.): *Science and Confucian statecraft in East Asia,* Brill: Leiden, forthcoming.
21. Michah Muscolini, *The Ecology of War in China: Henan Province, the Yellow River and beyond 1938–1950* (Cambridge University Press 2014), looks at destruction caused by war, disaster and engineering.
22. Pan Jixun: 'Hefang Yilan 河防一覽' in Wu Xiangxiang 吳湘湘 (compiler): *Zhongguo shixue congshu,* Vol. 33, Taipei: Taiwan Xuesheng Shudian, 1965, Chap. 8.23b, p. 658.
23. See also Miriam Seeger: *Zähmung der Flüsse: Staudämme und das Streben nach produktiven Landschaften,* Berlin: LIT Verlag Münster, 2014, p. 75; Randall A Dodgen: *Controlling the dragon: Confucian engineers and the yellow river in Late Imperial China,* Honolulu: University of Hawaii Press, 2001, p. 18.
24. Liu Heping (2012), 'Picturing Yu Controlling the Flood. Technology, Ecology and Emperorship in Northern Song China', in *Cultures of Knowledge: Technology in Chinese History*. edited by Dagmar Schäfer (Leiden: Brill), pp. 91–126.
25. Xie Xingpeng 謝興鵬, *Jiuzhou fangyuan hua Da Yu* 九州方圓話大禹. Sichuansheng Da Yu yanjiu hui, 2002, p. 50.
26. Hiromi Mizuni: 'Introduction: A Kula Ring for the Flying Geese: Japan's Technology Aid and Postwar Asia', in Hiromi Mizuni, Aaron Moore and John diMoia: *Engineering Asia. Technology, colonial Development and the Cold War order,* London: Bloomsbury Academic, forthcoming.
27. Judd Kinzley, Universities of Wisconsin Madison/Harvard Energy Conference from 2013/Dissertation: Staking Claims to China's Borderland: Oil, Ores and State-building in Xinjiang Province, 1893–1964. Conference website http://sites.fas.harvard.edu/~histecon/energy/Asia_History_Energy/participants.html /the project is ongoing at Harvard and MIT.
28. Cheng Li: *China's leaders: The new generation,* Lanham, MD: Rowman & Littlefield, 2001, p. 27. It is along those lines that Cheng Li identifies them as 'technocrats'.

3

Indigenous Resistance and the Technological Imperative: From Chemistry in Birmingham to Camphor Wars in Formosa, 1860s–1914

Ian Inkster

Introduction

This chapter will consider two main themes.[1] The first of these is how it was that an urban-based chemical programme spread at a global level under the force of a conjuncture of specific sites of local knowledge, academic research and intellectual property into the heart of a series of new industries associated with Europe and the USA in the later nineteenth century—from photography and artificial silk manufacture to gunpowder, craft and art nouveau products (in the latter case chemistry substituted for such expensive material as ivory or marble or mother-of-pearl), underground electrical cables and the even more important broad emergence of celluloid and plastic manufactures. Secondly, it will analyse how this new chemistry directly impacted on the politicisation of frontier

I. Inkster (✉)
School of Oriental and African Studies (SOAS), University of London,
London, UK
e-mail: ii1@soas.ac.uk

© The Author(s) 2018

D. Pretel, L. Camprubí (eds.), *Technology and Globalisation*, Palgrave Studies in Economic History, https://doi.org/10.1007/978-3-319-75450-5_3

warfare in Formosa (Taiwan) from the 1860s, escalating with the colonial invasion of Japan in the mid-1890s. This second task is designed to show how high-tech chemistry could, unbeknown to any of its commercial or expert agents in the 'West', and at the behest of no one evil-doer, radically upset the frontiers and checks and balances of non-Western ethnicities in relatively isolated locations of the world. Colonialism was a matter of commerce and useful knowledge in alliance with a process of cultural 'othering', which in the hands of Anglophone interests and Japanese colonial imperatives altered forever the entire history of a crucial part of maritime China at a time of its cultural degradation and commercial dismemberment. Finally, and ironically, it will be shown that such global connectivities or impacts might have been rendered either insignificant or turned to more positive account by the very chemistry that had created them in the first place. Chemical work from the First World War allowed the production of plastic material and product to be freed from its dependency upon several raw materials, including camphor. If such work had come earlier, cultural and political disaster in Formosa might well have been avoided.

The conclusion argues that older work on colonialism that visualises networks of technology and expertise as predominantly benign connectivities beyond or contra-cyclical to political colonialism cannot in itself replace the more basic forces at work in the global economy of the second industrial revolution—the technological imperatives of industrial capitalism served also to drive new frictions into old civilisations and in that process marginalised long-established indigenous communities. Many of these were spatially and culturally located on the very frontiers of colonialism, and at times persisted in their existence beyond colonialism itself.

Connectivities in a New Globalism

This chapter covers a period that witnessed a second industrial revolution based on a great surge of technological change, a massive growth in the international trade, investment and migration of the so-called Atlantic Economy, an almost world-wide entrenchment of colonialism and formal imperialism led by a small group of industrial powers who wielded those advancements in technology as tools of aggressive expansion. However much it is dressed up or nuanced by theories of cultural advancement or new cognitions, these basic elements interacted closely and cumulatively to form a dualism of winners and losers in the world system.[2] It is true that one nation broached this stark division—Meiji Japan from 1868 managed to react and industrialise just sufficiently to put her last amongst the dozen or so success cases. A few others appeared to benefit for a while from some supply advantage but then failed to sustain their trajectories (Argentina was an outstanding example), and a few lucky areas of recent settlement combined good natural resources, small census populations,[3] cultural proximity and frontier ingenuities to ensure their place amongst the suppliers of key natural or semi-processed commodities to the second industrial revolution (Australia and Canada leading this process). But for the huge populations of our planet, the years between the Great Exhibition of 1851 and the outbreak of world war in 1914 saw prosperity and optimism residing in winners, depression and resentment in all others. The two demarcation years used in my last sentence illustrate the absolute dominance of one side of this dualism, the terminal dates being defined entirely by events conjured up within the winner group. Of no other period in world history, including that of the classic industrial revolution, could this be said with the same sort of confidence.

Within winner nations the working classes began to share in some benefits of industrialism (a new urbanism, holidays, greater religious and political freedoms and in some cases even civil voice) whilst industrialists, bankers, insurers and shippers became the new projectors, employing in aid of their social comforts, their legal protection, good health and psychic well-being a new host of qualified professionals and experts. Most of the latter wielded income and status through their claims to expert knowledge and the civil application of that knowledge. Within many loser nations, especially in colonies, emerged in contrast an alienated elite, members of which moved their wealth from the land or from new commerce into the enterprises and exchanges of the second industrial revolution, often living in and sharing the cultural trappings of elite urban environs in Europe and the USA, whilst a host of millions became increasingly dependent on forms of lowly employment in agriculture, industry and commerce in their own countries that were irreversibly entwined with the dictates of industrial capitalism elsewhere.

Bearing in mind all such interrelationships we may speak clearly of a new globalism, and in this contribution we illustrate how technology was central to its connectivities, the densities of its networks and the character of its associations and institutions. We begin in Birmingham, in the heartland of that fusion of natural resources (minerals and energy densely arrayed) with the earned resources of skill, knowledge, technique and institutional felicity that in all represented both the formative ingredients of the first industrial revolution and the major building blocks of the second. In this later period the linkages between industrial growth, tacit knowledge in specific skill environs and systematic advancements of natural and experimental sciences became more formalised and identifiable, but still benefited decisively from continuing elements of spatial, social and cognitive proximity inherited from the earlier period of industrial enlightenment and artisanal skill.[4] Out of the knowledge networks of urbanism came Birmingham's lead in new processes of celluloid production, chains of experiment, skill and expert knowledge that in the period spawned major breakthroughs across the range of new industries.

In a world of machinofacture,[5] where overall economic growth in winners was spearheaded with a startling acceleration in both the output and range of manufacturing production centrally located in innovations in metal and chemical production processes, such urban locational advantages also created Birmingham's technological and commercial lead in the starkly contrasting fields of toy-making and arms manufacture. The arms manufacture that was integral to industrialism was central to the resulting global dualism. Chinese defeat by Great Britain in the so-called Opium Wars of 1839–1860 was not merely part and parcel of British industrialism's need to find a simple export product to China that would relieve the burden on the balance of trade of importing a valuable array of Chinese products, it was also an early aggressive expression of the technological synergies forged by Napoleonic warfare and the consequent great advance in British armaments and naval technologies.[6] Such combined commercial and technological imperatives ensured that industrial dominance was military dominance in the 150 or so years between the Napoleonic Wars and the Vietnam War. In the case of our study here, it was British commercial and territorial aggression after 1815 that eventually led a group of now Great Powers to undermine Chinese governance and commerce through a series of threats and wars, treaties and arrangements, which culminated in China's loss of commercial sovereignty, territorial erosions and failure of central authority. By the 1860s Chinese authority had been severely undermined along most of its land and sea borders, and in 1895 the loss in warfare to industrialising Japan and the resulting Treaty of Shimonoseki was exemplary of the global reach of the second industrial revolution.[7]

Similarly, the central importance of military and naval technologies and their frequent applications in the convoluted processes of expansive threat (e.g. to Japan after 1854), colonial formalisation and intensification (e.g. in India), or systematic partitions and dismemberment of nominally independent states (e.g. China) arose—in the felicitous terms of our two volume editors—ultimately from those energetic engineers defined 'as social actors with expertise and social authority in technology and science'.[8] We shall see how in Birmingham urban space and association yielded both expertise and authority just as it also broke down any stylised distinctions between 'technology' on one hand and 'science' on the other. But then, our case quite dramatically exemplifies how the prevailing civil society and urbanism of the second industrial revolution in England directly impinged upon the entirely contrasting political economy of the small island of Formosa, so many miles away, into forest and mountain regions and established indigenous communities that in all obvious respects inhabited an entirely different planet. That urban England and forest Taiwan were not in fact on effectively separate planets was because of the institutions and commercial competitions of late nineteenth-century capitalist expansionism, a central dynamic of which lay in the global networking of engineers, experts and their expertise.

Our case study embraces both sides of the resulting connectivity and confrontation. However benignly created, the celluloid and plastics technologies first generated in Birmingham had a direct and profound impact on the ethnic and economic history of Taiwan (Formosa), the small island situated on the Chinese southern coast, always marginal to mainland interests and entirely separated from them when Japanese colonial authorities took over in 1895. The global connection was complex, involving movements of knowledge and expertise, transfers through the expanding intellectual property systems (in this case patenting), encroachments by

Chinese settlers on the central and eastern territories of Taiwan, hitherto under the sway of non-Chinese indigenous groups, and the gunboat diplomacies of the Great Powers. Finally, we argue that just as this global confrontation was created in the process of technological advancement, so too it was to be relieved not by diplomatic or political solutions, but by further technological innovations stemming from the West.

Bustling Proximities: Out of Birmingham

By the 1850s, the city of James Watt and Matthew Boulton, of the Soho Works and the steam economy had become a central place of industrial capitalism. Birmingham, in the English West Midlands, was the stronghold of nineteenth-century industrial innovation. More intensively than anywhere else in the world at that time, Birmingham was crammed with artisans, workshop owners and small businessmen who competed avidly for the innovation guinea. A principal mechanism was cut-throat competition, but increasingly an improved system of intellectual property rights (the reformed patent system of 1851) was giving sufficient information and protection to allow for the emergence of what might be called technology research programmes, often as not dominated by one individual. As the engineer John Farey put it in 1835, the patented incremental improvements of the time were 'the origin of a number of considerable trades in Birmingham, Sheffield and in London'. A few years earlier he had emphasised 'the very expressive nature of the technical language that is used among all our artisans; also the established habit that the English have, more than any other people, of associating themselves into bodies and societies to act in concert to effect a common object'.[9] Here, then, was a fairly clear statement of the linkages between workshop culture, urban associative activities surrounding the diffusion and discussion of reliable knowledge, and the process of technological change.

Birmingham seems to illustrate the pertinent generalisations of this chapter. The city even at this time was predominantly a place of unmechanised workshops, whose toy and gun manufactures, brass and ironmongery, remained dominated by craftsmen benefiting from the fundamental Black Country technological advances of an earlier era—mineral fuel, puddling and rolling, and steam power.[10] In Birmingham during the 1830s and

1840s some 28% of patents were for metalworking processes, 11% were for machinery and motive power improvements, 14% for clothing and accessories, but a very large 43% belonged to a wide range of small metal products, and were related to advances in design, materials or use of equipment of such items as decanters and other glassware, firearms, lamps and gas-burners, wood-working and furniture-making, inkstands and ornamental items, and so on, the very stuff of small-scale Birmingham manufacturing until the early years of the twentieth century. Some 66% of patented inventions came from skilled tradesmen and manufacturers, and much of this was focused on product innovation, whereas improvements in metalworking, textile machinery, railway and other general machinery tended to come from self-styled engineers, often in partnerships—12% of over 400 patents were through partnerships, but this represented 24% of all Birmingham patentees, most of whom were small manufacturers, millwrights, machinists and skilled tradesmen engaged in a self-help industrial culture.[11]

As can be seen from Fig. 3.1, by the 1860s patenting was predominantly in the hands of a vast number of skilled tradesmen and small manufacturers, most of the latter of whom were upwardly mobile as they exploited the tacit skills and knowledge of their own earlier trade apprenticeships. There can be little doubt that the general and more formalised engineering skills of mechanical and civil engineers, mechanists and wrights created the advances in grain milling, steam motive-power, railway equipment, metal-cutting tools, metallic alloys and so on, whilst the brass founders, button manufacturers, Britannia-ware manufacturers, draftsmen, gunsmiths, japanners, tanners, stove-makers and silver operators devoted themselves to product improvements in their own fields.

The high technology that emerged from this crowded urban workshop milieu was clearly not any direct result of learned English university experimentation or of government interference or funding within Britain. Elite Britain had no impress on urban Birmingham.[12] We would claim the contrary. The camphor wars that would from this point escalate in Formosa were a far distant outcome of a culture of machinofacture that was absolutely inherent within the social processes of British industrialisation and which involved subtle informal linkages with many productive sites in Europe. Through such networks knowledge and intellect were brought to bear on technique in a myriad of ways. Of course, elite

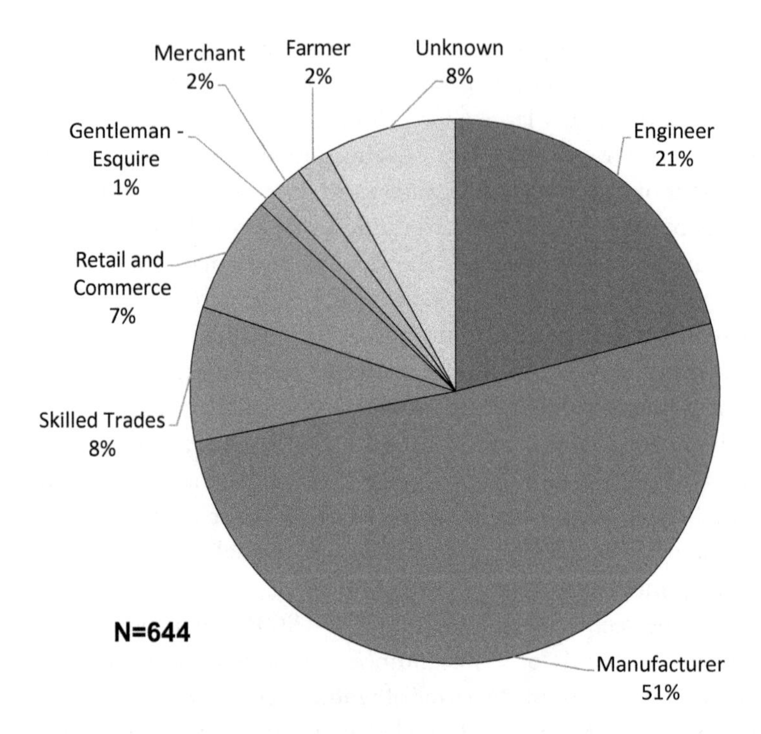

Fig. 3.1 The city of invention: occupations of patentees in Birmingham 1855–1870 (total for the years 1855, 1860, 1865 and 1870)

bureaucracy and colonial technocrats, experts in diplomacy, foreign languages and commercial stealth would soon enough take ownership of the expansion and application of such technologies during the following years, those of what is now often called the second industrial revolution.[13] As we shall see further, it was no coincidence that they were also the years of the New Imperialism.

One Birmingham inventor who was to know nothing of Taiwan was Alexander Parkes (1813–1890), who presented, on 12 December 1865, a lecture and exhibition to the members of the Royal Society of Arts in London concerning a series of objects made from plasticised cellulose nitrate utilising camphor as a key ingredient. Between 1841 and 1852 Parkes had successfully applied for several patents across the range of electro-depositing and metallic alloys, metal extraction and smelting, preparing gutta-percha and India-rubber solutions, and described himself variously as artist, experimental chemist, chemist and engineer.[14]

Inspection of trade directories shows that he was indeed moving upwards from the artistic use or decorative applications of new electrolytic processes to larger scale manufacturing as a metallurgist and chemist. From the perspective of world history, Parkes's inventive activity in Birmingham in and around 1865 was to have significant impact on the character of the second industrial revolution on the one hand and the social history of the small island of Formosa on the other.

Parkes styled himself both chemist and inventor during the long years in which he developed his practical experimental programme in Birmingham, issuing over eighty patents in the three main fields of electro-deposition of metals, desilverisation of lead and the series of varied inventions leading to the discovery and application of celluloid to a huge variety of industrial uses. After serving his articles as an art metal worker he entered and became head of the casting department of Elkington Mason and Company, a technological partnership focused on development of electroplating, and so at the heart of that Birmingham small company high-tech that was then directed mostly to design and art products.[15] During 1862–1865 he shifted to experiments on the application of camphor in the production of what was later to be called celluloid. This required the nitrating of cellulose fibres by soaking of cotton or wood or other cellulose sources in nitric acid with the addition of one of a number of feasible solvents.[16] From the 1830s European chemists had established most of the chemical framework.[17] In particular, C.F. Schönbein at the University of Basel reported to Michael Faraday at the Royal Institution in London in early 1846 that he had obtained cellulose nitrate [guncotton], this in turn encouraging T.J. Pelouze of Paris to return to his own earlier experiments on cotton and nitric acid, leading to his formulation of pyroxylin as a variety of cellulose.[18] In about 1847 or 1848 John Taylor, Schönbein's British patent agent, mentioned production of cellulose nitrate to Alexander Parkes and presumably directed him to Schönbein's British patent of 1846.

The link of this to Birmingham electroplating was through the increasing need for more efficient insulation, but an outstanding problem was that cellulose nitrate was dangerously inflammable, at times explosive. Parkes reported his experimental programme between 1855 and 1868 as made up of 'thousands of experiments' with plastic masses of nitrocellulose. In 1856 he had patented in England and France 'Parkesine', the first thermoplastic, a celluloid based upon nitrocellulose treated with a variety of solvents that could be moulded or pressed, turned, carved and rolled into sheets. On his own evidence given in 1878 he used camphor as a solvent and stabiliser from the outset, achieving real success in 1865.[19] In that year he presented to the Royal Society of Arts in London a huge range of objects made from plasticised cellulose nitrate using the new key ingredient—as both solvent and stabiliser—of camphor (Fig. 3.2).[20]

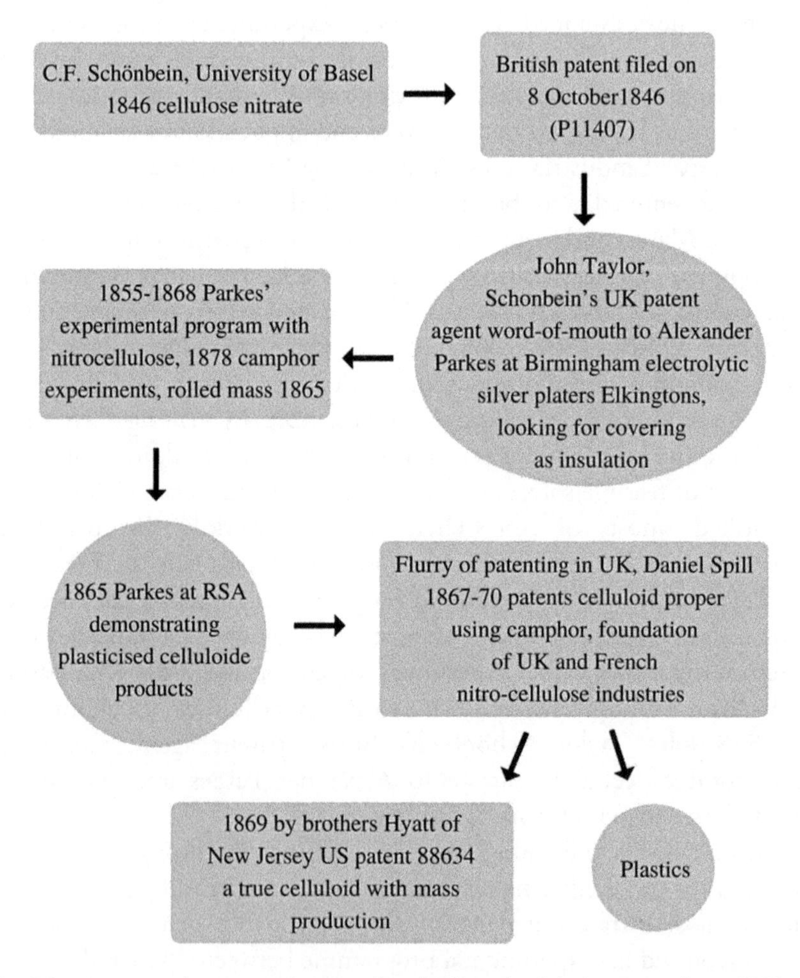

Fig. 3.2 An exemplary case: the camphor high-tech global pattern from 1846

Expert Networks and Technology Capture

Daniel Spill (1832–1887) was a factory general manager in his late twenties when he saw exhibits of Parkesine at the London Exhibition in 1862, the earliest public demonstration of the product capability of the new processes. He was not slow in coming forward, and wrote to Parkes enquiring if his brother's firm producing waterproofed fabrics, George Spill and Company, could use the patents under licence in their waterproofing processes. In 1864 Parkes travelled to Gloucestershire and entered into an agreement for the manufacture of Parkesine, which remained in force only until August 1866, when it was extended through to the new company, The Parkesine Company, with a subscribed capital of £9500. Alexander Parkes became its managing director and Daniel Spill its works manager, this lasting only until liquidation in 1868. Initially, in these early years Spill had no knowledge of nitrocellulose processing so he worked under Parkes's instructions throughout 1864–1868 and thereby received very specific tacit knowledge of how to manufacture. Very importantly, as works manager in the two Parkesine companies he was handling the bleached nitrocellulose, camphor, alcohol and other solvents that appeared in his own patents on xyloidine during 1867–1870, and that involved him in an extended patent litigation in the USA against the Celluloid Manufacturing Company and others during 1876–1888 that without doubt stimulated the wholesale global invention industry that developed around the new plastic materials. Much global patenting then emerged in order either to extend the basic applications or to circumnavigate possible challenges from aggressive patentees such as Spill.

The British xyloidine patents substituted alcohol as a solvent and developed compound solvents including camphor oil or 'liquid camphor', boiled or oxidised to produce soluble xyloidine (BPs 3984, 1868; 3102, 1869).[21] In particular Spill's British and US patents of 1869 and 1870 made a batch of claims for solvents, including those for alcohol and camphor, camphor and hydrocarbons, castor oil and camphor, carbon bisulphide and camphor as well as the bleaching of xyloidine 'after the removal of the nitration acids, with chloride of lime or a solution of chlorine, sulphurous acid or a soluble sulphite' (BPs 3102, 1869; 180, 1870:

USPs 97,454, 1869; 101,175, 1870), The Spill patents from 1868 have been identified as the invention of synthetic plastic, trademarked as xylonite, in effect a mixture of pyroxylin (a lower nitrated cellulose nitrate), alcohol and ether as a precursor to the similar plastic trademarked celluloid that emerged at around this time.

Spill's most important contribution was in bringing forward camphor as the principal solvent, but his early work was in effect a repatenting of several of Parkes's registered inventions, something that Parkes eventually addressed when he testified against his former manager during the very lengthy American legal proceedings.[22] When Daniel Spill brought a patent infringement lawsuit—ultimately unsuccessful—against John Wesley Hyatt, developer of celluloid in the USA in 1870, the judge ruled that it was in fact Parkes who was the true inventor owing to his original experiments.[23] As the USA became the centre for celluloid development, most new patents or court cases referred directly to the key Parkes patents as the true precursors.[24]

It was the inventive activity in the USA that created a camphor-based industrial regime. The British had not solved problems of volatility and the temporary character of many materials produced, so most success had been with minor trade products which together made an impact but disguised the future industrial potentials so proclaimed by Parkes; the American celluloid project created a composition which was at once decorative and hard at ordinary temperature but retaining indefinitely the property of plasticity at moderate heat. As Goldsmith summarised from his intimate knowledge of the chemistry, the industry and the persons concerned, although 'nitrocellulose, camphor and alcohol has been recognized as the essential ingredients of this composition, its correct mechanical and thermal treatment was not understood until Hyatt's inventions and years of experience had established the manufacture on a satisfactory basis'.[25] In 1869 John Wesley Hyatt (1837–1920) of New York developed the substance trademarked in the USA as celluloid, probably not independently of Parkes; a marked difference was that with his claim Hyatt won himself a $10,000 prize offered for a substitute for ivory in billiard balls, and later in 1888 developed thin films of celluloid as a replacement for glass as a base for plates in photography.[26] In a speech delivered much later, in 1914, J.W. Hyatt reported that he first learnt about the use of camphor when he was referred to 'some patents' which added a small amount of camphor to liquid solvents. When Hyatt added camphor to nitrocellulose with an alcoholic solvent this led him directly to celluloid.[27] But he, like several others, had little notion of the potential, many regarding the effects as temporary (USP2101, 1870). It was the focus on camphor as the solvent that was to lead to a whole range of celluloid-based products and industries, from photography to artificial silk to new materials of warfare.

Also important was a focus on plastic masses, as in Parkes's patents throughout the years 1843–1881, which as we have indicated ranged from vulcanised India-rubber to production of a compressed hydrate or 'fibre', and which to different degrees involved the use of camphor.[28] But in his 1856 patent he clearly anticipated the use of nitrocellulose films in photography either substituting for glass or layering the glass, in both cases with collodion. By 1865 it had become clear to him that production of large masses required volatile nitrobenzole or aniline as solvents, but that these could be rendered more suitable 'for use by the addition of camphor, by this means I obtain to some extent the same advantage as by the use of a less volatile solvent'. Although at that time in his English patent (1638 of 1854) J. Cutting had added camphor to ether and alcohol in the production of photographic collodion, Parkes's 1865 patent was the earliest to use nitrocellulose and camphor in the manufacture of a specifically plastic mass. In the same year, he used camphor for the production of electric telegraph insulation. Parkes repeated the same claims in a pamphlet for distribution at the Paris Exhibition of 1867, where he stated that a principal use of camphor was 'to assist in rolling sheets and in various mechanical arrangements, for the production of Parkesine in a cheap and more rapid manner'.

We should note that the decisive period for the rise of camphor as a crucial ingredient does seem to have been 1862–1872, moving from Parkes in Birmingham to Hyatt in Albany as the serious contenders, but that this was occurring within a very close international machinofacture environ. After the Journal of the Royal Society of Arts publication in 1865 (see above) of what was Parkesian celluloid, C.D. Abel and other British patent agents began releasing a clutch of related advances from other British investigators, involving the refining of camphor using temperatures up to

412 degrees Fahrenheit (BPs 2816, 1867; 1124, 1868). William McCartney was utilising camphor to dissolve India-rubber using alcohol to exhaust the surplus camphor, and a very large number of products under the titles of 'plastic compositions' as well as under 'India-rubber' began to appear in the patent literature, a major thrust being attempts to substitute for India-rubber with nitrocellulose compounds (Abridgements to Patents class 70, 1855–1896). Patentees throughout Britain were now using the term celluloid in their nitrocellulose advances, at times mixing pyroxylin with a great variety of substances using alcohol to thin, creating a huge number of products and embellishments, including insulation of telegraph wires. Attempts were being made to use nitrocellulose plastic compositions of semi-fluid collodion, with cotton and castor oil to manufacture textile-like fabrics at an early stage. (BPs 2775, 1860; 2675, 1864).

We have seen that amongst a host of new processes and structures, the second industrial revolution might be marked as strongly associated with plastics. Popularly, plastic is any material that may be moulded or shaped into different forms under pressure and/or heat. Chemically, plastics are polymers—substances composed of long chains of repeating molecules (monomers) made up predominantly of carbon and hydrogen atoms which, under the right conditions, join up into chain structures. What is very interesting from our account is how informed and systematic experiment stemming first from the industrial city of Birmingham then spread to other inventive centres, yet these early plastic pioneers could not yet be in a theoretical or instrumental position to fully understand the structure of their materials. More importantly, men such as Parkes clearly realised the basic working principles and the truly enormous potential technical and economic importance of their inventive work.

It is remarkable how many of the products and processes that were to be produced from camphor as a solvent were anticipated by Parkes prior to 1869. In 1856 the patents for collodion, waterproofing and its use in photography (BP1856, 1123: 1125) were covering a potentially huge industrial range. By 1865 he could add the production of electric telegraph conductors (BP1865, 2733), and in 1866 the manufacturing use of Parkesine for imitation of ivory and pearl (BP1866, 2709); followed by those in 1867 for coating metal tubes and 'rendering them suitable for ornamental purposes' and ornamenting paper, fabric and other surfaces for book binding (BP1867, 865, 1695); and in 1868 two for manufacture of varnishes and for billiard balls (BP1868, 1366, 1614). The work of Daniel Spill and his own continuing experimentation then led to new possibilities for Parkes in both product and process innovation. In 1879 the new nitrocellulose meant patents for billiard balls, combs and handles, varnishes, the coating and embellishing of surfaces, and in 1882 a composition of India-rubber for insulating electrical conductors (1879, BPs1865, 1866; 1882, BP5388) In 1884 a batch of patents covered compositions for the protection and insulation of wiring and compounds for coating surfaces (BPs for 1884; 10,054, 11,365, 12,280). Finally in 1885 Parkes completed his chain of institutional and technological logic with a patent for cartridge cases (BP1885, 6512). As his patent registration numbers indicate, Parkes was as heavily opportunistic as before, with visits to the patent office resulting in chains of near consecutive registrations as he presumably economised on the immediate costs of his travel and hotels to London.

The word 'celluloid' was coined by I.S. Hyatt in about 1868 or 1869 as a label for the substance that he and his brother J.W. Hyatt developed in Albany, New York and registered as a trademark in the US Patent Office on 14 January 1873.[29] As with Parkes and his customers before him, Hyatt had found that the substances he was experimenting with were dangerously flammable. He later wrote of his billiard balls that 'occasionally the violent contact of the balls would produce a mild explosion like a percussion gun-cap. We had a letter from a billiard saloon proprietor in Colorado, mentioning this fact and saying that he did not care so much about it but that instantly every man in the room pulled a gun.'[30] This is funny and entirely American, but also a good motive for further work. For Hyatt, a breakthrough came in about 1870 when he discovered that camphor was not merely an excellent plasticiser or a chemical that minimised shrinkage (generally Parkes's position) but made an optimum solvent for cellulose nitrate, producing a far more usable product. Stability was still a problem, this solved by Hyatt when he introduced ethyl alcohol as a solvent for camphor and dissolved the cellulose nitrate under pressure in the resulting solution.[31] It was this he christened celluloid, and it brought him the greatest commercial success of any of the early experimenters. Celluloid could more clearly and reliably be used to make decorative items such as dressing table sets, bags and low-cost collars and cuffs.

Although the term has now long replaced Spill's 'xylonite', in fact Hyatt's celluloid was all but identical to the collodion and Parkesine that Parkes was showing at the 1862 Exhibition. Writing as a contemporary celluloid expert, Goldsmith concluded that although not all authorities agreed 'that the Parkesine of 1862 is celluloid, yet the name cannot be denied to the Parkesine of 1865 which is fully specified as a rolled sheet prepared from nitrocellulose, wood naptha, castor oil, and camphor', which could be forced into rods or tubes or moulded into a grand variety of articles.[32] At the time that Goldsmith was writing, some eighty years ago, the Science Museum in South Kensington, London had mounted a permanent exhibition of the 1862–1865 Parkesine, which upon close observation Goldsmith argued should be accepted as 'celluloid' products. Certainly, the tubes and rods pressed out in the Parkes method were subjected to the same process as in the 1930s; and if not in strictly the terms of colloid physics, then in those of global industrial history and its products, the celluloid world does seem to have stemmed from the initial work of the artisans of Birmingham.[33]

It seems clear that in this period camphor was brought centrally into the world of the second industrial revolution by a complex and tortuous agency, at once highly individualised and dependent on key institutions, from the associations of European scientific investigators (themselves at times benefiting from pure accident),[34] to the patent networks and competitions of the USA. Standing well outside all this competitive humanity grew the noble camphor tree.

Indigenous Resistance. Camphor Wars in Formosa, 1860s–1914

The work of Parkes, competitively elaborated upon by an international band of patentees, had demonstrated that treating pyroxyline with camphor as the solvent yielded the greatest plasticity and stability, all other solvents being volatile. In the process, where pressure-rolling and steam-heating (to around 150 degrees Centigrade) yielded a basic paste, the product quality depended principally on the quantity of the camphor. By dry weight this was as high as one part camphor to two parts pyroxyline, where increased camphor afforded increased plasticity. It was this technical relationship that quite quickly raised the price of camphor on world markets, led by the demands of the fast-industrialising nations across several industries.

Until this technological change camphor had been a valuable and beautiful wood used for special ceremonies, perfumes and in parts of the world as an incense, preservative, or medicine.[35] By the 1920s, 80% of the vastly increased world supply of camphor was consumed by celluloid and film manufacturers, much of the remaining proportion for gunpowder or medicines. Celluloid was a major input to the manufacture of essential industries of the second industrial revolution, spreading initially from photography, paints and varnishes to mouldings and coverings and art products, electrical insulations, cementing for metals, artificial silk and electric light fittings, as well as providing a substitute for pearl, ivory and marble in the new middle-class art nouveau and crafts products of the industrial and capital cities.

The globally untested assumption was that the only worthwhile supply of camphor lay in the north-eastern region of Taiwan, as illustrated in the accompanying map. Until the early twentieth century Taiwan supplied between 80% and 90% of the world demand for camphor. Even prior to the colonisation of Taiwan in 1895, commercial expansion—especially that of Britain—ensured that the island was already linked to the Atlantic world through its exports of tea and sugar, as well as its growing import of opium, obtained both through the local junk trade via China as well as through British and other foreign suppliers. During the 1880s interest began to focus increasingly on the rich camphor supplies of Taiwan. The map shows clearly that the camphor trees were entirely located within the traditional lands of the Tayal people, one of several groups of indigenous peoples who in relatively small numbers thrived in the mountainous and forested regions, occupying over 50% of the total area of the island.[36] The much more populous Chinese settlers dominated the agriculture, commerce, industry and communications of the Western part of the island, in which Westerners (the British from the beginning of the 1860s) were settled as traders, missionaries, consul officials and adventurers of all kinds. The identification of Formosa as properly under Chinese governance tended to occur within the Empire only when danger loomed, as in the 1870s with the initial challenge of the Japanese, the 1880s with the wars against the French and in the 1890s with the full-scale war against Japan that resulted in its colonisation in 1895. Even then, China tended to see Taiwan only as its Western half, seldom pretending any sovereign control over the indigenous peoples of the East (Fig. 3.3).

Although the Formosan indigenous were small in number, they were large in fighting capability, and by the 1880s had secured through violence and frontier trading a supply of modern arms that amounted to little less than a veritable military revolution.[37] All the evidence is that by the time of the escalation in demand for camphor, around the 1870s or early 1880s, the Tayal people in particular, with strong hunting and head-hunting traditions, could and would resist forest encroachment. Earlier moves of the Chinese settlers eastwards, over a 200 year period, had been periodically repelled at great cost of life; guns were used more effectively by the indigenous than by the Chinese and later the Japanese. Hitherto, Western interests had honed in on the commercial and trading western portion of the island: the mountains were difficult, the eastern sea cliffs were amongst the most formidable in the world, the coastal seas were (and remain) violent and unpredictable. This was a setting for a sudden acceleration of violence directly linked to the technological imperatives of the second industrial revolution.

Late in 1863 a novelty of Western interest in the forests stimulated the Chinese imperial authorities to declare that the mountain forests, that is the camphor areas, were to be henceforth the property of the navy and that only government officials (not Chinese civilian settlers) could deal commercially in such regions. Camphor was now a monopoly in the hands of Chinese officialdom, and this prompted the British in 1868 to put direct naval pressure on the Chinese, with a brief skirmish (at times referred to as the Camphor War) resulting in the removal of the new regulations, whereby camphor prices fell, and in the hands of British and other foreign firms exports to Europe boomed.[38] With native Chinese forest producers selling on the free market there was an approximately twenty-five-fold increase in the volume of camphor exports in just the one year 1867–1868, with the local forest price halving.[39] But the longer-term

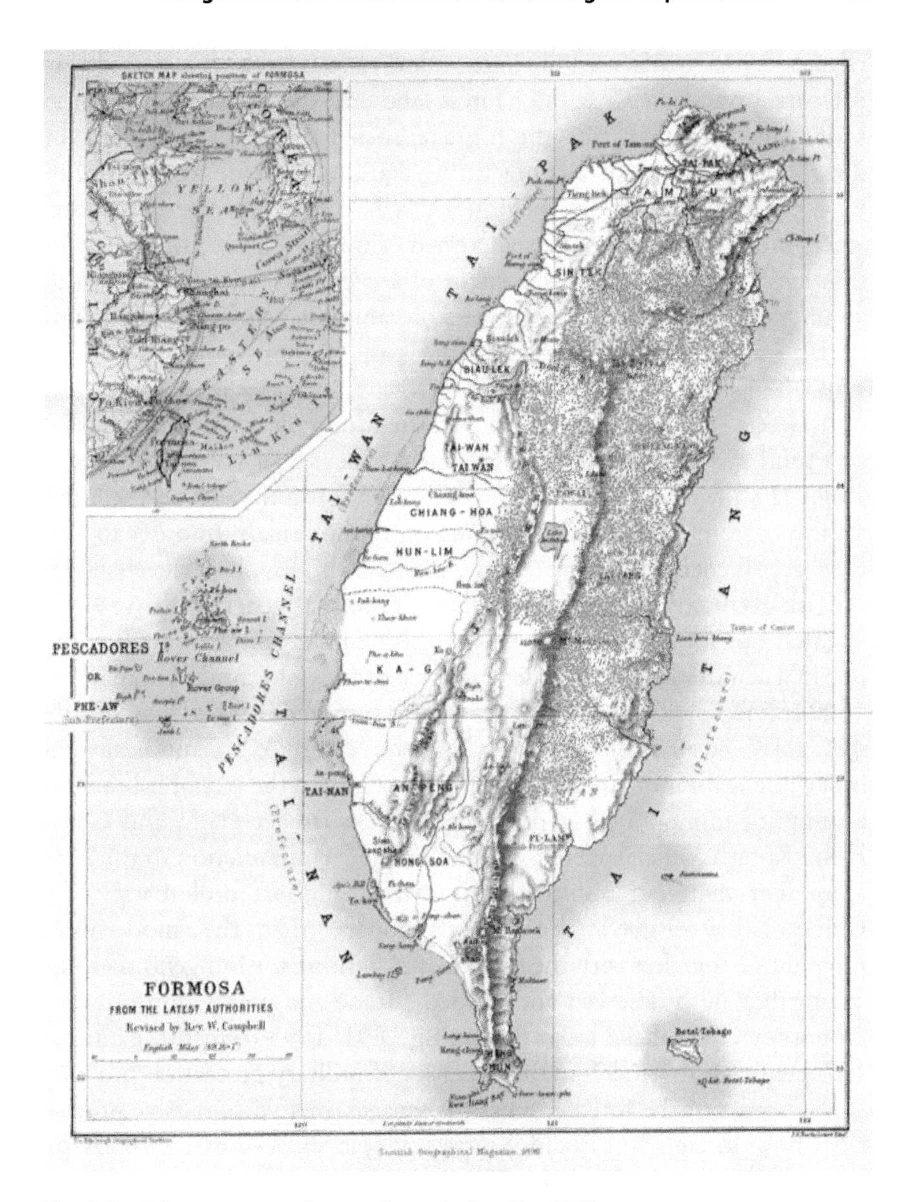

Fig. 3.3 Taiwanese camphor ecology during the 1880s

result was much-increased pressure on indigenous lands, compounded by the corrupt and illegal sales of forest land and the destruction of entire trees, which resulted from the mistaken belief that the crystalline camphor needed for global manufacturing could only be obtained by chipping of entire trunks in the forest itself. The result was the speedy emergence of forest Taiwan as an armed camp. By 1875, 'savage' opposition had stopped trade to the extent of a civil war in Formosa, so from around this time the global output of camphor was becoming conditioned—perhaps determined—by the cost of overcoming the 'savage wars'.

Between 1885 and 1891 the government monopoly was reestablished as a fund-raiser for the Chinese to meet expenses of the French war of 1884–1885, with officials now buying camphor in the east and selling it on the western coast at around three times the price.[40] In order to allay the direct French naval threat on the island itself, the Chinese were taking the risk of upsetting Western commercial interests generally. In 1887 the talented but somewhat extreme Liu Ming-Chuan was appointed governor of Formosa as a now independent province of China. He marched 1500 trained men against the tribes, with hundreds of lives lost, but the same forces ensured that the camphor wars continued spasmodically and broke out again more fully in 1891, this leading to the reestablishment of a camphor monopoly, but now with Reuter, Brockelmann and Co. of Hong Kong as official agents, this in complete contradiction to the 1869 agreement that had resulted from British gunboat diplomacy.[41] The Chinese now argued, reasonably enough, that the modernising programme together with the huge cost of fighting the indigenous groups meant they both deserved and needed a trade and production camphor monopoly. The official attempts during 1891–1895 to institute a tax of 13.5 cents per lb 'to defray the expenses of military protection provided at the border of the forest against the savages' met with massive protests from both foreign and island interests. It was replaced by an $8 tax per stove erected together with a 0.5 cent tax per lb transported from the forest.[42] Buyers of camphor were then compelled to buy at the ports, not in the forest. It is clear that violence together with confusions of changing regimes hampered the supply of camphor to global industry at a time

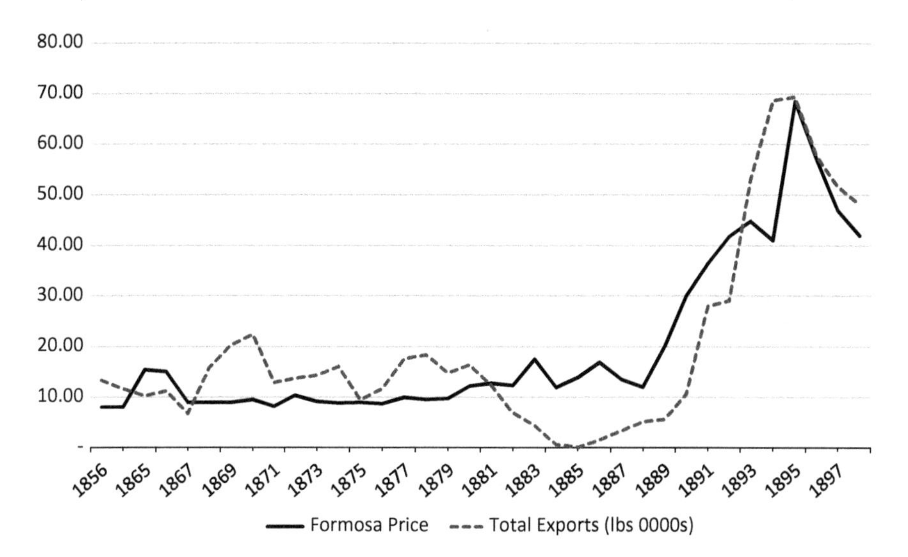

Fig. 3.4 Camphor prices and volume of production (Formosa price at Tamsui per picul [133 lbs] local currency)

when Taiwan had a virtual supply monopoly, and raised prices further through taxes, insecurity and other costs.

Figure 3.4 shows the impact of all this on camphor prices and volume of production between the 1850s and 1890s.[43] Prior to the new celluloid technologies both prices and volume fluctuated around a low level. From the mid-1880s both rose rapidly in roughly the same manner, the increase in production being matched by even larger increases in demand internationally and constrained by the high costs, delays and dangers of camphor forest expropriation.

With ethnically based civil unrest and warfare constraining supply, the entry of Japanese colonisation in 1895 politicised and internationalised the situation immediately. Indigenous groups had joined with bandits and with renegade Chinese soldiers after the defeat of the brief resistance of the Taiwanese Republican movement of 1895, which had been abetted by the imperial Chinese authorities in the face of the Treaty of Shimonoseki that had supposedly closed the Sino-Japanese War.[44] Determined to make their first colony pay its way, the Japanese had multiple reasons both to insist on civil obedience and to raise funds through monopolies over basic staple products, in particular camphor, tea, sugar and rice. The Japanese saw Taiwan as a test of the superiority of Japanese industrialising civilization over the moribund and fading culture of the Chinese empire. Where the Chinese had failed to govern their distant island, and where its Chinese settlers had equally been culpable in both mistreating the indigenous minorities of the island whilst failing to control the resulting unrest and warfare, the Japanese would firmly stamp clear marks of progress, stability and commerce. At least that was the plan.

The Japanese had shown more than passing interest in Taiwanese camphor even before they colonised the island. In the previous year, 1894, Japanese camphor-men set up a small community in Taiwan near Taiko to introduce new technology in the forest itself, using newly designed stoves for crystallisation of camphor chips at a time when the sheer demand in Europe had raised prices to $50 gold per cwt.[45] But the Japanese forces had little more control over frontier events than the Chinese: in 1898 'savages' attacked camphor men over 300 times during the year, killing 635 camphor workers and Japanese soldiers according to official Japanese figures. At that point, the Japanese were using some 2000 armed Chinese to protect border camphor workers. During 1900, indigenous groups attacked the camphor stills and cottages of camphor workers, with several hundred killed and more than 1000 driven from the camphor areas, captured the wealthier camphor manufacturers and demanded ransoms for their return. At the peak period of conflict, 1898–1901, deaths were averaging around 650 per annum, another high in 1907 producing a further 630 deaths.[46] From 1900 the guard-line that separated indigenous camphor areas from Chinese settlement was continually extended, fortified with Japanese guards and Chinese (and even indigenous) police, electrified and overlooked with modern field guns, the combination of which hardly altered the degree of resistance.[47] The rising costs of camphor extraction and processing to the Japanese government may well have made the enterprise less profitable, but two global forces counteracted this: the Japanese were determined to make colonies pay in terms of staple exploitation (as seen clearly in Hokkaido, Korea/Chosen and Manchuria/Manchukuo from the 1890s to the 1930s), and individual local and Western producers and traders could obtain a very high rate of return from such high-value product. Thus, the case became exemplary of the distinction between the return on colonialism for colonial governments (and their taxpayers), which could be modest, and the return on colonialism for individual traders, investors and projectors, which could be enormous given that most costs of transaction and governance were paid for by colonial authority and by taxpayers in the colonising nations rather than by individual enterprise.[48]

Technology and the Window of Destruction

Technology both opened and closed the window of destruction. The case abounds in sad ironies. The technologically determined global rush to camphor was met by the Japanese government's variable attempts into the 1920s to fix a monopoly price for camphor by starving the global market through dictating the price paid to producers by government, controlling the quantities of camphor produced per year or the quantities of camphor allotted to foreign sale, as against supplying the growing chemical manufacturing demands of Japan proper.[49] But in the meantime the chemical technology continued to set new parameters for commerce.

Three organisational changes began to close the window of destruction in Taiwan. First, new technologies of the forest, especially, allowed the obtaining of camphor without destroying the tree.[50] Felling and chipping were replaced by obtaining rich supplies from leaves and from bark-stripping, this representing directly a technological/ecological formulation to solve an ecological problem.[51] The camphor of Formosa could now be derived from the whole tree not just the trunk, this allowing coppicing and other protective measures. The Taiwan frontier was relieved also by Japanese reforestation activity both in Taiwan and in Japan, as well as through their own research programmes on camphor forestry. Secondly, there was a global move for the acclimatisation of camphor elsewhere. The earliest example, introducing camphor into Ceylon via the British Royal Botanic Gardens in 1852, was now emulated wherever possible, and from the beginning of the 1900s Japan was securing major supplies from Tosa.[52] Thirdly, technicians were increasingly saving on the use of camphor in the overall celluloid process, by making it generally more efficient (e.g. in the steeping process).

Three high-tech chemical advances further closed the window of destruction, opening opportunities elsewhere for profit across a range of products and industries. The USA was foremost in developing artificial camphor using turpentine (oxidation). Again, especially in Germany, camphor substitutes were formulated using a range of esters and naphthyls and products of petroleum distillation especially in the early 1900s. Finally, celluloid substitutes for imitating natural products such as horn and wood emerged with viscose and bakelite, especially led by innovation in France and Britain. Patents again captured most of such technological advance. This was especially true for developments in artificial camphor using terpenes, with major patents on camphoric ethers in 1898–1900 lodged from Germany and Britain.[53] Patents for camphor substitutes as an offshoot of petroleum distillation were initiated in the USA, followed by Germany.[54]

Conclusions

On the small island of Taiwan, as Western commercial invasions transformed into Japanese colonialism during the 1890s, high-tech advancements in new product areas provided the leading parameters of a new social, cultural and economic history. Table 3.1 shows how Taiwanese trade reflected the mix of traditions and modernities.

Despite the violence on the frontier of exchange, camphor exports under the Japanese were of strategic commercial importance, paying for

Table 3.1 Under the Japanese: camphor exports and opium imports 1896–1904 (Japanese sources, 000s yen)

Year / Product	1896	1897	1898	1899	1900	1901	1902	1903	1904
		War peak	War	War	War	War peak			
Camphor export	2248	1329	962	1733	1386	789	2849	2518	2199
Opium import	1165	1570	2044	2776	3393	2310	1477	1121	2866

the continued imports of opium that Taiwan inherited from British aggression in China via the Japanese after 1895.[55] Politically they demonstrated to other powers (rather poorly as it turned out) the command that Japan had over its new colony and linked that to new technological imperatives central to the flourishing of a second industrial revolution.[56] Colonialising modernity penetrated beyond hilltops into the mountain and forest 'fastnesses' eastwards, where tea and sugar did not go. In particular the pressure on the eastern lands of the Tayal politicised a resistance that had escalated during the brief Republican period in 1895, scaling up anarchic frontier tensions between dissidents, indigenous settlements, the Japanese colonists and the foreign traders and commercial agents. The increased demand for camphor, stemming from technological innovations of the Western world, initiated a 'colonial' problem for the Japanese in 1895 that had not arisen with their previous colonization of Hokkaido—how to exploit empire for basic raw materials without losing control on their own constructed 'frontiers of civilization'.

Notes

1. This chapter owes everything to my employment and my colleagues and friends in Taiwan since 1994, when I first took up a visiting Professorship at Foguang University in the graduate Institute of European Studies, Dalin, Chiayi. My subsequent research on several aspects of Taiwanese history was encouraged both by students and colleagues there and in the Department of International Studies, Wenzao Ursuline University of Languages, Kaohsiung, and by my doctoral students who have since become colleagues, especially Liu Chun-Yu Jerry 劉俊裕, and Pei-Hsi Lin Susan 林姵希. Earlier approaches to this subject were delivered as 17 November 2012, 'High Tech Europe and the Formosan Civilization Wars circa 1865–1900. A Study in Global Connectivity', *School of Oriental and African Studies*, University of London. November 2012, and as the Plenary Public Lecture 'Global Commodities as Connectivities. High Tech Chemistry in Birmingham and the Camphor Wars in Formosa, circa 1861–1914', *Cain Conference, Chemical Heritage Foundation*, Philadelphia, April 2014.

2. For further information see Inkster, I.: *Science and Technology in History, An Approach to Industrialisation*, London and Rutgers: Macmillan, 1991, and Princeton University Press, 1992, pp. 167–183.
3. This needs explanation. Especially in areas of recent white settlement, the subsequent years of rapid agricultural, commercial and (at times) industrial development were primarily centred in the émigré economy itself. It was the settlers that were counted. The seeming high rates of economic growth, and especially of per capita income growth, were predicated on an accounting process that basically excluded all indigenous elements. The growth centres in areas such as Australia or Canada would have had little or no impact on the measures of 'economic development' of India, China or Indonesia, areas of huge and dense population. Reversing this, if we measured the performance of China or India in terms only of their commercial and export/processing centres—often places crammed by and by with expatriate Western engineers and other 'experts'—then their 'growth' rates would have been amongst the highest in the world. As a final vision: what would have been the positive impact on the measured growth rate of India if Birmingham had been located there?
4. For the general and critical analysis of the earlier period of industrial enlightenment see Mokyr, J.: *The Enlightened Economy. Britain and the Industrial Revolution*, London: Penguin, 2011; For artisanal skills see Inkster, I.: 'Finding Artisans. British and International Patterns of Technological Innovation 1790–1914', *Cahiers d'Histoire et de Philosophie des Sciences*, No. 52, (2004), pp. 69–92, and the other papers in that volume by Pfister, Mottu-Weber, Soler, Chuan-Hui, and Woronoff.
5. A term used widely enough at that time, associated with both Marxist critique and engineering expertise, the establishment of mills, workshops and factories dominated by machinery and power based on metal and chemical manipulations and engineering. It is used here to focus analysis on the manufactures that were dominant in the world capitalist economy prior to the 1970s, strongly emerging in a small core from the 1830s. Most smartly and finely depicted by Joseph Whitworth in his *Miscellaneous Papers on Mechanical Subjects,* of 1858: the notion provides a backbone to my forthcoming work, *Prometheus Chained. Technology and the Victory of the Western World circa 1400–2017.*
6. In the years 1839–1842 and 1857–1860, whether the cause was primarily related to Britain's need to pursue imperial importing of opium into China in order to exchange for the great variety of high-value products imported into Britain from China is not of central concern here. Of

more importance is the resulting devastation of the Chinese economy, which had grown throughout the years of Europe's early economic expansion—fourfold between 1600 and 1820—but then ceased growth until 1913, falling more drastically by 1950; fuller recovery awaited the late 1970s. See Madison, A.: *Chinese Economic Performance in the Long Run, 960–2030*, Paris: OECD, 2007.

7. From the Meiji Restoration of 1868 Japan had benefited enormously from trading into the echoes of the first industrial revolution whilst also gaining industrially and militarily from the substantial technological advances of the second. Whilst several of the industrial Great Powers were apprehensive (and Tsarist Russia rightfully fearful), the commercial giant of the East, Great Britain, readily enough saw an openly trading Japan as a far better commercial ally than a disintegrating Chinese imperial system. See Inkster, I.: *Japanese Industrialisation. Historical and Cultural Perspectives,* London: Routledge, 2001, especially chapters 4, 7 and 11.
8. For an excellent most recent treatment on a large scale of military and civil technologies in world history see Andrade, T.: *The Gunpowder Age. China, Military Innovation and the Rise of the West in World History*, Princeton and Oxford: Princeton University Press, 2016, especially chapters 14–18. For earlier treatments see Black, J.: *War and the World: Military Power and the Fate of Continents 1450–2000*, New Haven: Yale University Press, 1998, and his 'Conclusion. Global military history, the Chinese dimension', in van de Ven, H. (ed.): *Warfare in Chinese History*, Leiden: Brill, 2000, pp. 428–42; and Headrick, Daniel R.: *The Tools of Empire, Technology and European Imperialism in the Nineteenth Century*, New York and Oxford: Oxford University Press, 1981. Mostly such work deals with the impact of overtly military, transport and communications technologies on major civilisations or colonised nations, rather than more insidious industrial process technologies at the indigenous frontier sites of such places. For any systematic treatment of the latter we must turn to the anthropologists and sociologists of technology; see as an excellent early treatment Hill, S.: *The Tragedy of Technology: Human Liberation versus Domination in the Late Twentieth Century*, London: Pluto, 1988.
9. Newton, W.: 'Letters and Suggestions Upon the Amendment of the Laws Relative to Patents for Invention', London, 1835, quote pp. 78–9; *Report of Select Committee on the Law Relative to Patents for Invention,* London, 1835, p. 132.

10. Hopkins, E.: *The Rise of the Manufacturing Town. Birmingham and the Industrial Revolution*, Stroud: Sutton Publishing, chapters 2–4. For the inventive context see Prosser, R.B.: *Birmingham Inventors and Inventions*, Birmingham, 1881; Berg, M.: 'Commerce and creativity in eighteenth century Birmingham' in Berg, M. (ed.): *Markets and Manufactures in Eighteenth Century England,* London: Routledge, 1991, pp. 173–204.
11. Patent data prior to 1855 in this chapter are calculations from the work compiled in the 1850s by patent officer Bennet Woodcroft, *Titles of Patents for Invention Chronologically Arranged 1617–1852*, London, 1854. Material from then to 1914, including Fig. 3.1, is derived from the original patent applications, and annual data includes a count of every patent applied for. All other patent data unless mentioned is from *Commissioner of Patents Journal* annually from 1852 to 1883.
12. Neglect of this simplicity was the initial and fatal mistake at the heart of Weiner, Martin J.: *English Culture and the Decline of the Industrial Spirit 1850–1980*, Cambridge: Cambridge University Press, 1981. For rebuff see Inkster, I.: 'Machinofacture and Technical Change: The Patent Evidence', in Inkster, I. et al. (eds.): *The Golden Age. Essays in British Social and Economic History 1850–1870*, Aldershot and Brookfield: Ashgate, 2000, pp. 121–142; For the earlier approach that is far more industrial and technological see Levine, A.L.: *Industrial Retardation in Britain 1880–1914*, New York: Basic Books, 1967.
13. Inkster, I.: 'Technology and Culture in the First Climacteric circa 1870–1914 and Beyond', *International History Review*, XXXI, 2 (June 2009), pp. 356–364.
14. MS Typed, J.N. Goldsmith, 'Alexander Parkes, Parkesine, Xylonite and Celluloid', London, 1934, (British Library 8233 d 6), especially p. 28; British Patent (henceforth BP) 313, 1865; Alexander Parkes, *Parkes' Evidence for the Defendant in Spill v The Celluloid Manufacturing Company*, New York, 1878, p.164.
15. Elkington and Co., *On the Application of Electro-Metallurgy to the Arts,* London: J. King, 1844; Watt, A.: *Electro-Deposition. A Practical Treatise, and Chapters on Electro-Metallurgy*, London: Crosby Lockwood and Co., 1886.
16. Masselon, Roberts and Cillard, *Celluloid, Its Manufacture, Applications and Substitutes*, London: Chas Griffin and Co., 1912.
17. Using nitric acid to convert cellulose into cellulose nitrate and water: $3HNO_3 + C_6H_{10}O_5 \rightarrow C_6H_7(NO_2)_3O_5 + 3H_2O$.

18. In summary, early euro-scientific experiments 1832–1846 were on nitric acid on cotton, paper or wood fibres, at Nancy, Paris, Basle, Frankfurt and Brunswick, this producing guncotton (also known as nitrocellulose or cellulose nitrate), at this stage highly unstable prior to the use of camphor.
19. Barnes, Robert F.: *The Dry Collodion Process*, London: G. Knight, 1856; Goldsmith, op.cit. 'Alexander Parkes', pp. 5–30.
20. In this presentation, he had the essentials: realistic plasticising required reducing the flammability of $C_6H_7(NO_2)_3O_5$ and the evaporative quality of solvents used to dissolve it, thus camphor. So Parkes was probably most important in producing patent recipes which saw camphor as a solvent, a stabiliser and as a plasticiser.
21. Henceforth individual patent references are sourced directly in this form: BP as British patent, USP as US patent and so on, with year and number given as archive references.
22. *Official Gazette of the US Patent Office*, Washington, Government Patent Office, 1872, vol. 2. July–December 1872.
23. *Annual report of the US Commissioner of Patents for 1869* vols. 1 and 2, Washington: GPO, 1871.
24. Bockmann, F.: *Celluloid. Its Raw Materials, Manufacture, Properties and Uses*, London: Scott and Greenwood, 1921, 2nd English edition trans. H. B. Stocks [1st ed. 1907].
25. Goldsmith, 'Alexander Parkes', p. 1.
26. In a 1870 patent, Hyatt describe the effect of camphor as temporary, the solidity of the substance from the mould only being improved 'by evaporation of the camphor until it becomes hard and tough like horn ... before the camphor has evaporated the material is easily softened by heat' (US2101 of 1870). The term Celluloid was coined during 1868–1869 by I.S. Hyatt for the substance he and his brother made in Albany and registered on 14 January 1873 [US Patent 133,229, 1872], but this was described by Parkes in 1862 in a leaflet describing his exhibit at the International Exhibition, London, class 4 as Parkesine.
27. *Journal of the Society of Chemical Industry* (1914), p. 227.
28. Goldsmith, 'Alexander Parkes', p. 12.
29. Schuetzenberger, Paul: *Technology of Cellulose Esters, A Theoretical and Practical Treatise*, New York: Van Nostrand, 1916; Goldsmith, 'Alexander Parkes', op.cit., pp. 17–26.
30. The production process itself was far more dangerous. At the outset the factory of the first British licensees of Schônbein, John Hall and Sons of

Faversham, blew up when using new material in 1847, and similar explosions occurred in France, Germany and Russia in the following years. G.E. Kahlbaum(ed.): *Letters of J. J. Berzelius and C.F. Schonbein* (transl. F.V. Darbishire), London: Williams and Norgate, 1900. The guided serendipity is fascinating in this early development. While he was experimenting on ozone—which he discovered—Schônbein found he could 'convert ordinary unsized paper into a substance having the following properties', coherence, resistance to most chemicals, firm and tough, especially as an improved paper (primarily that is waterproof) and strong, good for printing, packing and wallpaper (letter of 5 March 1846, pp. 81–85).

31. 'Landmarks of the Plastics Industry', London: Imperial Chemical Industries Ltd., 1962, p 10.
32. Goldsmith, 'Alexander Parkes', p 10.
33. For the detailed science–technology links see originally Worden, J.: *Nitrocellulose industry*, Vols. 1 and 2, New York: Van Nostrand, 1911; Sproxton, F.: 'The Celluloid Industry', *Journal of the Society of Chemical Industry*, 39, October 1920, pp. 351–368; Most recently, Reilly, Julie A.: 'Celluloid Objects: Their chemistry and preservation', *Journal of the American Institute of Conservation,* 1991, Volume 30, Number 2, 145 to 162.
34. Thus Schônbein's 1846 discovery was accidental but led to his flurry of entrepreneurial activity secured by his British patent filed on 8 October 1846, followed by British licence agreements in the next year.
35. In Britain at this time perhaps best known were Dr Fletcher's Quinine and Camphor Pills, and there was nothing at all that they could not cure! For an early advertisement that made its claims for many years see *Nottingham Journal* (11 April 1836)
36. For the indigenous people of Taiwan generally at this time, see Stainton, Michael: 'The Politics of Taiwanese Aboriginal Origins' in Rubinstein (ed.), *Taiwan: A New History,* op. cit., pp. 27–44, and the doctoral thesis by Susan Lin below (see endnote 37). For the Tayal it should be noted that by 1909 their population had dropped to 28,242 on official estimates but had risen to 87,794 by 1981 and then fallen to 74,321 (even including the 20,000 or more Truku in their number); these figures do seem unlikely so either census takers were too casual, tribal definitions/locations altered or individuals moved back and forth between the east and west of the island for a variety of reasons, this causing confused counts. See also, *Preliminary Statistical Analysis Report of 2000, Population*

and Housing Census, E., Taipei: Directorate General of Budget, Accounting and Statistics, Executive Yuan, R.O.C., National Statistics, 2000; Li, Jen-kuei, Cheng-hwa Tsang Tsang, Ying-kuei et al. (eds.): *Austronesian Studies Relating to Taiwan*, Taipei: Institute of History and Philology, Academia Sinica, 1995.

37. For full details of the military resistance of the indigenous peoples of Taiwan and a very valuable analysis of the way in which they embedded modern guns within their traditional values and cultural patterns, see Pei Linm, Susan 林姵希: *Firearms, Technology and Culture: Resistance of Taiwanese Indigenes to Chinese, European and Japanese Encroachment in a Global Context, circa 1860–1914*, PhD Thesis in History, Nottingham Trent University, March 2016. For an early Japanese appreciation of the military and cultural strength of the indigenous peoples see Shinji, Ishii: 'The Life of the Mountain People in Formosa', *Folklore*, 28, no. 2 1917, 115–132; 'The Silent War in Formosa', *The Asiatic Quarterly Review*, 2, October 1913, 77–92. For an introduction to the context and anthropology see Inkster, Ian: 'Anthropologies of Enthusiasm: Charlotte Salwey, Shinji Ishii, and Japanese Colonialism in Formosa circa 1913–1917', *Taiwan Journal of Anthropology*, 9, No.1 (June, 2011), pp.1–32.
38. The first two traders in Taiwan were Jardine Matheson and Company and Dent and Company, Hong Kong, British firms, who opened offices in 1860; see Grajdanzev, Andrew: *Formosa Today. An Analysis of the Economic Development and Strategic Importance of Japan's Tropical Colony*, New York: Institute of Pacific Relations, 1942.
39. For data on camphor production and prices see Davidson, James Wheeler: *The Island of Formosa, Past and Present: History, People, Resources, and Commercial Prospects. Tea, Camphor, Sugar, Gold, Coal, Sulphur, Economical Plants, and Other Productions*, London: Macmillan, 1903. The large appendix to this volume calculates prices, volumes and exports for camphor and is based on consular trade reporting—checked against the original sources the material is found to be accurate. See also Meskill, J.M: *A Chinese pioneer family. The Lins of Wu-Feng, Taiwan 1729–1895*, New Jersey: Princeton University Press, 1979; Pao-San Ho, Samuel: 'Colonialism and Development: Korea, Taiwan and Kwantung', in Myers, R.H. and Peattie, M.R. (eds.): *The Japanese Colonial Empire 1895–1945*, Princeton University Press, Princeton New Jersey 1984, pp. 347–398; Allee, A.A.: *Law and Local society in Late Imperial China: Tan-Shui Subprefecture and Hsin-Chu County, Taiwan, 1840–1895*, Unpublished PhD Thesis, University of Pennsylvania. 1994, especially pp. 190–240, on tea and camphor in the export economy.

40. Lin Manhong 林滿紅. Cha, Tang, Zhangnao Ye Yu Taiwan Zhi Shehui Jingji Bianqian 茶、糖、 樟腦業與臺灣之社會經濟變遷 (1860–1895) [The Tea, Sugar and Camphor Industries and Socio-Economic Change in Taiwan]. Taipei: Lianjing 聯經, 1997; Tavares, Antonio C.: *Crystals from the Savages Forest Imperialism and Capitalism in the Taiwan Camphor Industry 1800–1945*, PhD Thesis, Princeton University, 2004.
41. See Gardella, Robert: 'From Treaty Ports to Provincial Status, 1860–1894' in Murray A. Rubinstein (ed.): *Taiwan: A New History*, Armonk, New York: M.E. Sharpe, 2007.
42. Davidson, *The Island of Formosa,* op.cit, unpaginated appendix.
43. The sources for this figure are in accordance with data in endnote 39 as well as my calculations from the annual data of *The Japan Year Book*, 1906, pp. 526–531, and onwards, and are based on a kin weighing just over 1 lb or 600 grammes, the yen at around 2s in sterling.
44. Eskidsen, Robert (ed.): *Foreign Adventurers and the Aborigines of Southern Taiwan 1867–1874. Western Sources Related to Japan's Expedition to Taiwan*, Taipei: Institute of Taiwan History, Academia Sinica, 2005.
45. News of the Japanese invasion of Taiwan stimulated speculation over increasing prices, and in particular Colonel North, the speculator, established a London syndicate to buy up all supplies, prices jumping from 20 cents to 67 cents per lb., within days.
46. The death/wounded ratios of Japanese fighting on the Formosan frontier during 1896–1909 was around 7/3 according to Japanese data, the official ratio for casualties in the First World War for all sides was around 3/7, so the combination of camphor and colonialism was a particularly dangerous and costly one. May–August 1907 may have been the worst period, when a great advance of the Japanese frontier guard-line was attempted in the midst of camphor regions and insurgent activity. There ensued 107 days of fighting with large casualties. A precise Japanese estimate of 126,628 yen was required in construction of 277 guardhouses, casualties including ninety-three Chinese 'coolies', mostly unarmed servants of Japanese on the frontier, including some police. Calculations have been made from data in (Japanese) Government of Formosa, 'Report on the Control of the Aborigines in Formosa', Formosa: Bureau of Aboriginal Affairs, Taihoku (1911).
47. Compare this with conditions and instances of resistance—both physical and cultural—in Hill, *The Tragedy of Technology*, op.cit, and see also Berger, Peter: 'Four Faces of Global Culture', *National Interest*, 49 (1997), pp. 23–29 and Berger, Peter and Huntington, Samuel P.: *Many*

Globalizations: Cultural Diversity in the Contemporary World, Oxford: Oxford University Press, 2002.

48. Studies of cost and return on colonialism are usually marred by failing to distinguish between the costs to the colonising nation at large as against the returns to a small group of capitalists within such nations—colonialism represents a redistribution of incomes within the colonising nation—the costs of colonialism being paid by all taxpayers, the returns to colonialism being principally in the hands of commercial and industrial capitalists as well as some technocratic workers and experts. This is perhaps what explains colonialism continuing in the face of its supposed low return/cost ratios.
49. In 1913 exports from Japan of crude camphor were restricted to celluloid manufacture, this being applied to Formosa in 1918, effectively prohibiting exports for refining. By the 1920s, monopoly pricing of camphor by Japan was made ineffective through technological change and by increased world supplies of German camphor. By 1931 there was free competition between synthetic and natural camphor on world markets.
50. Mochizuki, Kotaro (ed.): *Japan Today. A Souvenir of the Anglo-Japanese Exhibition held in London 1910,* Tokyo: Liberal News Agency, 1910 and printed by Methodist Publishing House, Tokyo, pp. 160–65. In Japan private companies such as Japan Celluloid and Artificial Silk Co., in Aboshi, Hyogo, were hiring German technical experts in attempts to save the use of camphor. See also, Japan Monopoly Corporation, *Abstract of researches on Tobacco, Salt and Camphor,* Tokyo: JMC, 1958.
51. More recent investigations, that is, *c.* 1939, have proved that the leaves of the camphor tree yield more camphor and oil of camphor than the wood, and that their proper harvesting does no injury to the tree, so methods of distillation have now been greatly modified. If such a discovery had been made in about 1870, then this story would never have arisen.
52. Mitchell, Charles A.: *Camphor in Japan and in Formosa,* Chiswick Press for private circulation, 1900; Nock, J.K.: *Camphor Cultivation in Ceylon,* Colombo, AM and J Ferguson, 1905.
53. But see the early US patents also, as in the production of camphoric ethers, as a thick colourless oil, (USP10433 of 1898 G.W. Johnson, patent agent for Kalle and Co.) or production of camphor and wastes via the heating together of turpentine and anhydrous oxalic acid, this treated with caustic alkali and distilled with steam to separate out camphor (USP14754 of 1900 C.K. Mills, the patent agent for Ampere Electro-Chemical Co.).

54. Already in USA during the 1870s–1880s John H Stevens (and others at the Celluloid Manufacturing Co. of New Jersey, working Hyatt's patents) was inventing camphor substitutes, and from 1889 chemists throughout the world competed to produce alternative substitutes; but until the 1920s this tack was hindered by the cost of manufacture.
55. Calculated from Davidson Appendix, see endnote 39 above.
56. For some summary details see Inkster, 'Technology and Culture', op.cit, endnote No.13.

4

Global Engineers: Professional Trajectories of the Graduates of the École Centrale des Arts et Manufactures (1830s–1920s)

Darina Martykánová

The expansion of global capitalism in the decades around the turn of the twentieth century included growing investment in all kinds of projects and enterprises that required technical expertise. Railways were built crossing the territories of several countries, canals opened up new routes for ships by separating isthmuses and continents, irrigation systems enabled agricultural production on previously barren soil. The companies that carried out these works were often linked to particular national interests, but, at the same time, joint ventures abounded and the staff and workforce were often multiethnic and multinational. In this world, engineers carved out for themselves an expanding area of opportunities: as independent professionals offering their services, as private and public employees and as business owners.

The original version of this chapter title was revised to *Global Engineers: Professional Trajectories of the Graduates of the École Centrale des Arts et Manufactures (1830s–1920s).*
The correction to this chapter is available at
https://doi.org/10.1007/978-3-319-75450-5_15

D. Martykánová (✉)
Department of Contemporary History, Universidad Autónoma de Madrid,
Madrid, Spain
e-mail: darina.martykanova@uam.es

© The Author(s) 2018

D. Pretel, L. Camprubí (eds.), *Technology and Globalisation*, Palgrave Studies in Economic History, https://doi.org/10.1007/978-3-319-75450-5_4

The professional field of engineers was becoming truly worldwide in the period between the 1880s and 1920s. While some engineers accessed this global arena via nationally based companies or public institutions, others were required to mobilise their transnational networks and provide internationally recognisable credentials, such as a degree from an engineering school in a country that scored reasonably well on the international hierarchy of prestige. This chapter compares two heterogeneous groups of graduates from the prestigious Parisian engineering school, École centrale des arts et manufactures: alumni born in the Ottoman Empire and those born in Spain and Latin America, that is, coming from peripheral, but not colonial contexts.[1] These men were of different national, ethnic and religious origins and their life trajectories varied greatly. I pay special attention to the relatively small number of those who displayed a high degree of international mobility and developed a truly global career. Reconstructing their biographies allows me to ascertain the importance of different factors that shaped their professional mobility, such as ethno-religious networks, the links of friendship and expertise created during studies at the École centrale or the connection between the school and specific French companies operating beyond the frontiers of metropolitan France. Moreover, I discuss the hypothesis that these engineers played an important role in the creation of a globally recognisable figure of engineer and in the configuration of a transnational professional culture that made it possible to successfully carry out engineering works in different geographical and cultural settings.

École Centrale des Arts et Manufactures: A Global Engineering School

École centrale des arts et manufactures was founded in Paris in the late 1820s by a group of private investors who admired the dynamism of British industry and thought that, alongside with its prestigious schools for state engineers such as the École polytechnique, École des ponts et chaussées and École des mines, France needed a school that would train engineers for industry. While focusing on subjects related to construction and industrial production, the school did not actually adopt the most common kind of education among British engineers at that period, the one based on hands-on workshop training. Following the French tradition as it had come to existence during the eighteenth and early nineteenth centuries, sciences constituted a key part of the education of future engineers and manual work was cautiously framed as projects, fieldwork or labwork.[2] This standardised education that included sciences, contributed to making the school into a magnet for foreign students, accommodating their well-off family background and elite aspirations in a way workshop training could not. The British workshop model of engineering education, while praised all around the world and linked to the unparalleled British industrial development, proved much harder to export and was—in practice—less attractive for those whose families could afford to send their sons abroad to study engineering.

Unlike the French schools for state engineers that depended on the Ministry of War and where the access of foreigners depended on special permission, the École centrale was, in principle, open to all those who could pass the exams and pay the fee.[3] This policy continued once the school became public, after its founders transferred it to the French state in 1857, and shaped the students' profile: while at the schools for state engineers foreign students entered mostly thanks to the mediation of the French and/or foreign government, the École centrale was an attractive option for well-off families from all around the world that sought prestigious education for their children, but could not or did not wish to engage the authorities in providing access to it.

In order to enrol, the young men (and, from the beginning of the twentieth century, also women) had to pass an examination that required a thorough preparation. The studies lasted three years and the students were constantly evaluated. Besides dropping out (those who did not finish their studies were called *fruit sec* in the school's jargon), the less successful could also graduate receiving a mere certificate instead of a proper engineer's diploma. During the first year, the studies focused on general subjects such as descriptive geometry, geometrical analysis, general mechanics, general chemistry, natural history and movement transmission, but, following the ethos of the founders, practice-oriented subjects were also included, such as experiments in physics and chemistry, construction, machine design and drawing. The emphasis on 'application' increased during the second year: applied mechanics, machine construction, analytical chemistry, metallurgy, 'industrial physics', geology and exploitation, and public works and architecture. Furthermore, the students carried out practical exercises and projects, including special tasks for holidays. The last, third year, specialisation determined an important part of the curriculum: the students chose from mechanics, construction or metallurgy and their projects and practical tasks mostly fell within the chosen branch. New subjects such as steam engine machines or railways were added to applied mechanics, machine construction, analytical chemistry, metallurgy, geology and mineral exploitation, public works and architecture.[4] The final *concours* determined whether the student would receive a certificate or a full diploma, as well as the order in the yearly *promotion*.

Students from Spain and Latin America

Men from Spain and Latin America appear among the students soon after the École centrale was established, and their presence, though slightly diminishing, remains noteworthy until the end of the period analysed here. Between 1836 and 1921, approximately 220 students born in Spanish-speaking countries graduated from the school.[5] The great part of these students came from the domains of the Spanish Crown, both peninsular Spain (88) and the Spanish Antilles, Cuba (30) and Puerto Rico (6), that remained Spanish colonial possessions until 1898. Among the American republics, Mexico (25), Chile (14), Peru (12), Argentina (19), Uruguay (12) and New Granada/Colombia (6) feature prominently. The interpretation of the data is complex, because a student's nationality is mentioned neither in the register nor in the *annuaires* of the school's alumni association, only place of birth. Many students may not be citizens of the country in which they were born. In fact, the names and biographic information of some of the men included in the list strongly indicate that they were French. However, these indicators are not enough to ascertain this: foreigners often had their first names translated to French when in France and many French people were migrating to Latin America. Caution is encouraged by the fact that some of these men bearing French surnames later worked in Spain and Latin America. Vice versa, several graduates with Spanish surnames who lived and worked in a Spanish-speaking country are registered as born in Paris, Rome and so on.

Except for a few exceptions, we have little information on how these men got to study at the school. Was it their families that sponsored them or were some of them sent by their respective governments? The latter could be the case of some of the Latin American students. It is indeed the case of the first Spanish graduates, who were sent in early 1830s by the Spanish Crown to study at the École and help found a similar institution in Madrid when they returned.[6] Regarding the second half of the nineteenth century, though, it is highly improbable that the government of Spain would sponsor complete studies at the École centrale, as it founded and funded several schools for state and civil engineers in Spain.[7] Taking into consideration that graduating from a Spanish state engineering school was a precondition of becoming a state engineer in Spain, we may presume that the Spaniards who studied at the École centrale were supported by their families and planned for a career in the private sector, in Spain or elsewhere.

Though students came from all four corners of peninsular Spain, several cities and regions stand out: Madrid (15), Catalonia and the Balearics (14), the three Basque provinces (*Vascongadas*) and Navarra (13) and Malaga (12). This may indicate the economic dynamism of these regions, as men coming from well-off families willing and able to support them during their studies in Paris surely had expectations to make a good living as civil engineers or running technology-related businesses once they returned. While economic dynamism and successful industrialisation have traditionally been associated with Catalonia and the *Vascongadas* of the second half of the nineteenth and the beginning of the twentieth century, and Madrid, being the country's capital, is a priori in a privileged position, let us focus on the less-known case of Malaga.[8] In the mid-nineteenth century the city and the region of Malaga experienced a period of prosperity, based on commerce and booming industry (iron and steel, textile and food industries), competing with Catalonia in industrial production. Its further development was stymied by the difficulties in obtaining cheap and easy access to mineral coal, which gave comparative advantage to competing regions such as the *Vascongadas*. Several of the prominent families that featured in the industrial development of Malaga sent their members to study at the École centrale. This is the case of the Larios family. Richard (Ricardo) Larios, born in Gibraltar where one branch of the family settled in the first third of the nineteenth century, graduated in 1841. Though he died in Gibraltar in 1892, he had run his sugar industry business in Malaga. In 1857, two Malaga-born Larios brothers, sons of the influential businessman and politician Martín Larios

Herreros and his wife (and niece) Margarita, graduated from the École centrale.[9] Both went on to have remarkable careers in Spain. Their business activities ranged from sugar and textile industries to banking. Besides expanding his family business, Martín (1838–1889) was a Conservative deputy in the Spanish Parliament (Cortes) and shortly before his death was subject of a cause célèbre, an attempt by his mother and brother to declare him insane in order to take over the management of his property.[10] His brother and business partner, Manuel Domingo (1836–1895), who inherited the title of marquis that their father had achieved for the family, was also deputy and lifetime senator, active in the Moderate and later the Conservative Party.[11] The following generation is represented by Enrique Crooke-Larios (grad. 1877).[12] Product of a marriage between two prominent local families, Crooke (whose ancestor came from England in the mid-eighteenth century) and Larios, Enrique worked in the family businesses of banking, textile and sugar industry, and he was also a politician and deputy of the Cortes for the Conservative Party. Gibraltar-born Auguste Larios (grad. 1899), who was living in France in 1920, probably came from the Gibraltar branch of the family, though we may only speculate about the connection to the Spanish Larioses of Fernando Larios (grad. 1893), born in Belin, Nicaragua, who is probably the 'expert engineer' who built a dam in the Nicaraguan city of León in 1922 and appears as one of the nationalist leaders sympathising with the Sandinist movement in a document from 1931.[13]

Malaga-born graduates include two men bearing the surname of another prominent local family, the Heredia.[14] Ricardo Heredia (grad. 1852), whose grandfather Thomas Livermore Page was an English merchant settled in the region, worked as civil engineer in Malaga, while his sister Amalia, married to a local entrepreneur, engineer and Liberal politician Jorge Loring Oyarzábal (marquis de Casa-Loring), was a patroness of arts, philanthropist and woman of science, one of the founding members of the Spanish Royal Society of Natural History (1871).[15] Agustín Heredia (grad. 1874) was a master ironworker and owner of iron and lead mines in the region. Malaga-born Joaquín Almellones (grad. 1856) worked for the Sons of A. Heredia for thirty years, later establishing himself as engineer-counsellor. The city offered career opportunities for other *centraliens*, too. French-born Léon Polet (grad.1909) was working there in 1920 for the Unión española de fábricas de abonos, de productos químicos y de superfosfatos. David Veneziani, born in Constantinople into a distinguished Jewish family, was living in Malaga in 1897, working as a railroad and works inspector at the Compañía de ferrocarriles andaluces, a company whose general manager at that time was another *centralien*, Anatole Maegherman, born near the French town of Dunkirk, who graduated in 1861, just one year before Veneziani obtained his certificate. Before settling in Malaga, Maegherman had worked in Portugal for the Companhia Real dos Caminhos de Ferro Portugueses.

The following cases of *centraliens* with links to Andalusia illustrate the difficulty to establish the graduates' nationality. Antoine Germain (grad. 1920) was born in Malaga, though his name and surname are French. His links to the city were clearly stable and long lasting, as in 1930 we find him there working as engineer for the local railway company Ferrocarriles andaluces. 1876 graduate Charles Cuadra's surname is Spanish, but he was born in Paris. His career is closely linked to Andalusia: he managed the Colonia de San Pedro de Alcántara, an ambitious project of local industrial and agricultural development. He might be related to Luis Manuel de la Cuadra, marquis de Guadalmina, a well-known industrial and agricultural entrepreneur and banker with strong links to France, who bought the 'colony' in 1874. Later in his life, Charles/Carlos become the manager of the Casa López, Janer, Cuadra y cía, and worked as a civil engineer in Ronda. Eugène Poisson (grad. 1884), born in Valparaíso, Chile, is also linked to the *colonia*: in the 1890s, he was in Andalusia, working as engineer-representative of the Compagnie de Fives-Lille and managing the Compañía azucarera y agrícola de San Pedro Alcántara. Not all graduates with ties to Andalusia ended up contributing to industrial development. The professional career of Malaga-born Thomas Bryan (grad. 1849) took the most unusual turn: in the 1890s, he was the Bishop of Murcia and Cartagena.[16]

The Larios and the Heredia were not the only distinguished Spanish *centraliens*. Several of the graduates from the 1830s, who had come on goverment fellowship, had major impact on the institutionalisation of Spanish science and technology. Juan Cortázar (1809–1873) went on to be a mathematician and university professor. Eduardo Rodríguez (1815–1881) taught at the University in Madrid and at the Conservatorio de

Artes and when the Spanish Society of Industrial Engineers was founded in the early 1860s, he was its first president.[17] Joaquín Alfonso y Martí (1805–1868) became a director of the Conservatorio de Artes, founding member of the Academy of Exact Sciences in Madrid and he was put in charge of organising the Real Instituto Industrial, where he also taught. Deputy for Valencia at the Cortes Constituyentes (Constituent Assembly) in 1854, Alfonso was a radical democrat, blanquist and atheist, who chose to be buried under an olive tree in the countryside, instead of a cemetery.[18] Cipriano Segundo Montesino (1817–1901, grad. in 1837), from Valencia de Alcántara, went on to be a senator, member of the academies of science of Madrid and Lisbon as well as manager of the MZA (Compañía de los ferrocarriles de Madrid a Zaragoza y Alicante) railway company, linked to the French capital, particularly to the Rothschilds. Mobility, forced and voluntary, was part of his life. He came from a prominent Spanish liberal family. His father, a physician, had been deputy in the Cortes of 1812, the parliament responsible for the first Spanish constitution, and his family had to leave Spain for exile in London after the fall in 1823 of the constitutional regime known as the *Trienio liberal* (1820–1823). Montesino married Eladia, niece of Baldomero Espartero (1793–1879), Spanish liberal general and Progressive strongman (who was regent of Spain in 1840–1843). Montesino supported Espartero's rule, for which he had to spend another period of exile in London after Espartero's fall in 1843. When in Spain, he combined political activity with engineering. Taking into consideration that in the mid-nineteenth century Spanish engineers fought to limit access to public employment to Spanish males who held a diploma from Spanish engineering schools, Montesino revalidated his French title in the Spanish Real Instituto Industrial, so there would be no doubt that he could be officially acknowledged as *ingeniero industrial*.[19] He taught machine construction at the Conservatorio de Artes and later at the Real Instituto Industrial and compiled a textbook on this subject. He held the influential post of the Director of Public Works during the *Bienio Progresista* (1854–1856), when the Progressive Party he supported came to power. The Spanish Royal Academy of Sciences appointed him to the International Commission for piercing of the Isthmus of Suez, and he published a book on this subject in 1857. His lifelong commitment to progressive branch of Spanish liberalism led him to become one of the shareholders of the Institución Libre de Enseñanza, an ambitious project for education reform inspired by the values of German philosopher Karl Krause, whose teachings were highly influential among Spanish reformist intellectuals linked to the so-called *liberalismo avanzado*.[20] His son Luis Montesino y Fernández-Espartero (marquis de Morella, 1868–1957) followed in his father's steps, graduating from the École centrale in 1890. In 1891, Spanish engineers succeeded in banning those who graduated abroad from using their title in Spain, so Luis went on to study at the Escuela de ingenieros industriales in Barcelona. Like his father, he combined political activity with engineering. He was deputy and senator as well as municipal councillor of Madrid. He worked as engineer for the MZA railway company, he was director of the Escuela superior de artes e industrias in Madrid, vice-president of the Board for the Tramways of Granada, and held other posts in the public and private sector. He founded and directed the Escuela Nacional de Aviación (ENA, 1913).[21] During his life, he constantly renewed his links to France: working for the MZA, buying planes in France for the ENA, flying together with French pilots, representing the Societé des lignes Latécoère (Compagnie générale aéropostale) in Spain and living in Neuilly in 1920.

Spain offered plenty of opportunities for engineering graduates. Many Spanish-born *centraliens* went back to settle in their hometown and have a career in their field of expertise. Claudio Gil (grad. 1849), from a well-off family of Catalan entrepreneurs, worked as engineer for the Compañía catalana de gas in his hometown of Barcelona and he invested in several businesses linked to technological innovations.[22] Pamplona-born Aniceto Lagarde (grad. 1856) was engineer and director-in-chief of the roads of Navarra, while Mariano Planas (grad. 1863), born in Girona, established himself there as a machine-builder. Granada-born Francisco de Castro (grad. 1902) also earned his living in the private sector, working in his hometown as engineering consultant (electricity and hydraulics) and manager of industrial companies. Although his surname points to French origins, Louis Commeaux (grad. 1890) was born in the mining town of Mieres del Camino in Asturias, and he went back to work as an engineer at the Mieres Factory.

Few *centraliens* born in peninsular Spain had global careers. Those who worked abroad did so mostly in French-speaking Europe: Tolosa-born

Benigno Yrazusta (grad. 1878) became manufacturing manager at the paper mills of M. de Naeyet and Company in Willebroeck, Belgium; Charles de Tienda (grad. 1907), born in Santander, had a successful career in France, working as engineer at the Société Lorraine-Dietrich in Lunéville and at the Glassworks of Bousquet d'Orb (Hérault), and as manager of the Glassworks of Carmaux (Tarn). Some of the Spanish-born engineers with French surnames worked in Spain and France. François Saunier (grad. 1900), born in Santiago de Compostela, had worked as engineer specialising in gas and electricity in the Galician cities of La Coruña and Vigo, before moving to Paris where he ran a factory on polishing and abrasive material and presided over a business association of the producers of grinders and polishing products.

The surnames of the few Spanish-born *centraliens* who went to work beyond Europe indicate that they were probably French nationals. Charles Barbière (grad. 1885), born in the Basque city of San Sebastián, represented the French Société de construction des Batignolles in Cairo in the last years of the nineteenth century, when Egypt was de facto under British colonial rule. Madrid-born Léopold Collier de la Marlière (grad. 1895) was general manager of the Andrada Mines Ltd in Portuguese East Africa and engineering consultant at the Mines de la Falémé, iron mines in East Senegal. His experience enabled him to settle in Paris by 1920, working as consultant for mines and metallurgy. Galician-born Victor Meric (grad. 1903) worked as engineer on the railways in Syria and Yemen as well as on the construction of the railway line from Huelva to Ayamonte in Spain. In 1920, he lived in Paris, working as engineer for the Entreprise Dufour 'Constructions générales' that carried out works at the canal connecting Rhône to Rhine. Madrid-born Lucien Cocagne (grad. 1908) crossed the Atlantic to work as assistant engineer at the headquarters of the Compañía general de ferrocarriles en la provincia de Buenos Aires, in Argentina. The death on the battlefield in 1914 of Madrid-born Marcel Suss (grad. 1910), indicates that he was a French citizen, as Spain did not participate in the First World War. Before his untimely death, Suss had worked on the construction of a dry dock in Toulon and was commissioned to examine the possibilities of the exploitation of waterfalls in Finland. He also worked as engineer for the Sociedad española de ferrocarriles secundarios.

Peninsular Spain sometimes was a stop on the European or global career paths of other *centraliens*. This was the case of the French engineer Félix Ragon (grad. 1884), born in Curçay, Vienne. In the last decades of the nineteenth century, he worked in coal mines in the Silesian town of Dąbrowa Górnicza/Dombrowa (then controlled by Russia), together with his classmate from the École centrale, another Frenchman called Ferdinand Steenman. Later, Félix became engineer-director of the mines of Matallana in Spain, before moving back to the region of his birth. Even more global was the trajectory of Joseph Bastide (grad.1859), born in Rodez, Aveyron, France. He started relatively close to home, running a company specialising on public works in Bugeat, Corrèze. He then moved to Spain to manage the Company for the Construction of the Port of Santander. Afterwards, he was engineer-in-chief at the Compañía de minas y del ferrocarril Bacares—Almería. Crossing the Atlantic, he worked as a railway engineer for the government of Chile. In 1897, he was back in Europe, working as engineer-in-chief at the works of the port of Burgas, Bulgaria.

The relatively low mobility of the *centraliens* born in Spain becomes clear when we compare them to Latin Americans. Cubans and Puerto Ricans, who were subjects of the Spanish Crown until 1898, attended the École in great numbers. Men born in the Antilles could and did attend Spanish schools for state engineers, and some of them then went on to work for the Spanish state both in the Antilles and in peninsular Spain.[23] Nonetheless, it is clear that in the mid-nineteenth century, enterprising Antillean families that wished to sponsor their sons' technical education saw in the French École centrale a good option for an engineer heading to the private sector, maybe more prestigious and easier to access than the Spanish Escuela de ingenieros industriales in Barcelona. It seems that there was some coordination in sending out young men to the school, either based on family networks or municipal initiative. In the 1850 *promotion*, there are three Cubans from the city of Matanzas. Cuban graduates display a high return rate to the island, which indicates plenty of opportunities in Cuba for qualified engineers with high expectations in terms of monetary gain, either working for others or in family businesses (considering that only well-off families could fund their sons' studies at the Parisian school). My analysis confirms that Cuban graduates from foreign engineering schools either used their knowledge in family businesses or worked as managers or in 'prestigious' branches of engineering, such as railways and urban infrastructures. As others have pointed out, few Cuban or Spanish engineers worked as mechanics or machinery operators in Cuba in the mid-nineteenth century.[24] One of the 1850 graduates from Matanzas, Eugenio Pimienta, managed a sugar refinery in Cárdenas, Cuba. Manuel Montejo (grad. 1853), born in the Cuban town of Puerto Príncipe (Camagüey), managed the gas company of his hometown. Havana-born Charles Theye (grad. 1875) was chemistry teacher and manager of the Santísima Trinidad sugar factory in San Marcos (Cuba), while his fellow countryman Luis de Arozarena (grad. 1877) worked as an engineer at the Havana railway, was associate professor at the Faculty of Sciences in Havana and general manager of the salt mines of Cayo-Romano in Nuevitas. Puerto Ricans, too, had a high return rate. German Mooyer (grad. 1896), born in Mayaguez, was engineer at the railway company of Puerto Rico and in 1920, after Puerto Rico had been taken by the USA, he worked as assistant engineer of public works in his home town. The number of Antillean students decreased towards the end of the nineteenth century. We may only speculate that by then ambitious young men headed to the engineering schools in the USA.

Some of the Cubans and Puerto Ricans settled in peninsular Spain: Matanzas-born Juan de la Torriente (grad. 1862) lived in Santander in 1897, Henry Berrocal (grad. 1886) in Barcelona. 1872 graduate Julio Apezteguía y Tarafa (1843–1902), born in Trinidad, Cuba, moved across the Atlantic several times. Member of a slave-owning family that owned a sugar plantation, he modernised it, turning it into one of the biggest and most productive *ingenios* on the island, even in the world.[25] He was a deputy for Santa Clara and for Havana in the Spanish Cortes for six terms (1879–1890, 1893, 1896–1898), representing the Unión Constitucional. After Cuban independence, he lost his Spanish citizenship (although he tried to get it back), and he died in 1902 in New York, struggling with financial difficulties, as he had lost great part of his possessions and wealth to his US business partners and creditors.[26]

Other Antillean *centraliens* (from less privileged families?) had international careers. Havana-born John Knight (grad. 1857) was a manager in the French Compagnie des chemins de fer Bône-Guelma, before dying in

Paris in 1884. This company was founded in 1875 to build railways in Algeria and Tunis (Algeria having been a French colony since 1830; Tunis was to become a French protectorate in 1881) and was a daughter company of the French Société de construction des Batignolles that employed many *centraliens*. Matanzas-born Emilio del Monte (grad. 1874) worked as an engineer at the railways in Girardot and Antioquia, Colombia. Later, he returned to Cuba to work as assistant engineer at the railway in Matanzas, as engineer-in-chief of the Havana railway and of the railway between Cienfuegos and Villa-Clara. Others settled in France. Cuban-born *centralien* with a French surname Ernest Raoulx-Jay co-owned the Compagnie de navigation E. Roaulx-Fournier et Cie and was the head of technical services of La Rochelle municipality, France, while Frédéric Henrique, born in Havana, was an engineer of the electrical substation of the Cité Bergère, engineer-director of the electric lighting of Perpignan (Entreprise Bartissol), engineer at the Société pour la transmission de la force par l'éléctricité and at the Société du gaz de l'électricité of Marseille, where he lived. Blas Batanero de Montenegro from Cuba worked as examiner for physical sciences at the Parisian establishment Collège Stanislas, before dying in Switzerland in 1917.

In a sense, professional trajectories of the graduates born in Latin American republics were global from the instant they crossed the Atlantic to study engineering in France. Nonetheless, many of them simply went back to their country of birth and developed a career there, in the public and/or private sector (unlike in Spain, engineers who studied abroad could work for the government in most Latin American republics). The very first *centralien* from a Spanish-speaking country, Juan Cano (grad. 1836), born in Mérida, Mexico, became coronel of the Mexican corps of engineers, which indicates that his studies might have been sponsored by the Mexican government. Manuel Echegaray (grad. 1855), born in Cuzco, Peru, worked as a civil engineer in Lima. Another Peruvian, Francisco Paz Soldan (grad. 1861), worked as an engineer for the Peruvian government and managed private companies in Lima. Alfred Puelma Tup(p)er (grad. 1882) from Santiago de Chile became sub-chief of the section of architecture at the Direction of Public Works of Chile and authored a treatise on the Chilean sugar industry.[27] Jorge Hayton (grad. 1887) from Buenos Aires worked as electrical engineer for the Maison

Sautter-Harlé et Cie, but also for the Argentinian government. Juan González (grad. 1860) from Bogotá, New Granada, went back to work as civil engineer in his hometown, now the capital of the United States of Colombia. His classmate Ignacio Pedralbes y Capúa from Montevideo, son of a mathematics professor, went on to have a prolific career as engineer–architect in his country, while he also taught at and served as dean of the Faculty of Mathematics of Montevideo. He was director general of the Uruguayan public works administration and one of the founders of the Ateneo.[28] Another Uruguayan, Juan Lamolle, worked at the design office of the Northern Railways to later become engineer-in-chief of the Montevideo Municipality. Born in the Argentinian capital, Jules Lacroze (grad. 1863) founded the Tramway Company of Buenos Aires. His company employed a French *centralien* (grad. 1896) with strong family links to the Ibero-American world, Charles d'Ornellas (1874–1961), born in Paris. In the early twentieth century, d'Ornellas was general manager of the Lacroze Tramway Company and of the Central Railways of Buenos Aires, also representing the interests of MM. J. G. Withe et Cie in South America.[29]

The most outstanding of those Latin American *centraliens* who established themselves successfully in their country of birth is probably Ernesto Madero Farías (1872–1958), born in Parras de la Fuente, Mexico. Before studying at the École centrale (grad. 1893), he had graduated from the Johns Hopkins University in the USA. He was minister of finance of the Mexican government in 1911–1913, in the tumultuous period when the revolution broke out. Coming from a rich family involved in politics and in diverse industries and agriculture, he expanded the family business and founded new companies (Compañía metalúrgica de Torreón, Compañía carbonífera de Sabinas, Compañía Harinera del Golfo Nuevo-León, etc.). He spent some time in exile in Cuba and the USA.[30]

Other Latin American *centraliens* had international careers. Some moved within Latin America. Felipe Victora (grad. 1875), born in Montevideo, taught at the Escuela nacional de Ingenieros in San Juan de la Frontera, Argentina, before becoming director general of the Uruguayan public works administration. Marie de Lapeyrouse, was born in Piura, Peru, but after he graduated in 1877, he set up business in public works in Iquique (a province that switched hands from Peru to Chile in the

Pacific War, 1879–1884), where he later managed sodium nitrate factory San Donato. Several of the graduates settled and worked in France or Spain. Some of them might have been French citizens. Louis Mosès (grad. 1893), born in Lima, was living in Paris in 1920, working as an engineering consultant specialising in patenting inventions. Maxime Heurtematte, born in Panama, graduated in 1892; in 1920 we find him in Paris as a partner in the Maison Raguet et Heurtematte, entrepreneurs in public works, but the Heurtematte family is still present in Panama nowadays.[31] Mexico-born Emile Pinson (grad. 1895) moved between Mexico and France. He had been director general of the Compañía de San Ildefonso, chief executive of the Compañía hidroeléctrica del Chapala; vice-president of the Compañía de papel San Rafael. In 1920, he was living in Paris, working as engineer for the Compagnie générale de la électricité, and partner of the Mexico-based Maison E. Pinson et L. Matty. Gabriel Foullioux (grad. 1921), born in Rauquén in Chile, was working in 1930 as engineer for the Cristalería Española, in Arija, Línea de la Robla, Spain. Joseph Lopes-Dias (grad. 1878), born in Vallenar, Chile, had a successful career in France.

While some moved between two countries, others had a truly global career. Juan Caubios (grad. 1897), born in Montevideo, worked, as several other *centraliens* of different origins, for the French Société de construction des Batignolles (involved in railway construction in the Mediterranean) and for the Maison Hennebique (trading in reinforced concrete), another French company linked to several *centraliens*. He had been director of the company that managed the port of his Uruguayan hometown, before becoming director of the port of Mar del Plata, Argentina, where he lived in 1920. Léopold Cassou (grad. 1885), born in Guanajuato, Mexico, had worked at the construction of steel mills and at the headquarters of the Creusot factories (a company that employed many *centraliens*), before he settled in 1890s in Bilbao, Spain, working as engineer at the Talleres de Deusto. Eugène Renaud (grad. 1885), born in Mexico, worked as engineer for Argentinian government and managed the distillery L. Poiron et compagnie in Saint-André.[32] In 1897, he was back in Mexico as subdirector of the silver mines of Quintera, in Álamos, Sonora. Félix Courras (grad. 1880), born in Montevideo, was working in the Ottoman Syria for the Compagnie des chemins de fer de Beyrouth-

Damas-Hauran et Biredjik sur l'Euphrate in 1897. Marcel Landormy (grad. 1910), born in Gatun, Colombia, had worked in the French colony of Algeria and in metropolitan France; in 1930 he managed Gelatinous White Co. Ltd in Great Britain.

Mobility sometimes hints at social disadvantage and instability (less well-off family, recent migrants, ethno-religious minority, etc.); in other cases, it may be attributed to individual choice or unknown circumstances. The case of two brothers born in Santiago de Chile who graduated together in 1874 is telling. Between 1874 and 1897, Charles (Carlos) Robert de la Mahotière developed his career in Chile. He worked as engineer at a foundry and mining company of Chañaral, as chemical engineer of a saltpetre company of Antofagasta, as engineer-in-chief of the railway from Talca to Constitución, and he ran a family business, Maison Mahotière fr. et Sainte-Anne, a carpentry workshop and mechanical sawmill, in Temuco.[33] His brother Louis (Luis) also worked in Chile, producing soap, working as engineer-in-chief at the construction of the railway from Victoria to Temuco and as Charles's partner in the family firm. Nonetheless, he also worked as engineer in the silver mines of Huanchaca, Bolivia, and by 1897 he was working as a civil engineer in Paris.

Several of the *centraliens* born in Latin America died in battle. Cornelio Borda (grad. 1854) from Bogotá, New Granada, died in the service of Peruvian government in the Battle of Callao (1866), during the Chincha Islands War between Peru and Spain. The other two fallen, Edouard Challé (born in Santiago de Chile, grad. 1912) and his Mexican-born classmate Albert Huguenin, died in the First World War (1915), which indicates they were French citizens.[34]

The *centraliens* born in Spanish-speaking countries are a highly diverse group. Some of them were French citizens born abroad. Others held the citizenship of or had strong links to their country of birth and developed their career there. The École centrale clearly enjoyed great prestige among the Spanish (and Cuban) enterprising bourgeoisie in the 1850s–1870s. For many, their studies at this Parisian school led to more than just acquiring a polish of French-style modernity: several high-profile graduates undertook technological modernisation of their family businesses. The fact that hardly any *centraliens* from Spain worked for the Spanish public sector proves the efficiency with which the engineers trained in

Spanish state engineering schools established exclusive access to public employment. In Latin America, the governments tended to be willing to employ not only their citizens who graduated abroad but also foreign nationals. The engineers who did develop their career abroad, entirely or partially, tended to work in other Latin American countries, in Spain, in France or for a French company anywhere in the world (mostly North Africa and Middle East). A diploma from the École centrale and the contacts they established there helped them fashion themselves as representatives of French engineering knowledge and skills, and get acknowledged as such by French companies in France and abroad and by potential business partners and clients.

Students Born in the Ottoman Empire

The Ottoman Empire was a vast, highly heterogeneous monarchy that extended over three continents. During the period covered in this chapter, it experienced more or less successful separatist movements as well imperial/colonial predation by European powers, and, eventually, its disintegration in the aftermath of the First World War. Facing growing internal and external challenges, Ottoman governing elites reacted by enacting a long series of reforms, but the growing interconnection between 'Europe' and the Ottoman Empire led to the deepening of Ottoman economic, technological, even political dependence on foreign powers.[35] The first Ottoman-born students appear at the École centrale later than the Spanish and Latin American ones, in the early 1850s. Between 1853 and 1923, approximately one hundred students born in the Ottoman domains graduated from this school. Their names indicate great diversity in terms of nationality, ethnicity and religion. There is an additional complication because the Empire was losing territory throughout the period and even if some regions were officially considered as the domains of the Ottoman sultan, they had their own policies regarding public works, sending students abroad, management of religious diversity and so on. Previously, I analysed two groups of Ottoman *centraliens*: Armenians and Jews. Particularly the latter were disproportionately represented, constituting approximately 20% of Ottoman-born graduates.[36]

Here I summarise the information on the students' backgrounds, before focusing on return rates and mobility patterns.

The Ottoman-born *centraliens* were an extremely diverse group in terms of regional and ethno-religious origin. First, several were probably French, taking to consideration their name(s), surname and career patterns. We have to bear in mind that 'Western Christian' communities had lived in the Ottoman Empire for centuries or had settled there recently.[37] Thus it is hard to ascertain who could be a member of such family and who was just born in the sultan's domains because of his father's temporary employment there, for instance working for the French government or a French company. The place of residence of the student's 'head-of-the-family', noted in the register, provides hints, though we need to consider that a family that had been in the Ottoman Empire for centuries or decades could have moved to France sometime between the future student's birth and his enrolment in the École (this was the case of several Ottoman Jews).

Some students were born in regions that had a great degree of autonomy and were separated de facto or de jure from the Ottoman Empire during the period covered in this chapter. This would be of particular concern when examining the careers of the large number of students born in Moldavia and Wallachia (later Romania) and in Egypt. Ottoman students of Greek ethnicity display specific mobility patterns: besides staying abroad or heading home, a few settled and worked in the independent Greece; in some cases, their Ottoman hometown (i.e. Salonica) actually became part of Greece during their studies or later in their life. Another complex issue is classifying the Ottoman-born students bearing Italian surnames. They could be Levantines, that is Western Christians or Jews settled in the Ottoman Empire, or children of men from the Italian lands working in the Empire.[38] In the former case, there would be a remarkable difference between those Christians and Jews living in the Empire for centuries and those whose families had settled there more recently, for instance as merchants, without ever interrupting their links with Italian cities and states. Ottoman Armenians (about a dozen) and Muslims (besides the Egyptians, there were only two Ottoman Muslim graduates) are easy to identify owing to their naming patterns.

Analysing the mobility patterns of the Ottoman-born students, it soon becomes obvious that Egypt and particularly Romania were lands of successful returns. The great part of the students born in the Ottoman Wallachia and Moldavia returned to their homeland, which was a de jure Ottoman vassal state of United Principalities in 1859–1877, before becoming fully independent. It is clear that the École centrale was popular among the people from Moldavia and Wallachia, lands whose elites cherished cultural and political links to France, which, in turn, was happy to patronise them and their project of an independent Romania. In an article that shows the impact that the École centrale, its staff and its graduates had on engineering and science in Romania, Horia Colan outlines the careers of the most successful *centraliens* of Romanian origin, including those born under the Ottoman rule.[39] They enjoyed great opportunities in the expanding public works administration of an independent country in the making that was in the midst of a state-building process. The 1869-graduate Gheorghe Duca (1847–1899), born in the Moldavian town of Galati, directed and reformed the Scoala Nationala de Poduri si Sosele in Bucharest, the Romanian version of the French École des ponts et chaussées, was general manager of the Romanian Railways and directed construction works of the port of Constanta. Several of these *centraliens* engaged in politics: 1839 graduate Alexandru G. Golescu (1819–1881) fought in the revolution of Walachia in 1848, went to France to gather support against the Ottomans and Russians and later served as minister (once as prime minister) in several governments of the United Principalities. The 1870 graduate Constantin P. Olănescu (1844–1928) was minister of public works and shaped the legislation on public employment of technicians.

Egypt, an Ottoman territory with a highly autonomous government that became a de facto British colony in 1882, was another land of fruitful returns, for Muslim and non-Muslim graduates. Furthermore, it provided opportunities for French and foreign *centraliens*. Muslims from Egypt were the only noteworthy group of Muslim students at the École centrale. Two Egyptian Armenians who graduated before the British occupation had prominent careers after their return. Cairo-born Joseph Manouk Bey (grad. 1855) was general subinspector of the Egyptian Telegraphs in 1897. Coming from a prominent Egyptian Armenian family (his father was Egypt's prime minister), Boghos Nubar, though born in Constantinople, developed his professional career in Egypt, where he managed the Société anonyme d'irrigation in the Behera and was executive manager of the Egyptian Railways in Cairo. He moved between Egypt and the core regions of the Ottoman Empire, becoming a prominent voice for the Ottoman Armenians. Until the Armenian genocide during the First World War, he understood that the future of Ottoman Armenians lay within the Empire, where he tried to improve their legal standing, education and life conditions (founding the still-existent Armenian General Benevolent Union in 1905 in Cairo). After the war, he protested at the lack of interest of the victorious powers in helping the Armenians in their dire situation, in promoting their national interests and in rewarding them for their support of the war efforts.[40]

Jewish Egyptians could also go on to have a successful career in their homeland. From a Sephardim family, 1882 graduate Joseph Aslan Cattaui (1861–1942) served in the public works administration and ran businesses linked to technological innovation, including a prosperous sugar refinery in Hawandiah, near Cairo. He was a member of the Egyptian delegation sent to London in 1905 to negotiate the independence of the country. Later, Cattaui served as minister and senator, and was active in the Egyptian Jewish community.[41] Two Egyptian Muslims who graduated together in 1883, Farid Abdelaziz of Alexandria and Ismail Sirri of Menia, went on to work as engineers, the first of the Egyptian railways, the second for the administration of irrigations. Interestingly enough, their classmate Auguste Souter, born in the French Val-et-Chatillon, worked as an engineer for the Egyptian state in Korachia. Another Egyptian Muslim, Mahmud Fayed (grad. 1904), born in Cairo, worked as engineer-in-chief at the Ministry of Public Works of Egypt, but in 1920 he was living in Constantinople, where he established himself as a private engineer, specialising in surveys and concessions of public works, agriculture, mines and industry. He might have left the Egyptian civil service for political reasons. Egyptian institutions employed French *centraliens*, too, as was the case of Eugène Villard, born in Vallon, Ardèche. Until his death in 1890, he worked as engineer at the engineering school in Bulaq in Cairo.

Regarding the graduates born in the core regions of the Ottoman Empire, particularly in Constantinople, Adrianople and Salonica, many stayed in France or went on to work elsewhere in the world, but others went back and a few had successful careers. The Armenians who studied together at the École centrale in the late 1870s were elite young men with family links to the Ottoman civil service. Gabriel Servicen (grad.1880) was the son of a distinguished physician who had studied in Pisa and Paris and worked at the Hospital of the [Ottoman] Ministry of War. Gabriel spent some time working as an engineer in colonial Algeria, before moving back to Constantinople when he had a successful career at the Ministry of Public Works. Grégoire (Kirkor) Humruzian (grad. 1878) rose to be imperial commissioner of the Oriental Railways at Salonica. Others did well in business, such as the sugar merchant Manouk Aslan or the famous architect Sarkis Balyan (1831–1899) from a prominent Armenian family of architects and builders, who dropped out of the École in the early 1850s. Ottoman Jews who returned tended to work in the private sector. Salonica-born Isaac Fernandez lived in Constantinople during the first two decades of the twentieth century, managing several mining companies operating in Macedonia and Asia Minor. Gabriel Tedeschi (grad. 1872), born in the Ottoman town of Varna (today's Bulgaria), managed a brick factory in Constantinople, owned by the Camondos, a prominent family of Ottoman Jewish bankers that had recently settled in Paris.[42] Few Ottoman Greek *centraliens* had a career in the Ottoman lands. Alexandre Antoniades (grad. 1893), born in Adrianople, managed chrome mines in Dağardı (Fethiye region of Asia Minor) and a charcoal furnace in Kolzu (Zonguldak region of Asia Minor), before he established himself in Constantinople as representative of MM. Hersent, Schneider et cie.

Despite these stories of successful returns, migration was a sharply growing trend among the Ottoman-born graduates, particularly Greeks and Armenians, but also among the Jews. While many settled in France, set up businesses or worked for local companies, several Ottoman-born Jewish and Greek *centraliens* display a high degree of mobility. The 1900 *promotion* is striking: all three graduates born in the Ottoman domains engaged in international mobility. Salonica-born Ernest Farraggi-Vitalis worked as manager of the mines in Ticapampa, Peru, before establishing himself in Paris as a construction engineer. Demosthenes Lykiardopoulo, born in the Anatolian town of Mersin, worked as an engineer for the Metallurgical Society of Taganrog, in the Russian town of the same name, before returning to his hometown. Maurice de Toledo, born in Adrianople, was secretary general of the Société ottomane d'Héraclée, a mining company operating in the Zonguldak basin, exploited mostly by the French. Later, he moved to France, where he worked as a chief engineer of the Société Tréport-Terasse Ltd (most probably the company operating the famous local funicular) in Normandy and then at the Bureau Technique Hispano-Français in Paris.

Occasionally, friendship ties created in the school led to international business partnerships: Démétrius Papa (grad. 1895), born in Constantinople, set up a business with his classmate André Richaud from Basse-Terre in Guadeloupe. Démétrius was executive manager of their Société cochinchinoise de béton armé in Paris, while André ran the company branch in Saigon, in the French colony Cochinchina (in today's Vietnam). They dealt in reinforced concrete, being agents of one of the most important companies in the sector, the *Maison* Hennebique pour les travaux de béton armé, founded by the French–Belgian inventor and entrepreneur François Hennebique.

Transnational mobility was stimulated and mediated by French companies or companies owned partially by French investors. Thus Samuel Léon (grad. 1892) from Smyrna worked as an engineer for MM. Schneider et Cie in Transvaal, before settling in Paris as executive manager of the Société des Forces motrices d'Auvergne. It worked the other way round for Émile Foucart (grad. 1879), born in Oiron, France, who ended up managing a factory on ceramic products near Constantinople, owned by the Camondo family. The pattern of studies in France, high professional mobility and settling in Paris was not displayed exclusively by Jewish engineers from the Ottoman Empire. Miron Goldberg (grad. 1887) was born in Pruzhany (Russian Empire, today in Belarus), a town with an important Jewish community. He worked as an engineer in oil explorations of Kertch and Crimea, and later in Romania, Galicia (most probably the Eastern European region), Peru, Egypt, Italy and Algeria. By 1920, he had established himself in Paris as a consultant-engineer specialising in oil fields, geology and exploitations.

The Ottoman-born *centraliens* who tried their luck on the other side of the Atlantic all seem to have been Jewish. The above-mentioned Ernest Farraggi-Vitalis managed mines in Peru; Constantinople-born Leon Lévy (grad. 1888) ran a business in port works in Chile; and Richard Viterbo from the same city (grad.1872) was general manager of the Tramways de Constantinople, chief engineer in New York, Texas and Mexican Railways and head manager of a refinery in the USA, before going back to the Ottoman Empire in the late nineteenth century to be general manager of the Mersina-Tarsus-Adana Railway Company Ltd in the Anatolian city of Mersin. Thirty years after Richard, another Constantinople-born Viterbo, Lionel, graduated from the school. If Lionel, who was living in Chicago in 1920, was Richard's son, the Viterbo family probably settled in the USA, while still keeping their ties to France.

Career patterns of the Ottoman graduates were heavily shaped by ethno-religious origins and political circumstances. The Ottoman Empire was indeed a land of opportunity for the *centraliens*, in terms of doing business and finding public or private employment; their links to France often played in their favour. At the same time, however, the core lands of the Empire proved less and less welcoming for many of those who were born there.[43]

I do not wish to argue that global mobility was particularly typical of foreign *centraliens*; there were striking cases of highly mobile French graduates too. Parisian Louis Cugnin (grad. 1891) worked as a railroad engineer in Tunis, Madagascar and China. He managed mines in Brazil, worked as an irrigation expert in Baghdad; and was *en missions* (appointed by the French government or a French company?) in Turkey, Morocco, Russia (in Turkestan and Caucasus), before settling in France by 1920. Joseph Volay (grad.1897) from Lyon worked as railway engineer in China, Ottoman Empire (the Soma–Bandırma railway employed several *centraliens*), Spain, Cuba, Syria and France, before becoming (by 1920) the engineer-in-chief of the Régie général de chemins de fer et travaux publics in Beirut. While the Ottoman Empire was sometimes a stop on a mobile career of French and other foreign *centraliens*, some settled there for a long period of time. Louis Chenut (grad. 1879), born in Choisy-le-Roi, France, worked as an engineer of the Serbian state railways and later as subdirector of the works at the railway connection from Salonica to Constantinople. By the end of the nineteenth century, he was a representative of the Société ottomane du chemin de fer de Smyrne–Cassaba in Smyrna. Paris-born Ernest Michot (grad. 1891) worked as a mining engineer in the Ottoman Empire, including the company managed by the Ottoman Jewish *centralien* Isaac Fernandez. By 1920, Michot lived in Paris, where he was general manager of the Compagnie du Boléo, French mining company founded in 1885 by the bank Mirabaud et Cie to explore the rich mines of Boleo, in Baja California, Mexico.

Conclusions

The École centrale des arts et manufactures was a school that acted both as a French and as a global institution of education. Its global dimension was underlined by the sheer number of students from different corners of the world who sought a diploma from this engineering school and thus helped perpetuate its international standing and worldwide renown. At the same time, the École centrale was deeply rooted in France and linked to French companies. Besides being a magnet for young Frenchmen from families linked to industry and commerce, it attracted important numbers of foreign-born French students; it fostered links among future engineers of different origins; and, through its graduates, it served to expand and strengthen French economic interests in many regions of the globe during a period of intensive capitalist expansion, sometimes called the First Age of Economic Globalisation. It also supported French colonial enterprise by supplying qualified men, French and foreign (who often had additional useful knowledge, skills and contacts), for companies operating in colonies, in the areas envisioned for colonial expansion and in the regions where intense competition among global powers for direct and indirect domination was taking place.

To its foreign students, the school offered the quality and prestige of French engineering education as well as useful contacts, all of which helped their professional careers in their countries of origin, in France itself and in many regions of the globe, particularly—but not exclusively—in regions where French capital and companies were active. While foreign students generally came from well-off families that could afford to fund their trip to and costly stay in Paris, it provided them the possibility to reproduce (or improve) their social status not only in their country of origin, but also elsewhere, which became particularly important for those whose origins limited their professional options or/and exposed them to persecution. In the turmoil in the first three decades of the twentieth century, the school might have played a life-saving role for some men, in particular the Armenian (and other Christian) students of Ottoman birth.

The school was part and parcel of global capitalism and it fuelled it, supplying companies that did business on a transnational basis with highly qualified technical staff and management, fostering the creation of new business partnerships and companies. Its graduates were engaged in enterprises that were innovative at the levels of technology, business and workforce management. It might be too bold to affirm that they also participated in the configuration of a truly global engineering culture, but the analysis of their trajectories does indeed indicate that the combination of the graduates' education, mobility and professional practice promoted the mutual acknowledgement of men as engineers across national, ethnoreligious and to a certain degree racial boundaries (gender boundaries resisted until much later). The *centraliens* did not limit their activities to industry, infrastructures, agriculture and commerce; many became involved in politics, either directly or contributing to public debate. Beyond their words and deeds, their very presence as engineering graduates in the public sphere legitimised all kinds of political action via implicit and explicit references to progress, to expert knowledge and to specifically French claims to modernity and universal civilisation.

Notes

1. I thank Feza Günergun, Nelson Arellano, Antoni Roca Rosell, Guillermo Lusa and Jesús Sánchez Miñana for sharing with me several documents as well as for their generous advice. For my article on the Ottoman Jewish and Armenian graduates of the *École centrale*, see Martykánová, Darina: 'A Gateway to the World: Jewish and Armenian Engineers of Ottoman Background at the École centrale des arts et manufactures (1853–1923)', *Diasporas. Circulations, migrations, histoires*, Vol. 29 (2017), pp. 29–51.
2. Grelon, Amdré: 'Du bon usage du modèle étranger: la mise en place de l'École centrale des arts et manufactures', in Grelon, A., Gouzévitch, I.,

Karvar, A. (ed.), *La formation des ingénieurs en perspective: Modèles de référence et réseaux de médiation: XVIIIe – XXe siècles*, Rennes: Presses universitaires de Rennes, 2004, pp. 17–21. Belhoste, Bruno: 'Ingénieurs civils contre ingénieurs de l'État: la création de l'École centrale des arts et manufactures et le tournant de 1830', in Nicolaïdis, E., Chatzis, K.: *Science, Technology and the 19th-Century State*, Athens: Institut de recherches néohelléniques/Fondation nationale de la recherche scientifique, 2000, pp. 45–56.

3. I discuss this issue in Martykánová, Darina: 'Les ingénieurs entre la France et l'Empire ottoman (XVIIIe-XXe siècles): un regard mosaïque pour une histoire croisée', *Quaderns d'Historia de l'Enginyeria*, Vol. 15 (2016), pp. 297–319.
4. Programme of the year 1850, *École centrale des arts et manufactures*.
5. This chapter is mostly based on two primary sources: the bulletins (*annuaires*) published by the Association amicale de l'École centrale des arts et manufactures, the alumni society of the École centrale, and the student registers of the school.
6. Ramón Teijelo, Pío-Javier: *El Real Conservatorio de Artes (1824–1887): un intento de fomento e innovación industrial en la España del XIX*, PhD thesis, Universitat Autònoma de Barcelona, 2011, pp. 117–118.
7. For the multiple meanings of civil engineer, see Lundgreen, Peter: 'Engineering Education in Europe and in the USA, 1750–1930: the Rise to Dominance of Schools Culture and the Engineering Profession,' *Annals of Science*, 47 (1990), pp. 33–75. Here I use it to distinguish engineers working in the private sector from state engineers. On engineering schools in nineteenth-century Spain see Silva Suárez, Manuel (ed.): *Técnica e Ingeniería en España*, vol. 5, *El ochocientos, Profesiones e instituciones civiles*, Zaragoza: Real Academia de Ingeniería/ Institución 'Fernando el Católico'/Prensas Universitarias de Zaragoza, 2007.
8. Nadal, Jordi: 'Industrialización y desindustrialización en el sudeste español, 1817–1913', *Moneda y Crédito*, Vol. 120 (1972), pp. 3–80. Nadal, J. (ed.), *Atlas de la industrialización de España: 1750–2000*, Barcelona: Crítica, 2003.
9. On their father, see in Parejo Barranco, J.A.: 'Martín Larios Herreros (1798–1873)', in Parejo Barranco, J.A. (ed.): *Cien empresarios andaluces*, Madrid: LID Editorial empresarial, 2011, pp. 74–80.
10. García García, Emilio: 'El caso Larios (1888): diagnósticos médicos contrapuestos e intereses económicos,' *Revista de Historia de la Psicología*, Vol. 32, 1 (2011), pp. 33–54.

11. Fernández-Paradas, Mercedes: 'Manuel Domingo Larios y Larios (1836–1895)', in Parejo Barranco, J. (ed.): *Cien empresarios andaluces*, Madrid: LID Editorial empresarial, 2011, pp. 283–285.
12. Bulpes Fernández, Carmen: 'Enrique Crooke y Larios: hombre, político e industrial. Una mirada desde la prensa de la Restauración', Universidad de Málaga.
    https://www.academia.edu/31198652/Enrique_Crooke_y_Larios_hombre_pol%C3%ADtico_e_industrial._Una_mirada_desde_la_prensa_de_la_Restauraci%C3%B3n.
13. Transcripts from the local press about the dam in https://eduardoperez-valle.blogspot.com.es/2015/03/a-guisa-de-cronica-advenimientode-la.html. Regarding the political activity of Nicaraguan engineer Fernando Larios, see Selser, G. (ed.): *Cronología de las intervenciones extranjeras en América Latina: 1899–1945*, vol. III, 1899–1945, Mexico: UNAM, 2001, p. 1931.
14. García Montoro, Cristóbal: *Málaga en los comienzos de la industrialización. Manuel Agustín Heredia (1786–1846)*, Córdoba: Universidad de Córdoba, 1978.
15. Fernández-Paradas, Mercedes: 'Jorge Enrique Loring Oyarzábal (1822–1900)', in Parejo Barranco, J. (ed.): *Cien empresarios andaluces*, Madrid: LID Editorial Empresarial, 2011, pp. 177–179; Ramos Frendo, Eva: *Maria Amalia Heredia Livermore, Marquesa de Casa-Loring*, Málaga: Servicio de publicaciones de la Universidad de Málaga, 2000.
16. As stated in the *annuaire* of 1897.
17. Ramón Teijelo, ibid.; Cardoso de Matos, Ana and Roca Rosell, Antoni: L'Ecole centrale, les centraliens et la péninsule ibérique: des intérêts réciproques (in press); Cano Pavón, José M.: "El Real Instituto Industrial de Madrid (1850–1867): medios humanos y materiales", *LLULL*, Vol. 21 (1998), p. 42.
18. Maluquer de Motes, Jordi: *El socialismo en España: 1833–1868*, Barcelona: Crítica, 1977, p. 301.
19. Nonetheless, as there was no state corps of industrial engineers, this step brought little immediate advantage to Montesino.
20. Montesino, Cipriano S.: *Resumen de las lecciones del curso de construcción de máquinas*, Madrid (?): Real Instituto Industrial, 1853?; id. *Rompimiento del Istmo de Suez*, Madrid: Imprenta nacional, 1857. For his political activity, see:
    http://www.senado.es/web/conocersenado/senadohistoria/senado18341923/senadores/fichasenador/index.html?id1=3114

Regarding the Institución Libre de Enseñanza: https://www.ensayistas.org/critica/generales/krausismo/textos/estatutos-ILE.htm

Echávarri Otero, J. et al.: '"El curso de construcción de máquinas" del profesor Cipriano Segundo Montesino', in Emilio Velasco Sánchez et al. (eds.), *XXI Congreso Nacional de Ingeniería Mecánica: Libro de Artículos*, Elche: UMH, 2016, pp. 428–435. The volume 85 (2017) of *Alcántara*, journal published by the Institución Cultural El Brocense, focuses on Montesino.

21. On its opening, see the newspaper ABC: http://hemeroteca.abc.es/nav/Navigate.exe/hemeroteca/madrid/abc/1913/02/18/009.html.
22. Rodrigo y Alharilla, Martín: *La familia Gil. Empresarios catalanes en la Europa del siglo XIX*, Barcelona: Fundación Gas Natural, 2010, pp. 149–177.
23. The life of Eduardo Cabello Ebrentz, Philadelphia-born Spanish state engineer from Cuba, in Montenegro López, Amador (ed.): *Memorias de un ingeniero del siglo XIX: Eduardo Cabello Ebrentz*, Madrid: Colegio de Ingenieros de Caminos, Canales y Puertos, 1991.
24. Fernández de Pinedo, Nadia; Pretel, David: 'Circuits of Knowledge: foreign technology and transnational expertise in nineteenth-century Cuba', in A.B. Leonard and David Pretel (eds.), *The Caribbean and the Atlantic World Economy*, Basingstoke: Palgrave Macmillan, 2015, pp. 263–289.
25. Fernández Prieto, Leida: *Espacio de poder, ciencia y agricultura en Cuba: el Círculo de Hacendados, 1878–1917*, Madrid: CSIC, 2008, pp. 76–78 and others.
26. Ibid. Other Apezteguías appear in the documents: Emilio Apezteguía (grad. 1859), born in La Trinidad, Cuba, is listed as civil engineer living in Paris in 1897. In the school register of 1869, a Monsieur Apezteguia, engineer, appears as Parisian contact person for Richard Viterbo, son of a banker from Constantinople. I was unable to confirm whether this is the same person as or, more probably, a relative of an Emilio Apezteguía of Cienfuegos, merchant and landowner (*hacendado*) who was member of an influential association of Cuban landowners and plantation-owners the Círculo de Hacendados de la Isla de Cuba in 1887–1888. See Fernández Prieto, *Espacio de poder*, p. 95.
27. Puelma Tuper, Alfredo: *La industria azucarera en Chile: y establecimiento de una nueva fábrica nacional de azúcar de betarraga en Santa Fé*, Chile: Imprenta Gutenberg, 1887. http://www.libros.uchile.cl/557.

28. Loustau, César J.: *Influencia de Francia en la architectura de Uruguay*, Montevideo: Trilce, 1995, pp. 52–60. Regarding the founders of the Ateneo: http://www.oocities.org/webateneo/historia/fundadores/fundadores0023.htm.
29. His father Antonio Evaristo Ornellas was a physician born in Portugal and his mother María Dolores Heeren was born in Spain. His wife's name, Ana Ponce de León, indicates further links to Spanish-speaking countries.
    https://gw.geneanet.org/wikifrat?lang=en&n=d+ornellas&nz=de+riqueti+de+mirabeau&ocz=0&p=charles&pz=honore+gabriel.
30. *Diccionario histórico y biográfico de la Revolución Mexicana (primera edición)*. México, D.F.: *Instituto Nacional de Estudios Históricos de la Revolución Mexicana, Secretaría de Gobernación*, 1994.
31. http://biblioteca.clacso.edu.ar/Panama/cela/20120717021132/consideraciones.pdf.
32. It is unclear whether the distillery was in Saint-André on the island of Réunion or in the peninsular France.
33. Working as engineer-chemist for the Compañía de Antofagasta, Carlos de la Mahotière published an article in the French journal *La Génie Civil* in which he put forward the formulae describing the precipitation reaction of iodine with sodium bisulphite. Another Chilean centralien, Alfredo Puelma, published an article critically assessing de la Mahotière's formulae and method. See: Puelma Tupper, Alfredo: *Estudio comparativo de los diversos sistemas del beneficio del yodo en Chile*, 19th January 1892. https://nuevosfoliosbioetica.uchile.cl/index.php/AICH/article/view/30538/32306.
34. Marcos Rodrigue (grad.1913), born in Tucumán (Argentina), who 'died of his wounds' in August 1918, may be another example.
35. Pamuk, Şevket: *The Ottoman Empire and European Capitalism, 1820–1913. Trade, Investment and Production*, Cambridge: Cambridge University Press, 1987; Quataert, Donald: 'The age of reforms, 1812–1914', in Halil Inalcık and Donald Quataert (eds.), *An Economic and Social History of the Ottoman Empire, vol. 2, 1600–1914*, Cambridge: Cambridge University Press, 1997, pp. 759–943; Thobie, Jacques: *Intérêts et impérialisme français dans l'Empire ottoman (1895–1914)*, Paris: Sorbonne, 1977.
36. The biographic data is analysed in detail in Martykánová, 'A Gateway to the World'.
37. Schmitt, Oliver Jens: *Les Levantins. Cadres de vie et identités d'un groupe ethno-confessionnel de l'Empire ottoman au « long » 19e siècle*, Istanbul: Isis Press, 2007.

38. Panutti, Alessandro: *Les Italiens d'Istanbul au XXe siècle: entre préservation identitaire et effacement*, Paris: Université de Paris III – Sorbonne Nouvelle, 2004.
39. Colan, Horia: 'Léon Guillet, l'École centrale de Paris et la Roumanie', *Noesis*, 2003, pp. 91–95.
    http://noesis.crifst.ro/wp-content/uploads/revista/2003/2003_2_01.pdf.
40. George, Joan: *Merchants in Exile: The Armenians of Manchester, England, 1835–1935*, Princeton/London: Gomidas Institute, 2002, pp. 184–185.
41. http://www.blackwellreference.com/public/tocnode?id=g9780631187288_chunk_g97806311872888_ss1-85.
42. Şeni, Nora; Le Tarnec, Sophie: *Les Camondo ou l'éclipse d'une fortune*, Paris: Actes Sud, 1997.
43. This was not a gradual deterioration of intercommunity relations, rather a process of traumatic transformation marked of moments of hope for a new, successful arrangement adapted to modern times. For such moments, see Campos, Michelle U.: *Ottoman Brothers. Muslims, Christians, and Jews in Early 20th Century Palestine*, Stanford: Stanford University Press, 2010.

5

Re-Designing Africa: Railways and Globalization in the Era of the New Imperialism

Maria Paula Diogo and Bruno J. Navarro

Crafting New Anthropogenic Landscapes in Africa: Colonial Infrastructure and Territorial Management[1]

The Berlin Conference (1884–1885) consecrated a new colonial order that fitted perfectly the imperial agendas of Great Britain, Belgium, France, and Germany.[2] As historical rights over colonial territories gave place to the policy of effective occupation of African territories, the rivalries among European colonizing powers became increasingly critical. Portugal, a peripheral country in Europe, responded to this new international framework by launching a set of "civilizing outposts,"[3] to secure Portuguese sovereignty over the hinterland of both

M. P. Diogo (✉) • B. J. Navarro
CIUHCT, Faculty of Science and Technology, NOVA University of Lisbon, Lisbon, Portugal
e-mail: mpd@fct.unl.pt; bjnavarro@gmail.com

© The Author(s) 2018

D. Pretel, L. Camprubí (eds.), *Technology and Globalisation*, Palgrave Studies in Economic History, https://doi.org/10.1007/978-3-319-75450-5_5

Angola and Mozambique and ideally linking the western and the eastern African coasts.

Colonial strategies, however, echoed European political balance, thus limiting the way countries were able to enact their specific agendas. Diplomatic and military events often disguise clashes between opposing technological projects, and technological superiority often acts a soft power, influencing diplomatic negotiations and shaping international treaties. Concepts such as techno-diplomacy, techno-politics and techno-economics helpfully portray the complex entanglements between technology and society at large.[4]

The centerpiece of the Portuguese strategy was, however, not different from the British, the French, or the German ones. Building railways was the most efficient way to "domesticate" and exploit the African natural and human landscape in situ and, at the same time, to establish an economic hierarchy among geographical spaces both within the colonies and in a world-wide context.

As with all colonial agendas, the Portuguese one was designed to exploit the colonies, imposing a European world view that implied altering the physical, social, and economic structure of the colonized territory under the label of the "civilizing mission."[5] Railways and other infrastructure—roads, harbors, dams—designed by engineers transformed both the African landscape per se, by molding it to the needs of building the railway lines (earthmoving, changes in river beds, tunnels), and its use, by carving the way to plantations, mining, and trade outposts in the land formerly used by indigenous as pastures or hunting territories, and by establishing white settlements.

Under the banner of progress, Portuguese engineers crafted technological anthropogenic landscapes in Angola and Mozambique based on the European concept of "resource" and "commodity," displacing and dispossessing native people of their land, changing their representations and uses of the territory, and imposing a new rationale of development. This active intrusion disrupted the native human and natural landscapes, forcing them to integrate the capitalist economic world system.

Portuguese Survival Manual in Africa: Building Railways to Secure the Colonies

In 1877, already in the context of the Scramble for Africa, Portugal mounted two public works expeditions to its African colonies.[6] Headed by two engineers, Rafael Gorjão (Angola) and Joaquim Machado (Mozambique), their purpose was to assess the viability of building two railway lines, one in Angola linking Luanda to Ambaca and the other in Mozambique linking Lourenço Marques (now Maputo) to the Transvaal. By putting forward the South Portuguese Africa project (the so-called Pink Map), which aimed at linking by train the two coasts of the major Portuguese African colonies (western coast of Angola to the eastern coast of Mozambique) and claiming sovereignty over this land corridor, the Portuguese government tried to recover its status as a colonial power and to resist the expansionist imperial policies of other European countries.[7]

Although the Portuguese west–east railway axis never came to life, as it collided with Cecil Rhodes's railway project of connecting Cairo to Cape Town,[8] the building of the African railway system remained critical to the Portuguese colonial agenda and was perceived as the main landmark of the country's presence in Angola and Mozambique.

These two expeditions—which ran in parallel with two much more publicized expeditions led by the Lisbon Geographical Society (Sociedade de Geografia de Lisboa) to secure Portuguese sovereignty across the territories of Zambezi, Bié (Angola), Niassa (Mozambique), and the area today known as Shire River—were the first of various surveys associated with the construction of the railway network both in Angola and Mozambique.[9]

Coal, copper, iron, cobalt, gold, diamonds, timber, rubber, coffee, and cocoa were valuable commodities that demanded strict management rules across the territory, thus granting infrastructure a vital role. Railways played a strategic role in the promotion of commerce, industry, and agriculture, bringing Africa to the heart of the nineteenth-century world economy. Decisions on which route to choose were, therefore, extremely important as they should balance, on the one hand, the economic status quo and, on the other, the leading economic areas of the future.

The detailed reports carried out by engineers reveal how their expertise shaped the Portuguese techno-political agenda and legitimized the governmental elites' political action. Their work established the guidelines for roads, railways, and harbors, the backbone of an integrated economic flow of raw materials, commodities, industrial goods, and labor force. As experts, engineers were responsible for designing the economic geography of the colonies, building an intelligible representation of the space, and decisively influencing central and local governance, the development of an administrative network, the creation of new political and economic centers, and the consolidation or depletion of existing ones.

The absence of basic colonial structures in the Portuguese colonies worried politicians and engineers. As stated by the engineer Joaquim Machado, "to bring civilisation to the natives" was an ideological imperative,[10] and the best way to enhance their desire for progress was to dazzle them with technological devices.[11] However, building a railroad in Africa was a difficult task for Portugal. The effective implementation of railway lines resulted in major political, economic, and diplomatic constraints.

Portugal was a small country unable to assert its sovereignty unilaterally in the concert of nations, often tangled in economic and financial crises, and with few entrepreneurs willing to invest in projects of uncertain profitability and without sound governmental guarantees. The alternative of using private and frequently foreign funds was perhaps inevitable, but it went hand in hand with conflicts concerning different interests and tensions among stakeholders. Even in such a difficult financial context, the Portuguese government was quite committed to build a railway network in Angola and Mozambique. Two examples make this clear. In 1877, from the total national budget for building infrastructure, 41% was allocated to the preliminary studies concerning the railway line Luanda–Ambaca.[12] And in 1878 engineer Joaquim José Machado was shipped off to Lourenço Marques in order to study the railway line linking Lourenço Marques to the Transvaal border, in spite of the fact that he would have to work during "the worse period of the year [...] The season of heavy rain, of strong heat and dangerous fevers."[13]

The year 1878 marked the beginning of railways in African territories under Portuguese administration. The railway from Lourenço Marques to the Transvaal border was the first ordered by the government (Royal

Decree 144, August 10, 1878). The engineer Joaquim Machado was in charge of this project. Four years later, he published in the *Journal of Mines and Public Works* (*Revista de Obras Públicas e Minas*), the official journal of the Portuguese Association of Civil Engineers (Associação dos Engenheiros Civis Portugueses),[14] a memoir about the project titled "Memoir on the Railways from Lourenço Marques to the Transvaal Border" ("Memória ácerca do caminho de ferro de Lourenço Marques à fronteira do Transvaal"). The first chapter was named "The need to build up the railway to civilise Africa." Its subtitle is a quite clear statement of the author's view of the role to be played by railways in Africa. In addition to a technical assessment, four main problems are raised in this memoir: the potential profits of the coveted and extensive African colonies; the backwardness and slavery of the native peoples; new destinations for the Portuguese emigration; and the effective occupation of the territories. After analyzing all four topics, Machado advocates replacing the slave-trader with the entrepreneur, the missionary with the technician (that is, the engineer), and to build up fast means of communication, such as railways, roads, and seaports.

The memoir on the railway from Lourenço Marques to Transvaal and the memoir on the railway from Mossamedes to Bié (1889),[15] both by Joaquim Machado, show a deep concern for the absence of basic colonial structures on which the new project of colonization stemming from the Berlin Conference should be built. The international context, namely the growing interests of England and Germany in the Portuguese African colonies, forced the Portuguese government to reassess its priorities. The main concern was to occupy the inland regions of Angola and Mozambique and to link the two colonies by railways and roads. The presence of Portugal in its African territories had to be made visible to the European great powers.

It is in this context that Joaquim Machado presented his sketch of the first part of the railway from Mossamedes to Bihé, linking Mossamedes to Alto da Chela along 178 km, which provided the necessary starting point for the effective occupation of Angola and was the first step to attain prosperity in Angola and Mozambique. The engineer argued that the prosperity of the English territories of Cape and Natal and of the states of Transvaal and Orange, inhabited by Dutch descendants, vis-à-vis the poverty of Angola and Mozambique, was due to the lack of communication with the good inner plains, especially those of the district of Mossamedes where the climate was similar to that of southern Europe and, therefore, able to welcome countless settlers.[16]

Although some doubts were raised concerning the railway connecting Mossamedes to Bihé—namely concerning the profits of the investment, the appropriateness of its configuration (with considerable slopes, tight curves, and the use of reversions), as well as the appropriateness of the alternative Benguela route—the Portuguese government supported the project.

In the Field: Building the Railways in Angola and in Mozambique[17]

In Angola, the building of the first railway line was a long-time cherished project, postponed for approximately forty years. The aim of the Luanda line was to overcome the constraints of navigation on the Quanza River, establishing a link between the coast and the hinterland, where "the most productive lands and the most valuable elements of wealth" were believed to lie.[18] Portuguese engineers such as Ângelo Sárrea de Sousa Prado and Arnaldo de Novais Guedes Rebelo dealt with the guidelines of the railway line that first connected Luanda to Ambaca, continued deep into the African unexploited wealthy wilderness, and finally reached the Mozambican coast, and particularly its harbors also under construction. The project was commissioned to a Portuguese business consortium, led by Alexandre Peres, who, even with a state subsidy in 1885, was unable to find Portuguese investors and had to call upon the intervention of English trustees to secure the investment using mortgage bonds.

On September 7, 1899, twelve years after the launch of the project, the first 364 km of the railway line was inaugurated. The difficulties of the building site, the African climate, and the 1892 bankruptcy justified the long delay. The dubious financial contours of the venture and the fear that in case of default by the company the English trustees would take over the mortgage and assume ownership and management of the line, thus leading to international diplomatic conflicts, also contributed to this very slow pace. In fact, secret negotiations between England and Germany for the division of the Portuguese empire into spheres of influence between 1898 and 1899 revealed the German desire to control this railway line, a centerpiece of its great colonial design of Mittelafrika.[19]

Initially designed to be a trans-African railway, rivaling Cecil Rhodes's Cape to Cairo dream,[20] the Ambaca line turned into a political, financial, economic, and judicial *imbroglio*—dubious financial contours concerning investments and allegations of promiscuity between political and economic interests. It was only solved in the 1930s, in a completely different world context. The first railway project to be launched in Mozambique linked Lourenço Marques (now Maputo) to the border of the South African Republic of the Transvaal. It was a project discussed and negotiated over several decades. Designed by Portuguese, South African, and British engineers, the Lourenço Marques–Transvaal line had to face several political and diplomatic disputes, such as the judicial suit opposing Portugal and Great Britain on the property at the Lourenço Marques bay or the Anglo-Boer wars (1880–1881, 1899–1902) that significantly altered the international political balance in the area, mostly by limiting the British ambition of dominion over the southern part of Africa.

The concession contract was signed on December 14, 1883 by the Portuguese government with a private company, led by Eduard McMurdo, a colonel in the US Army, known for his interest in speculative business but with the advantage of being neutral as far as European interests in the South African region were concerned. As in Angola, an endless series of misunderstandings compromised the viability of the entire venture, leading to political tensions between Portugal, Great Britain, and the USA that resulted in delays in the construction of the line.

It was an African gauge line (106.7 cm wide) of extraordinary geo- and techno-political relevance to South Africa, a "landlocked country," for securing access to the sea without passing through British-influenced territory.[21] McMurdo saw the economic potential of the line and therefore agreed to settle the route without any public aid. However, he was not able to secure a tariff agreement with British South Africa—and consequently the traffic to the Lourenço Marques harbor—and left the line unfinished 8 km from the border with the Transvaal. The rest of the railroad was completed in 1888 by the Portuguese state, which ensured its exploitation until the 1970s.

The Portuguese grand plan to build a trans-African line from coast to coast, which already faced huge opposition by Cecil Rhodes, had yet another technical obstacle to overcome—the diversity of gauges. Portuguese engineers, most probably because of their French training and interest in the French African lines, favored the meter gauge (100 cm), which was used in the Ambaca line in Angola; on the contrary, the Lourenço Marques railroad used the 106.7 cm distance between rails, which became the standard as English colonial engineering became the leading reference in the African continent.

As political tensions grew in Africa, mostly after the British ultimatum to Portugal and the Fachoda incident between Great Britain and France, both caused by railroad rivalries, railways became increasingly more important as a visible symbol of imperial dominance.[22] Despite financial difficulties—Portugal went bankrupt in 1892—the Portuguese government commissioned in 1891 the Beira railroad (in Mozambique), to Henry Theodore Van Laun, who later passed on the project to the Beira Railway Company Limited. Open in 1898, this line was one of the English impositions on Portugal, following the ultimatum of the 1890s. It was indeed a techno-diplomatic decision that nevertheless served Cecil Rhodes's interests; through his British South African Company he was one of the main shareholders of the Beira Railway Company.

Engineers, Colonial Expertise, and Professional Empowerment

The discussion on the railway system in the Portuguese African territories continued to fill the pages of the *Journal of Mines and Public Works* as one of the major subjects of discussion among Portuguese engineers until the 1930s. The paper by João Pereira Dias on a global plan for the Portuguese railway network in Africa and on the line from Mossamedes to Alto da Chela (1891),[23] the lecture by Henrique Lima e Cunha on the Lobito line (1897),[24] and the paper by Manuel Costa Serrão on the Angola line (1898),[25] are excellent examples of this long-lasting interest.

A general overview of the articles on public works in Africa published in the *Journal of Mines and Public Works* shows a sustained increase in this topic only disturbed during the period of the First World War, when the periodicity of the journal itself is irregular (Fig. 5.1).

In thematic terms, the railways are clearly the majority, reaching 75% of all articles on African topics. The depth of these articles is, of course, variable, from simple enumerative news to articles presenting intervention projects in Africa (Fig. 5.2).

As we have seen, since the mid-nineteenth century the Portuguese colonial agenda placed technology at its core. Technology and engineering provided the metrics of development and progress, and technical objects and systems became the visible face of modernity. The relationship established with non-industrial cultures was shaped by a technological matrix. The civilizing mission was, above all, the integration of Africans into the techno-scientific European rationality and, by extension, in the worldwide economy. Technology and engineers were fundamental pillars of Portuguese colonizing ideology and politics.

Portuguese engineers were quick to realize this new status. The architecture of the African empire was in fact in their hands: communication infrastructure (railways, roads, ports) was the instrument for the economic

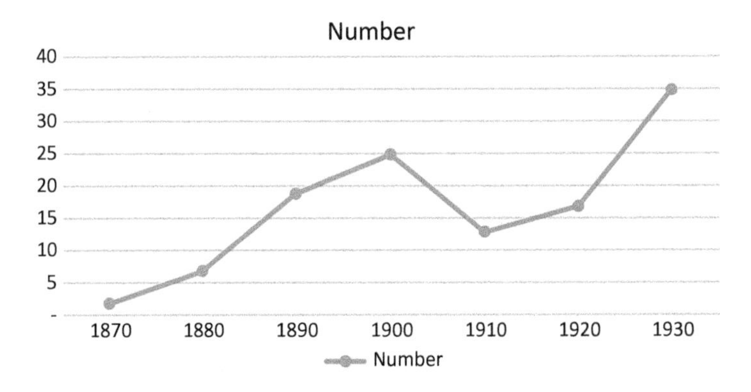

Fig. 5.1 Articles on public works in Africa in the *Journal of Mines and Public Works* (1869–1930)

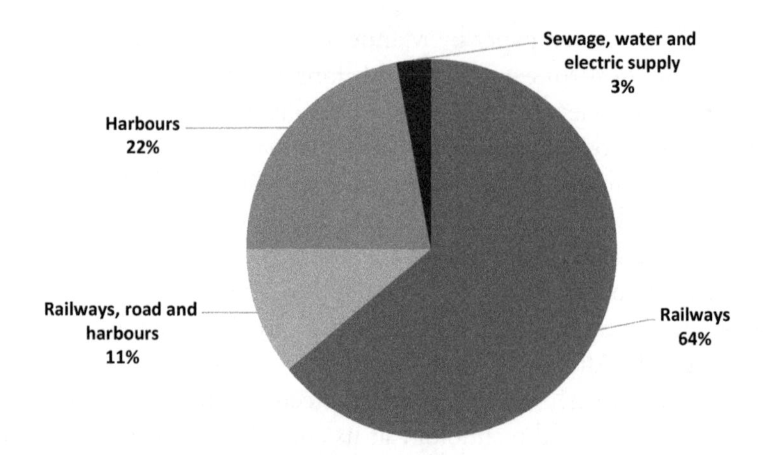

Fig. 5.2 Articles on public works in Africa in the *Journal of Mines and Public Works* by topic (1869–1930)

homogenization of Angola and Mozambique, overcoming the gap between coast and the hinterland, linking the agricultural and mining areas to the neuralgic centers of commerce. Engineers knew they were an indispensable and unavoidable part of the intervention in Africa: they designed the infrastructure that ensured the flux of raw materials, commodities, goods, and people, critical to the Portuguese colonial project, and were thus active actors in the decision-making process concerning Africa's economic geography.

The role of engineers in the African colonies also provided a seductive job market for young engineers, who easily had the chance to work on major infrastructure and to acquire specific and valuable training that later on could be reinvested in their careers back home.[26] In 1899, the *Journal of Public Works and Mining* published an important document known as the *Alvitres* (*Suggestions*),[27] in which Portuguese engineers discuss the main topics concerning their profession and propose to the government a set of measures that should be undertaken. Concerning the problems of employment (suggestions 9 and 10), the document emphasizes both the importance of the African territories as a promising job market and the role played by engineers as the artisans of progress and sentinels of national pride in the overseas territories.

Prestige was, undoubtedly, a commodity that was available in the colonial territories; both individual prestige and professional status. Engineers as a professional group were changing the face of the Portuguese African territories by building infrastructure that secured the empire. Conscious that their intervention in the colonies had a translatable value in terms of know-how, prestige, and capacity for political intervention, Portuguese engineers remained extremely active in this market, constituting one of the fundamental tools of the colonial agenda. One of the many articles published in the pages of the *Journal of Mines and Public Works* stated that "When colonial nations want to take real possession of their territories they send their engineers overseas."[28]

"Portugal Is Not a Small Country": Building Portuguese Landscapes in Africa

In 1934, Henrique Galvão, an army officer and supporter of the Portuguese dictatorship,[29] prepared the map *Portugal não é um país pequeno* (*Portugal is not a small country*) to be presented at the Portuguese Colonial Exhibition, held in Porto in 1934. As general commissioner of the exhibition, Galvão wanted to showcase Portugal as a strong colonial power (Fig. 5.3).

The English and the French versions of the map are slightly but significantly different: while in the Portuguese version Portugal is painted in yellow, as all other European countries (image a), thus bringing to the forefront its position as colonizer, the international versions present Portugal and the colonies both in red (image b), highlighting the message of an unified strong and large country. These two different messages embodied the Portuguese colonial agenda: the overseas territories were used as a strategic asset when Portugal had to negotiate its place in the new worldwide economy and as territories to be exploited and used in a typical colonizer/colonized relationship. The success of both these strategies heavily depended on the efficient management of the territory, both concerning its natural and human landscapes. Technology, and particularly the building of infrastructure that secured the flows of raw materials,

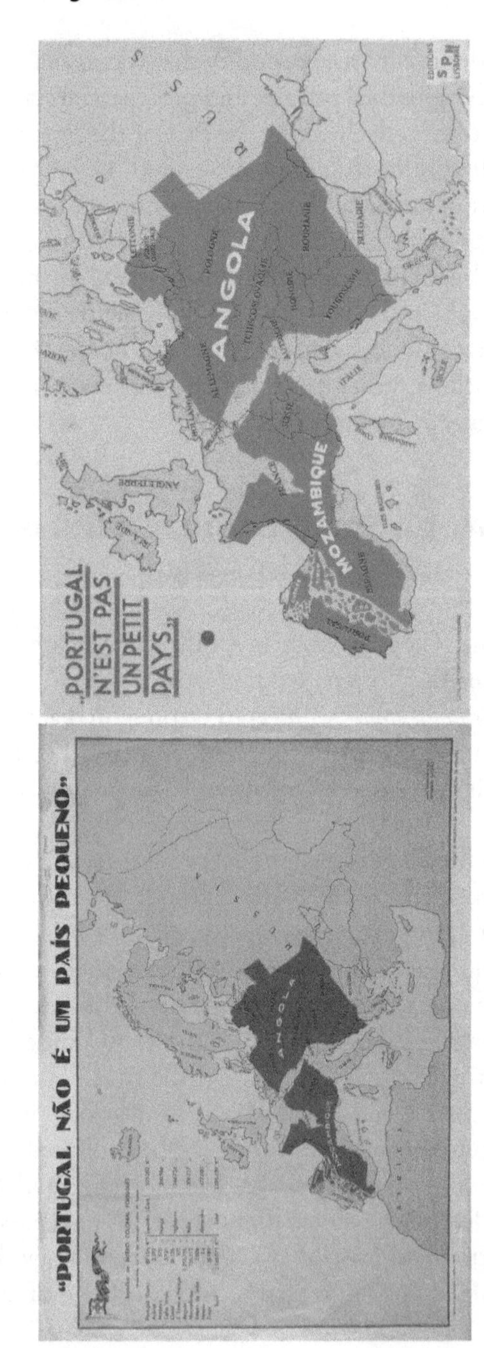

Fig. 5.3 **(a)** Portugal is not a small county (Portugal não é um país pequeno)—national version. **(b)** Portugal is not a small county (Portugal não é um país pequeno)—international version

commodities, goods, and people, was at the core of the colonial agenda, carrying with it the profound change of local traditional appropriation and use of the territory.

In fact, Portuguese colonial engineers played a decisive role in mapping and changing the landscape of overseas territories. The reports and projects on public works, particularly the technical and management guidelines and procedures for building railways, are critical to understanding the scale of changes introduced both in urban and rural landscapes in the African colonies. The Portuguese colonial policy of "material improvements" shaped the territory in order to mirror the European mainland, thus creating a new geography and hierarchy of spaces that was imposed on the traditional and local ways of appropriating and using the land.

Moreover, the creation of European-driven anthropogenic landscapes in Africa was at the core of the colonial mythology, which was largely based on the building of infrastructure. An important part of that mythology was to present Africa as a *terra nullius*, as if African landscapes had not already been transformed by human populations for millennia but were "virgin" for the Europeans to turn them into human landscapes. Engineers such as Joaquim José Machado pioneered the building of the Portuguese colonial imaginary, namely by keeping a systematic recording of detailed observations, which went far beyond the immediate purposes of engineering as a science. Both in journals and in conferences addressing audiences engaged in Portuguese colonial policy, he portrayed the territory, the life of the indigenous peoples, and, mostly, the difficulties white men experienced when trying to transplant the European technological framework to the African context.[30] During the building of the Lourenço Marques–Transvaal railway, he witnessed the precarious working conditions, owing to the limitations imposed by the wilderness and the adverse climate. The thick bush—virtually unknown to national and international cartography—was a recurrent obstacle that compromised the rigor and reliability of technical studies, as perceived by European engineers; the climate, particularly during the season of rains, forced frequent stops and hosted dangerous insects that caused disease outbreaks (sleeping sickness, marsh fevers).[31] These difficult and insalubrious conditions, which Machado tried to soften in his descriptions, gave rise to a regional black legend, widely exploited by the neighboring British colonies eager to reduce the geographical and hydrographical potential of the Lourenço Marques harbor and its most advantageous tariffs. By encouraging these rumors, Great Britain defended its own interests in the southern part of Africa, by inhibiting the hiring of workers for the Portuguese railway and enforcing a change in local commercial routes.

In 1924, Francisco Pinto Teixeira, a colonial engineer who would play a decisive role in the reorganization of the railways in Mozambique, also recorded crucial information for a better understanding of the difficulties that Portuguese engineers faced in their missions in Africa. At a conference held at the Association of Portuguese Civil Engineers, Teixeira reminded the audience of the new axiom established in the Treaty of Versailles: it was only worth the countries that had the resources to rapidly develop colonies having them.[32] Having been commissioned to scout the area around the Bay of Tigers, he argued that despite the public work missions in Africa being already a long tradition, working conditions had not changed significantly. At the end of the day, and as Machado also mentioned, engineers were forced to dismiss their traditional training and often improvise to solve unexpected challenges that were not listed in their French textbooks. As in the *Heart of Darkness* by Joseph Conrad, it is clear in these accounts that there was a deep gap between the triumphalist colonial rhetoric and the conditions in the field to truly implement it.

Two years later, in 1926, Alfredo Augusto de Lima addressed the question of colonial engineers as a specific group within the civil service demanding an urgent requalification of the overseas engineering career, not only in terms of professional promotions, but also concerning expertise. To assert this specificity within the colonial administration framework, Lima proposed the creation of a specific course on colonial engineering in the curriculum of the Colonial Higher Education School, drawing government and public attention to the well-known difficulties and complexity of exercising the profession of engineer in an unpredictable and dangerous environment (torrential rains, floods, cyclones, corrosion of building materials, instability of river flows, severe drought in some regions, absence of technical and scientific instruments, and inaccuracy of cartographic and meteorological information).

The fourteenth course on Notions of Civil Constructions, Colonial Constructions and Roads responded to this need for a specific training and career. Alfredo Augusto de Lima was responsible for the course and in its presentation to academia he stressed again the heavy responsibilities of colonial engineers and the difficulties they had to face every time they had to work in wild regions that were not yet domesticated by European infrastructure.

Engineers were in fact the eyes of the state in Africa: their reports showed how their expertise was critical to carry forward the Portuguese colonial agenda.[33] Political power, as well as members of the intellectual elite, gathered around the Geographical Society of Lisbon and the Association of Portuguese Civil Engineers to praise enthusiastically the work of these experts as active builders of new landscapes that allowed Portugal to take possession of its colonies. Their specialized opinion, regarded as politically independent, was regularly used to legitimize colonial development policies, namely supporting laws and diplomatic negotiations for the delimitation of overseas borders and participation in ministerial and parliamentary decisions.[34] It was their work in the field that allowed Portugal to claim its vast African empire as portrayed in Galvão's maps.

Heroism and Criticism

The colonial engineer was thus part of a powerful imagery, playing the role of an adventurous explorer with an aura of heroism and patriotism, quickly matching and replacing the missionary as the centerpiece of colonization. Colonial engineers nourished this image by consolidating a collective memory of engagement in the service of Portugal, which would have positive repercussions in terms of professional careers. Lopes Galvão was one of the most outstanding champions of the merits of Portuguese engineering and its "brilliant action towards the development of our overseas lands," which was at the core of the economic resurgence of the colonial empire after the First World War.[35]

António Vicente Ferreira, professor at the Technical Institute of Lisbon, the main engineering school in Portugal at the time, went even further in considering colonial engineers as "social leaders" in the process of colonization of the African provinces. Being "first rate personalities" in the context of the white minority, given their "real or presumed" capacities as organizers, one should consider them as naturally fitted to lead the economic and social policy in emerging countries. Technical leadership was also relevant concerning their privileged contact with the natives, with whom they often interacted, thus embodying "the prestige of the white race" by affirming its qualities of leadership, "steadiness of character, human sympathy and ability to understand the indigenous mentality."[36]

He emphasized that engineers were an important piece of a larger elite of "agents of civilization," which also encompassed doctors, missionaries, and administrative authorities, who carried the heavy burden of "civilizing" both nature and humans in the African continent by building anthropogenic landscapes. They created large-scale infrastructure that made a new territorial order to which the old local ways had to adapt. Although this idea was not original (Joaquim José Machado had already mentioned it in several of his reports, writings, and presentations), Vicente Ferreira used it as an instrumental concept to assign engineers the task of promoting the "human geography of the colony for the benefit of all who inhabited it."[37] Vicente Ferreira considered that colonial engineers had the moral and professional duty of knowing extensively the "physical, biological and human environment" upon which they acted in order to change it through scientific and technological approaches and interventions.

After years of investment in infrastructure, the balance of the Portuguese colonizing effort was, however, ambivalent. On the one hand, it was clear that "progress" had made its way through the wilderness, particularly by domesticating the territory and exploiting its resources within the new imperial post-Berlin Conference agenda. This clearly utilitarian-driven agenda was designed specifically to counteract the British (and in some extend the German) predatory agendas that looked at the large Portuguese colonial empire as easy prey.

The official triumphalist narrative admittedly shared by many engineers was clearly part of a motivational discourse based on the concept of "moral victories"—for instance the transcontinental railway linking the west coast of Angola to the eastern coast of Mozambique—in which

Portuguese engineering, although in the vanguard of African colonization, was prevented from assuming a leading role because Portugal was a small country in the European arena.

At the same time, however, this patriotic rhetoric was seasoned by a much more critical assessment of the real limitations of the Portuguese colonial action. In this parallel approach engineers pointed out the entropy of developmental projects, which were set in sparsely populated overseas territories, subject to the inclement climate of the tropics and deprived of the immense mineral wealth that had triggered the economic development of the neighboring colonies. Pressured by the urgent need to affirm their rights of sovereignty and with limited possibilities concerning the economic use of the territory—agriculture was the only viable option available and only possible in the most fertile lands—dependent on the migratory flows from Portugal to Africa and facing a slow process of "civilization" of the indigenous peoples, Portugal anchored its colonial and imperial hopes in railways. Railroad lines supported new commercial dynamics by linking the rich agricultural and mining hinterland to the seaports and decisively influencing the attractiveness of new settlers, which in turn would give rise to new colonialized lands.

Soon it became clear that the rational of this agenda was compromised both by the clash of international imperial interests and by the impotence of Portugal to play the game on equal terms. Being a peripheral country with little international agency and a subaltern status concerning diplomatic relationships and marked by political instability, economic and financial crisis, Portugal was not able to cement its presence in Africa through a coherent and credible plan of colonization. On a regular basis and particularly in Africanist forums, colonial engineers addressed the erratic profile of the Portuguese colonial administration, regretting the lack of a clear long-term plan, previously discussed and agreed between experts and politicians that would allow for a continuous implementation regardless of partisan rivalries.

Portuguese foreign policy, and particularly colonial policy, was thus subordinated to opportunities and not to a previously set plan, most difficulties being solved case by case and not as part of a more global picture based on a rigorous assessment of the consequences of both political and technical decisions. It was this drifting approach that explains, for instance, the policy of granting royalties or the construction and exploitation of strategic railways to foreign companies that, more or less disguisedly, were at the service of other nations' deliberate and systematic imperialist agendas, It was this same case-by-case appraisal that explains the choice of transnational railways, much more in the service of the interests of the neighboring colonies than of the real needs of the Portuguese colonization agenda, and the chaotic multiplication of gauges. The lack of coordination and scattering of responsibilities and competences—often disguised as autonomy—was particularly striking in Mozambique's railway management structure, where for 656 km of railroad there were four general directions and several commissions for improvements, in stark contrast to the centralizing administrative profile that had already been adopted in South Africa.

After the military coup of 1926 that eventually led to the Estado Novo dictatorship,[38] Portuguese colonial engineers and administrators demanded urgent reforms that, respecting the nationalist criteria in vogue, would centralize the railroad administration of the colonies and enforce unity of action by putting an end to a desolate scenario of total inertia and slouch. Despite the "highly qualified" Portuguese engineering corps, the truth is that Portuguese railways in Africa and mostly in Mozambique were still insignificant in relation to the area they covered. In the words of one of these critics, wittily summarizing the situation despite its seriousness, "When there are engineers to make the required technical studies, there is no money to pay them; when there is money there are no engineers available and when there is money and engineers there is no material. And when you open a tender to purchase the material in need, there are protests that take years to solve."[39] No wonder, then, that in the early 1930s the colonization of Portuguese Africa was still incipient and considered by a large part of the ruling elite as a marginal topic. The colonial question was thus a niche for colonial experts—engineers, scientists, and physicians—and very few politicians. The Lisbon Geographical Society, together with the Portuguese Association of Civil Engineers, continued to champion the colonial agenda although poorly supported by the public authorities. During the 1930s, reclaiming the Portuguese colonial tradition, new public works brigades were formed and sent to Angola and Mozambique. Although these expeditions presented a large set of studies concerning colonial infrastructure, most of them were forgotten on the shelves of the ministry. The only result of this last breath was the construction of the Tete railroad, which began in 1938 and was completed in 1949, connecting the Moatize coal region to the port of Beira. This was served, as we saw, by the Transzambian railway, owned by foreign capital.

Conclusion

Portuguese engineers perceived their work in colonial settings as a way of developing technical expertise and acquiring prestige, both in national and professional contexts, and as a new labor market for professionals who could use their African experience as a trampoline to their future European careers. One should not take their corporate ideology at face value, and we have pointed to several instances in which their plans failed or produced unintended consequences. However, we do take seriously their status as agents of empire. The technological landscape of the Portuguese empire in Africa was decisive in keeping Portugal as an active, even if fragile, actor in the European and international scene, engineering ways to translate colonial power into power in the European arena, and thus avoiding being completely overrun by the English, French, and German colonial agendas.

On the other hand, as in all other imperial agendas, this process of domestication of overseas territories—which accounts for the asymmetrical relationship between colonizers and colonized, built on domination and coercive forms of disciplining the territory and its inhabitants—is critical to understanding and discussing the complex entanglements between land management and ecological, social, cultural, and economic systems. Engineers were at the heart of the technologies of the land and of the technologies of the state,[40] which redesigned the African landscape and affirmed Portuguese power in Africa. Large technological systems—railways, roads, dams, urbanization, sanitation—allowed for the successful implementation of new agricultural, commercial, and demographic maps in colonial settings, thus designing unbalanced landscapes of prosperity and poverty on a global scale that are still visible (and maintained) today in post-colonial contexts.

Notes

1. Anthropogenic Landscapes or "Human Landscapes" are areas where direct human alteration of ecological patterns and processes is significant. We use this concept to stress that European colonial infrastructures are not just instrumental in managing the territory, but they create and support specific entanglements between ecological, social, cultural, and economic systems.
2. The Berlin Conference (1884–1885) rewove the fabric of the traditional colonial order. The principle of the "effective occupation" of colonial territories displaced the traditional historical rights, clearly favoring the British, Belgian, and German colonial agendas that aimed at securing African new sources of raw materials and markets.
3. The term was very much in use during the nineteenth century to identify niches of European civilization that contrasted with the "primitivism" of local traditional ways of life. These outposts were perceived as strongholds of civilization in hostile territories and acted as a basis for future military actions. In the context of European rivalries over imperial spaces, these outposts were national signs of dominance.
4. The concept of techno-diplomacy was first used by Der Derian, J.: *On Diplomacy: A Genealogy of Western Estrangement*, Oxford: Basil Blackwell, 1987. It accommodates a cascade of other terms such as digital diplomacy and railways diplomacy. The term techno-politics was used by Hecht, G.: *The Radiance of France: Nuclear Power and National Identity*, Cambridge (MA): MIT Press, 1998; the term techno-economics is largely used by economists and historians of technology.
5. The civilizing mission, also known as the *mission civilisatrice*, is the nineteenth- and twentieth-century colonial rationale that considers Europe has the moral duty to spread civilization by Westernizing indigenous peoples in accordance with the colonial ideology of assimilation. The "white man's burden" was, thus, to bring European civilization to what were perceived as backward peoples. This rationale is particularly relevant for the French and the Portuguese colonial agendas.
6. The Scramble for Africa was the occupation, division, and colonization of Africa by European powers between 1881 and 1914.
7. Portugal lost a lot with the Berlin Conference: the effective occupation rule was imposed against the historical rights defended by Portugal, the free navigation of the African rivers was also imposed (Portugal lost its

rights concerning the rivers Congo, Zambezi, and Rovuma), and Portugal lost the territories of the estuary of the Congo.

8. Cecil Rhodes's Cape to Cairo railway project and his concept of an "all red" (only British) railway line is a classic of colonial historiography. Conflicting Portuguese and French railway projects are much less known. Portugal proposed the so-called Pink Map project that aimed at linking the west coast colony of Angola with its east coast colony of Mozambique, more specifically at connecting Luanda and Lourenço Marques. France planned to build a railway line across its colonies from west to east across the continent, from Senegal to Djibouti. See Diogo, M. P. and van Laak, D.: *Europeans Globalizing: Mapping, Exploiting, Exchanging*, London and New York: Palgrave Macmillan, 2016.
9. It is the expedition to the river Shire, headed by Serpa Pinto, that was the pretext for the British Ultimatum, a critical event in Portuguese history that eventually led to the Republican Revolution in 1910. The British government claimed that the Portuguese explorer had lowered the British flag and taken over by force the territory of the pro-English tribe of the Makololos.
10. Letter of the Chief Engineer of the Companhia Real dos Caminhos de Ferro atravez d'Africa (Royal Railway Company across Africa), 1888. AHU, 2678, Room 3, Bookxase16, Shelf 17, n.13420.
11. Machado, Joaquim José: 'Memória ácerca do caminho de ferro de Lourenço Marques à fronteira do Transvaal', *Revista de Obras Públicas e Minas*, No. 12 (1882), pp. 1–57; David Nye's concept of technological sublime fully applies to this case. See Nye, D.: *American Technological Sublime*, Cambridge (Mass.): MIT Press,1996.
12. AHU, 866, DGU, 3ªRep. 1874–78.
13. Letter of J.J. Machado, AHU, 2678, Room 3, Bookcase 16, Shelf 17, n.119.
14. The Portuguese Association of Civil Engineers (Associação dos Engenheiros Civis Portugueses) was created in 1869. It was the first professional association of engineers in Portugal and aimed at bringing together all engineers that although having a military training (the only one existing in Portugal at the time) worked as civil engineers following the Ponts et Chaussées' spirit. See Diogo, Maria Paula: 'In search of a professional identity – The Associação dos Engenheiros Civis Portuguezes', *ICON*, No. 2 (1996), pp. 123–137.
15. Machado, Joaquim José: 'Caminho de ferro de Mossamedes ao Bihé', *Revista de Obras Públicas e Minas*, No. 21 (1890), pp. 219–296.

16. Between 1879 and 1889, 100,000 Portuguese from mainland Portugal, the Azores, and Madeira left Portugal for Brazil. Machado, Caminho de Ferro, pp. 236–7.
17. For a detailed account on the building of Portuguese African railways see Navarro, B. J.: *Um império projéctado pelo "silvo da locomotiva" O papel da engenharia portuguesa na apropriação do espaço colonial africano. Angola e Moçambique (1869–1930)*, Lisboa: CIUHCT/Colibri, 2018; See also Pereira, Hugo Silveira: 'Especulação, tecnodiplomacia e os caminhos de ferro coloniais entre 1857 e 1881', *História. Revista da FLUP*, IV, No. 7 (2017), pp. 137–162.
18. *Legislação e disposições regulamentares sobre caminhos-de-ferro ultramarinos*, Vol. 1 (1857–1894), p. 17.
19. Mittelafrika was the name of the German project to build a continuous strip of German dominions in Africa, from the Atlantic to the Indian Ocean, by annexing the Portuguese territories of Angola and Mozambique to the German colonies of East Africa (Rwanda, Burundi, and Tanzania), South-West Africa (Namibia), and Cameroon.
20. Diogo and van Laak, *Europeans Globalizing*, pp. 148–165.
21. Faye, Michael L. et al.: 'The challenges facing landlocked developing countries', *Journal of Human Development* Vol. 5, No.1 (2004), pp. 31–69.
22. Both incidents result from the conflict between Portuguese and French railways and the Cairo to Cape railway of Cecil Rhodes. See Diogo and van Laak, *Europeans Globalizing*, pp. 148–165.
23. Dias, João José Pereira: 'O Caminho de Ferro de Mossamedes', *Revista de Obras Públicas e Minas*, Vol. XXII, 255–256 (1891), pp. 62–75.
24. Cunha, Henrique Lima: 'Caminhos de Ferro de Benguella a Mossamedes', *Revista de Obras Públicas e Minas*, Vol. XXVIII, 329–330 (1897), pp. 257–273.
25. Serrão Manuel Costa: 'Systema Ferro-Viário de Penetração em África – Linha do Sul de Angola', *Revista de Obras Públicas e Minas*, Vol. XXXI, 367–369 (1900), pp. 211–351.
26. This pattern continued throughout the twentieth century (until 1974) during the Portuguese dictatorship (*Estado Novo*). See Saraiva, Tiago (eds.): 'Science, Technology and Fascism' (special issue), *HoST, Journal of History of Science and Technology*, Vol. 3 (2003).
27. Diogo, Maria Paula: 'Industria e engenheiros no Portugal de fins do século XIX: o caso de uma relação difícil', *Scripta Nova*, Vol. 69, No. 6 (2000).

28. "Exposição" *Revista de Obras Públicas e Minas*, 1899, pp. 353–354, 382–383.
29. Portugal was ruled by a dictatorship from 1926 until the Carnation Revolution in 1974: first by the military National Dictatorship (1926–1933) and afterwards by the *Estado Novo*, an authoritarian, autocratic and corporatist regime led by Oliveira Salazar (1933–1974). Henrique Galvão participated in the military coup of 1926 and was a fierce supporter of Salazar until the 1950s, when he became critical of the regime. Accused of conspiring against Salazar, he was imprisoned and expelled from the army but he managed to escape in1959, taking refuge in the embassy of Argentina and having obtained political exile in Venezuela. It was during the exile that Galvão performed a spectacular action against the Portuguese dictator—Operation Dulcineia—hijacking the Portuguese ship *Santa Maria*, full of passengers.
30. Actas das Sessões da Sociedade de Geografia de Lisboa, Vol. I (1876–1881), p. 274.
31. Machado, J.J.: *Relatório acerca dos trabalhos para a fixação da directriz do caminho-de-ferro projectado entre Lourenço Marques e a fronteira do Transvaal*, Lisboa: Imprensa Nacional, 1884.
32. Francisco Pinto Teixeira: 'Gazeta dos Caminhos de Ferro', No. 1 (February to July 16, 1925). The Treaty of Versailles ended the state of war between Germany and the Allied Powers; Article 119 of the treaty required Germany to renounce sovereignty over former colonies.
33. Diogo, M. P.: 'Um olhar introspectivo: a Revista de Obras Públicas e Minas e a Engenharia Colonial' in Diogo, M. P. and Amaral, I. (eds.): *A outra face do Império*, pp. 81–82.
34. Cardoso de Matos, A.; Santos, M. L.; Diogo, M. P.: 'Obra, engenho e arte nas raízes da engenharia em Portugal' in Heitor, M., Brito, J. M. B. and Rollo, M. F. (eds.): *Momentos de Inovação em Engenharia em Portugal no Século XX*, Lisboa: Dom Quixote, 2004, pp. 25–27.
35. Galvão, J. A. L.: *A engenharia portuguesa na moderna obra de colonização*, Lisboa: Agência Geral das Colónias, 1940.
36. Ferreira, A.F.: *Estudos Ultramarinos*, Vol. III: *Angola e os seus problemas* (part 2), Lisboa: Agência Geral do Ultramar, 1954, pp. 198–200.
37. Ferreira, *Estudos Ultramarinos*, pp. 181–203.
38. The May 28, 1926 *coup d'état* was a military nationalist coup that put an end to the unstable Portuguese First Republic (1910–1926) and initiated the Ditadura Nacional (National Dictatorship), which later became

the Estado Novo, an authoritarian dictatorship led by Oliveira Salazar, that lasted until the Carnation Revolution in 1974.

39. Galvão, J. A. L.: 'Rede ferroviária de Moçambique em relação com as possibilidades da colónia', *Boletim da Sociedade de Geografia de Lisboa*, 7–8 (1929), pp. 278–279; 'Importância dos caminhos-de-ferro no desenvolvimento económico das Colónias', *Boletim da Sociedade de Geografia de Lisboa*, 7–8, (1929), pp. 299–301.
40. Arnold, David: 'Europe, technology, and colonialism in the 20th century', *History and Technology* Vol. 21, No. 1 (2005), pp. 85–106.

6

The Global Rise of Patent Expertise During the Late Nineteenth Century

David Pretel

Introduction

Ever since the institutionalisation of modern innovation systems in industrial and industrialising countries, diverse social actors—from lawyers to engineers, from intermediaries to consultants—have been active participants in the regulation and operation of patent institutions. The period from about the 1870s to the early twentieth century marked a pivotal moment in the nature and level of participation of specialised experts in patent rights in several national systems. Since then, these 'invisible' agents have shaped the direction and transmission of technical innovation worldwide. Their particular kind of expertise has consistently contributed to the making of new technologies and the management of property rights over such technologies. It has also, however, been a source

D. Pretel (✉)
Centro de Estudios Históricos (CEH), College of Mexico,
Mexico City, Mexico
e-mail: dpretel@colmex.mx

© The Author(s) 2018

D. Pretel, L. Camprubí (eds.), *Technology and Globalisation*, Palgrave Studies in Economic History, https://doi.org/10.1007/978-3-319-75450-5_6

of controversy insofar as patent experts have been regarded as actors that support an excessive concentration of power over technologies, a problem that remains endemic in modern societies.

The diffusion of science and technology is, as Steven Shapin observes, 'an active process that is undertaken by specifiable social groups for their particular purposes'.[1] This is especially true for the production and transmission of patent rights and patented technologies within and among societies. Patenting and patent management are complex activities in which a varied range of experts participate. This expertise is a central component of the modern institutional arrangements by which patents are granted. An array of social actors are active mediators in the process of transforming an idea or a material device into a piece of tradable private property in the form of a patent.

Most of the studies on patent professionals hitherto available have adopted a national perspective that rarely examines the rise of patent agents beyond the country of study.[2] The historiography has tended to focus on case studies of agents assisting specific inventors or firms. The question of the international extension of this professional activity has not been thoroughly analysed thus far. The role of agents in international dynamics, such as international patenting, has likewise been overlooked.

This chapter adopts an international perspective. The interplay between patenting activity and globalisation is a subject of growing interest.[3] With the late nineteenth-century wave of globalisation, industrial property regimes achieved an international reach. Patenting turned out to be increasingly transnational, although patent systems remained nation-state institutions. In this context, the active role of patent experts and their expertise during a period of accelerating globalisation is not well understood. Nor does the historiography tell us much about the globalisation of the patent profession.

The purpose of the first section of this chapter is to shed light on the global institutionalisation of patent agents during the late nineteenth century. This section takes a broad comparative perspective, offering some generalisations about the growing centrality of patent experts in several national systems. Here I reflect on the diverse forms of patent expertise in industrial and industrialising countries. The essential insight of this section is that patent professionals of the late nineteenth century had varied roles, reflecting heterogeneous innovation cultures, institutional environments and degrees of industrialisation across countries.

The second section considers patent practitioners as links among national systems. The question here is whether these specialists contributed to international technology transfer and, if so, how? I argue that patent experts should be seen not only as system-builders but also as agents of globalisation. Beginning in the 1870s, not only was patent expertise becoming globalised, but patent experts were acting as globalising agents. In other words, experts became central to the burgeoning connections among diverse patent institutions worldwide. The question of which actors legitimised patent institutions is another concern of this second section. Did these professionals participate in the making, regulation and legal shaping of the international patent system from the 1870s? To address all these questions, I examine the activities of professional associations, international institutions and transnational networks of patent agents. Particularly relevant to this discussion are agents' publications and international directories.

Much of the research analysing patent expertise has concentrated on the business strategies of multinational corporations. Scholars have focused on patent management in high-technological industries and on the in-house agents of European and American companies. Building on this work, I discuss in the third section the controversial role of patent experts as agents of corporate globalism. Neither the problematic aspects of the rise of corporate patenting nor the expansion of agents to peripheral economies should go uncritiqued. Multinational activities in the outer and distant peripheries—that is, Latin America, South Africa and Australia—remain much less well understood. Corporate patent strategies connected to global networks of commodity production, such as in agro-industry and mining, reflect the limits and contradictions of late nineteenth-century technological globalisation.

A Global Profession

Because of the multiple roles of patent practitioners in various historical contexts, the concept of 'patent expert' can be both vague and highly complex. There is no standard definition for patent expertise valid for all countries during the late nineteenth century. Patent expertise was, in Anna Guagnini's words, a 'hybrid occupational activity' at the interface of legal, economic and technological realms.[4] In spite of the ambiguity that characterised their profession, patent practitioners at the turn of the twentieth century had already gained professional 'jurisdiction', to use the term coined by sociologist Andrew Abbott.[5] By the 1880s, so-called patent agents in most European countries and the United States as well as peripheral and colonised countries began to be recognised as a professional body with expertise in legal procedures and technical issues. This professional expansion was not uniform across nations. In many countries, especially those that were less industrialised, patent agencies were less developed. Sizeable professional communities of patent experts were concentrated in just a few countries, with the United States, France and Britain hosting the overwhelming majority. The number of agents in each national system correlated with their level of patent activity. Early industrial countries, such as Britain and France, had more informal agencies, whereas latecomers such as Germany and the United States had larger numbers of formal agencies.

Studying the institutional organisation of national patent systems during the late nineteenth century means going beyond the narrow definition of the present-day profession of 'patent agent'; it means investigating all the social actors who, as Ian Inkster puts it, 'organised relationships and information flows within systems'.[6] In this chapter, the term 'patent expert' refers to the work of patent attorneys (the American term) and patent agents (the British, French and Spanish term), as well as other actors who occupied some intermediate position between inventors, manufacturers, investors and patent officers. I use the terms 'patent practitioner' and 'patent professional' as synonyms for 'patent expert' throughout this chapter. The definition used here is thus broad in scope and transhistorical so as to encompass the various kinds of professionals who operated in different national patent systems and their relationships to these systems. In short, I use 'patent experts' to refer to the professionals who assisted patentees—whether as solicitors, attorneys, intermediaries, legal advisers, property rights merchants or technical consultants within or through intellectual property institutions.

From the early nineteenth century, patent practitioners could be found in Britain, France and the United States. During these foundational years, members of this diffuse community began describing themselves as 'patent agents', although there is little evidence of their early activities.[7] It was not until the expansion of the specialised patent business in the second half of the nineteenth century that agents became a community of full-time professionals, with their own regulations and associations. In both early industrial countries and latecomers, a larger market for patenting services emerged in the second half of the nineteenth century. During the final decades of the century, professional patent agents also developed in industrial latecomers, including Japan, India, Sweden, Germany, Spain and Australia.[8]

The activities of patent practitioners varied geographically, in line with the industrial and institutional disparities from country to country, even region to region. Likewise, the training and background of these practitioners, their influence on governmental regulations and their degree of participation in international patenting differed significantly. The degrees of specialisation, as well as the legal and technical standards required of agents, differed across countries. What is significant here is not so much the national variations themselves but what these variations tell us about the heterogeneity of patent cultures and legal structures during these years.

From the mid-nineteenth century, there was an increasing division of labour between inventors and those who registered and commercialised patents. Professional patent practitioners in both core and peripheral systems participated in various related activities and offered clients a range of services. The primary role of agents was to support patentees in the application process with regard to administrative procedures, patent writing and the preparation of models and drawings. The writing of patent specifications was the core endeavour of this process.[9] The presentation of claims had to be accurate enough to secure a patent, yet vague enough to preclude imitation. It was through the process of writing or translating patent specifications that inventors effectively became owners of their own practical knowledge in different national jurisdictions.

Only the inventive idea embodied in a text could generate rights.[10] The patent specification did not just convey the inventive idea, but it translated and transformed its essence. After being granted a patent, not only did the inventor become the owner of a property right but also the author of the text that embodied the idea. In 1882, A.V. Newton, professional patent agent and fellow of the British Institute of Patent Agents, made this point, stating that a main agent's duties were:

> To collect the inventor's ideas, to arrange them in an intelligible form, and ultimately to embody them in a specification, which will not only stand the scrutiny of the Law Court, but which will effectually prevent any rival manufacturer from doing anything in the direction of the patent […] At the same time, he must be careful that the boundary is not so indefinitely drawn as to overlap existing rights, or to interfere with rights of new-comers.[11]

Marconi's patent on wireless telegraphy, granted in 1897 in Britain, is a particularly telling example of the central importance of patent lawyers and agents in the making of a patent at the turn of the nineteenth century. In a groundbreaking article, Anna Guagnini showed how dependent the Nobel Prize-winning Italian inventor and industrialist was on London-trained agents for the successful writing of the patent specification of his famous invention.[12] Given Marconi's scant formal scientific background and ignorance of British patent application procedures, he could not have succeeded in obtaining the patent on his own. With the help of expert professionals, however, Marconi's specification resulted in a model of patent-writing and subsequent commercial success. The patent experts with whom Marconi was associated included the prestigious London firm of patent agents Carpmael & Co. and the barrister and technical advisor Sir John Fletcher Moulton, a distinguished lawyer with outstanding scientific training. Interestingly, Marconi's agents, especially Moulton, wrote the final specification of the patent. Moulton's contribution was more than a simple adaptation of the text; he introduced real changes in the new technology—he was making technical knowledge—and, more importantly, he transformed a scientific idea into a valuable piece of property. In short, Marconi's agents were connecting different fields and transforming the status of technical knowledge.

One important issue to consider is how specialised patent lawyers provided inventors with vital legal support, negotiating authorship, power and authority in the background.[13] As experts on patent legislation, patent lawyers, attorneys and barristers guided patentees in the preparation of their applications so as to avoid litigation trials, and represented them in courts of law in the event of infringement proceedings. The increase in litigation that occurred during the late nineteenth century reflected the need for patentees to enforce their patents in court in order to ensure the successful commercial exploitation of their inventions.[14] The rise of legal assistants and expert witnesses in patent disputes, meanwhile, indicated that the capitalist appropriation of invention was a collaborative process in which various actors were interacting. This argument connects to a long-running debate in technology studies. Indeed, the legal process of making a patent largely negates the traditional heroic accounts of technological progress.[15] The participation of lawyers and expert witnesses in patent appropriation has serious implications that continue to threaten the very foundations of intellectual property institutions.[16]

Primarily in core industrial countries, some experts acted as patent brokers, assisting inventors in the commercial exploitation of their patented technologies. These intermediaries helped patentees to assign and license their property rights and to facilitate external investment. These professionals connected two realms, invention and the market, that were already becoming closely linked by the mid-nineteenth century. Being a valuable commercial asset, a patent—or the prospect of one—could be sold and licensed. Finding some commercial sponsor or partnership could be an essential way of minimising manufacturing risks as well as attracting resources to finance the invention process.[17] Several large firms, such as the 'hub company' Brush Electric, and professional inventors such as Thomas Edison, Lord Kelvin and George Westinghouse, participated intensively in the American and British markets for technology through patent rights licensing and partnerships. The practice of collecting patent royalties as an alternative to monopoly made agents indispensable in the market for invention. It was, however, in the United States where expertise in the patent trade, systematic licensing and partnerships developed most rapidly.[18] In countries with less-developed markets for patents, such as Spain and Mexico, the occupational activity of brokerages was less professionalised than in the United States.[19]

In the period 1870–1900, agents placed themselves at the centre of patent systems; as a result their services became indispensable for successful patent application. They monopolised the process of obtaining patents in European countries such as Britain, Spain, Sweden and France. A small number of high-priced agents with technical and legal backgrounds, concentrated in European capital cities (such as London, Paris and Madrid), dominated the management of economically valuable patents through influence and close personal contact with patent officials.[20] By contrast, in the United States, patent agencies were not concentrated in a single city but distributed among several commercial and industrial cities, namely New York, Boston, Washington DC and Philadelphia. American agents were more involved in the commercialisation of intellectual property rights and the financing of inventions, owing to the larger size and diversified structure of that country's market for technologies. The higher rate of patent litigation in the United States necessitated a higher number of agents who specialised in patent trials.

British, Spanish and French agents focused primarily on the patent application process owing to the highly bureaucratic, costly and complex administrative processing of patent applications. The British and French systems, in contrast to the American one, have been accused of concentrating patent applications among a restricted elite community of patentees and agents, thereby making access to intellectual protection less democratic. The American patent system appears to have encouraged patent applications across a wider range of individuals and regions.[21] During the 1880s and 1890s, patent intermediaries in industrial latecomers and colonised countries acted mainly as agents of technology diffusion, working throughout patent systems designed not only to protect invention but also to promote industrial emulation. In countries with reduced markets for patents, such as Spain, agents' activities overlapped with other occupations, for example engineering consultancy or commercial endeavours. Even during the 1890s, many Spanish lawyers and engineers were still not yet professionally engaged as full-time patent professionals and did not recognise themselves as such.[22]

During the second half of the nineteenth century, patent agents published their own specialised technical and trade journals that provided detailed information about patent procedures and descriptions of patented technology. The most relevant examples of patent journals include the weekly *Scientific American*, edited by the leading American patent agency Munn & Co., the *Patent Journal and Inventors' Magazine*, edited by the London patent business of Barlow, Payne and Parker, and the monthly publication *Le Génie Industrielle*, published by the French *ingénieur-conseils* from the Armengaud family.[23] For Latin America and the Caribbean, the agency Munn & Co. published, from 1890, its first international edition, *La América Científica e Industrial*, a mechanical magazine written in Spanish. These widely circulated agents' journals contained lists of patents, reports about patent issues and information on the necessary specialised training that agents required. Meanwhile, patent agents published manuals, pamphlets and doctrinal treatises on how to invent and patent, national and international patent laws, practical mechanics and draught, and other issues of interest for the development of the patent profession.[24] An early example of a patent agent's manual is Armengaud's *Practical Draughtsman's Book of Industrial Design* (1853), translated into English by the British agent William Johnson.[25]

Another critical dimension of patent experts' activity throughout the late nineteenth century was their participation in the regulation of patenting activity.[26] In Europe and the United States, patent lawyers and patent practitioners in general were, alongside judges, the most highly regarded specialists in the interpretation and drafting of legislation. Their body of expertise gave agents the authority not only to shape the interpretation of industrial patent law but also to influence legal reforms. For instance, agents participated in parliamentary debates leading to the reform of national patent laws in countries such as France, Britain and Spain.

The major technical journals edited by agents were an important instrument by which to demand reforms in patent laws and to advance the interests of agents, inventors and patentees. For example, London-based agents such as Moses Poole, Alfred Carpmael and William

Johnson—the latter the editor of the *Practical Mechanics' Journal*—were major players in the British patent system reform of 1852. In Spain, the weekly journal *Industria e Invenciones*, edited by the engineer and agent Gerónimo Bolivar, lobbied for the legal recognition of the patent profession that was eventually attained through a 1902 law.

After the patent controversy that swept Europe between 1850 and 1875, agents emerged as an empowered group that actively pushed for national reforms and the international sanctioning of intellectual property rights. Agents were among those who campaigned for worldwide patent protection and were active in drafting multilateral agreements in patent law in the 1880s. As I will discuss in the next section, professional associations were critical to patent agents' international lobbying efforts and an instrument that promoted the development of an international network of agents.

International Expert Networks

Despite the increasing incentives for international patenting in the second half of the nineteenth century, there remained important barriers that inhibited or retarded the flow of applications beyond national borders. The variations in national patent laws and the lack of reliable information about market opportunities for patent rights in different countries were seen by inventors, firms and agents as significant constraints in their efforts to extend their property rights to multiple national jurisdictions. In the 1870s, after two decades of controversy on patents, inventors and industrialists became increasingly interested in establishing a set of international regulations that would protect patents globally.[27] As the leading British mathematical physicist and engineer Lord Kelvin pointed out in an 1869 letter to the Glasgow Philosophical Society, inventors had 'the grand object of obtaining a common patent law among all civilised nations'.[28] Similarly, some years later, in a public address delivered in November 1882, the vice-president of the British Institute of Patent Agents, the renowned engineer John Imray, called for 'something like an International Patent Law', describing the current situation as 'a horrible abuse' for inventors patenting in foreign countries'.[29] Demands for an international patent law, or at least a greater degree of uniformity among national models, came as well from free-trade supporters who saw fragmented national regulations as constraints to the liberalisation of international trade.[30]

The growing trend toward transnational patenting, along with the pressures of large companies and agents, set in motion a process of relative institutional convergence during the 1880s. Yet this process had its direct origin in the special international Congress on Industrial Property held in 1873 on the occasion of the Vienna World's Fair.[31] The Vienna Congress served mostly to resolve the patent controversy. Patent experts, such as the civil engineer Carl Pieper, a German who served as secretary of the event, played a prominent role in the conference.[32] The patent lawyer George Haseltine, a British representative, declared that if a preliminary international agreement on patent protection was attained, delegates 'shall have contributed more to the material interests of mankind than any congress of modern times'.[33]

The discussions initiated in Vienna continued in 1878 in Paris with a second International Congress on the Protection of Industrial Property, once again coinciding with a world's fair.[34] At this conference, with French civil engineer and agent Charles Thirion acting as conference secretary, industrialists and patent experts enthusiastically supported the international harmonisation of industrial property rights.[35] Finally, in 1883, at the International Convention for the Protection of Industrial Property in Paris, a formalised international patent system was established. The result was an early intergovernmental treaty for the gradual convergence of national patent laws signed by, among others, France, Spain, Portugal, Italy, Belgium and Brazil. The United Kingdom joined in 1884, the United States in 1887, Japan in 1899 and Germany and Mexico in 1903.[36] Regulations were established for a range of fundamental issues, such as priority rights, compulsory working and temporary protection in international expositions.[37] However, the impact of the agreement on international legal convergence was limited and, as Sam Ricketson notes, 'institutionally the new Union was quite unstable'.[38] The Paris Union implicitly accepted international institutional diversity and consequently had the effect of reinforcing national patent models.

This progressive process of legal convergence contributed to the emergence of an international patent system, or in the words of Peter Stearns, a 'global political institution'.[39] There remained, however, significant institutional and bureaucratic differences among national systems, a diversity that presented difficulties for patentees who wished to register their inventions in several countries. The international patent system was more than a group of patentees engaging in transactions across national borders. International patenting continued to be unfeasible without agents, who facilitated the global exchange and registration of property rights of invention by writing and translating patent specifications.

Patent experts likewise were essential in the establishment of the International Association for the Protection of Industrial Property (AIPPI), a non-governmental association established in 1897.[40] Under the auspices of this interest group, industrialists, lawyers and engineers pursued the advancement of international agreements on patent protection and the effective implementation of the provisions of Paris Convention. Agents such as the British Edward Carpmael, the French Armengaud Jeune and the Spanish Francisco Elzaburu were part of the association's first executive committee.[41] In 1902 AIPPI had 530 members, mostly from European countries, with Germany (111), France (110) and Britain (110) with the largest number of affiliates. Countries from the European periphery, such as Italy (26), Russia (4) Spain (5) and Sweden (2), were less represented. The United States had nine members and Canada only one.[42] In 1906 came the founding of another non-governmental professional association, the International Federation of Patent Agents (FICPI), which consisted of patent lawyers and attorneys.

Experts were essential actors in the global race for patents. Heterogeneous patent bureaucracies and the complexity of science-based industrial technologies led patentees to seek out specialised professionals who could mediate in the international arena. In a context of growing interconnectedness among national systems, experts helped build an international network in which knowledge, practices and information were exchanged. We can think of the international patent system as a series of interlocking and overlapping networks of national offices, companies, engineers, capitalists and inventors that transcended the boundaries of Western Europe and the United States. In these networks, agents were linkers or networkers. The relationships between patentees and their agents transcended the borders of national jurisdictions long before the emergence of international agreements on patent law in the 1880s. However, it was during the 1880s that these patenting networks became truly cross-national, to the extent that they began to include the extra-European and colonial worlds.

The various actors who operated across national patent systems constituted a social and informational network mediated by intermediaries. Networks of late nineteenth-century patent experts were not merely social networks; they were socio-technical networks intertwined with material circuits. Patent experts were mediators between different realms. These experts were not only networked with other social actors but also with texts (patent specifications in different languages) and artefacts (specific technical devices).[43]

Agents networks had hubs in New York, Washington, London and Paris, cities where the central offices of both larger international patent agencies and professional associations of agents were located. These cities, which received information from all over the world, were the centres that oversaw the patenting occurring in many diverse countries. In the industrial periphery, professional patent experts worked primarily as correspondents or representatives of foreign firms and inventors, acting as subagents of agencies based in core industrial nations. From this perspective, agents could be considered consultants in international patenting. When the services provided by agents were limited to the administrative communication of property rights, these specialists could be considered interactional experts insofar as they were performing an activity of an inherently bureaucratic nature. Most often, however, the communication services performed by agents involved the translation of specifications and the adaptation of drawings, in which case their activity was not merely administrative but substantially influential in the international making and circulation of technological knowledge on a global scale.[44]

The international interconnectedness of the various national systems is best understood if we look at the professional associations of patent agents established during the last decades of the century.[45] These organisations, which first appeared in industrial countries in the 1880s and then in peripheral ones, created a space of transnational socialisation that gave rise to an international community of experts. Good examples of this trend are the creation of the French Syndicat des Ingenieurs-Conseils en Matiére de Proprieté Industrielle (1884), Chicago's American Patent Law Association (1884), London's Chartered Institute of Patent Agents (1882), the Australasian Institute of Patent Agents (1890) and the Spanish Association of Commercial and Industrial Property Agents (1907). Agents' associations in major industrial nations had foreign as well as local members, thus facilitating the activities of international networks of agents in the periphery. These associations served the twofold mission of connecting professionals and circulating information about patent laws, markets for inventions and professional practices worldwide. As an article in the *London Journal of the Society of Patent Agents* put it: 'No other profession has so constantly to deal with legal matters in all countries, or needs the co-operation of its members in so many and so distant cities.'[46]

During its first decade of existence, one of these associations, the Chartered Institute of Patent Agents (CIPA),[47] was described by its secretary Henry Howgrave Graham as being 'of the most cosmopolitan and international character'.[48] One of the earliest committees appointed by the CIPA, in 1882, was for foreign laws. This committee read and discussed papers concerning foreign and colonial legislation with the objective of promoting improvements in patent laws and regulations for foreign patents. In 1890, this professional association had fifty-seven foreign members and sixty-nine fellows practising in Britain. Foreign fellows were based in twenty-six cities—mostly in Western Europe (Paris, Vienna) and North America (New York, Washington DC, Montreal, Toronto) but also in the European periphery (Barcelona, Lisbon, Madrid, Milan, St. Petersburg, Turin) and colonial or post-colonial settings (Mexico City, Jamaica, Rio de Janeiro, Cape Town, Hong Kong, Calcutta, Melbourne, Sydney). Fellows based in Asia and Latin America were often British or American citizens working primarily for foreign companies. These included the American Richard E. Chism, a lawyer based in Mexico City with expertise in the mining sector, and the solicitor Henry L. Dennys, born in England and from 1874 based in Hong Kong.

The criteria for the selection of qualified local professionals by foreign patentees included not just past records and accomplishments but also membership in one or more associations of patent agents. Agents' associations were, in short, instruments for establishing long-term collaborations based on mutual trust. Owing to the lack of official registration or exams in many countries, patent experts garnered international reputations through their participation in professional associations in a variety of countries. For example, in 1901 the civil engineer Louis Bordes, based in Buenos Aires, was a member of professional associations in Britain, France and Australia. Bordes had in 1870 established the International South American Patent and Trade Mark Agency, which specialised in patent matters for all of Latin America. In Calcutta, Henry H. Remfry and Maurice Remfry, of the agency Remfry & Son (which exists to this day), were members of professional associations in Britain, France and Australia. Remfry & Son was also active in Borneo, Ceylon, Hong Kong and Japan, where it had correspondents. Henry Remfry was the author of several treatises and pamphlets on intellectual property, including *Inventions likely to 'take' and 'pay' in India and the East* (1892).[49]

International networks of agents grew rapidly during the 1890s. One indicator of this trend was the number of agents registered in the *International Directory of Patent Agents*, published in London from 1893. This directory, organised by countries, provided information about agents' locations and in some cases their credentials, services, costs, educational backgrounds and memberships in professional bodies. The number of agents registered in this directory increased from about 2200 in 1893 to more than 4000 in 1901.[50] In other words, there was an increase in the nodes connecting the agents' networks. The 1893 edition of the directory lists patent agents in fifty-nine countries and reveals that more than 70% of the world patent agent offices were located in France, Britain and the United States.[51] There were also dozens of agents in places as diverse and far-flung as Japan, India, Mexico, South Africa, Argentina, Australia and Hong Kong. Interestingly, in the preface of the 1893 edition, the editor explained the reason for its publication:

> The enormous growth of patent agencies (more particularly what may be termed its international developments) has suggested the publication of this volume, which will enable patent agents to find readily the addresses of members of their profession with whom they may wish to communicate, or to ascertain as regards any particular country or colony, whether patent agency is there practised, and by whom and at what address.

Despite international agreements, professional associations and directories, agents still faced uncertainty in international patenting and by the turn of the century would continue to be advocating for a more integrated international patent system. This was made clear in a 1902 editorial in the British *Journal of the Society of Patent Agents*, where agents demanded more uniformity among national laws so as to reduce 'the great risks an agent runs through the negligence or inability on the part of foreign agents employed by him'.[52]

Agents of Corporate Globalisation

During the late nineteenth century, American and European companies began a movement towards international patenting.[53] Through the management of intellectual property rights in various national jurisdictions, large firms acquired valuable patents in countries not only in the fast lane of industrial development (countries such as France, Germany, Britain and the United States), but also in latecomers (such as Australia, Spain, Italy and Mexico). The transition to a modern corporate business model and the growing demand for trained experts in patent issues were closely related.

The industrial enterprise described by the business historian Alfred Chandler was a prototype of the kind of company that would go on to develop a professional management of patent rights.[54] Patents became the foundation of a new business model.[55] For the research industry, patent rights served as an indispensable means of controlling the market and preserving its commercial interests—often through patent pools and patent-based cartels—as well as a source of considerable capital amassed through royalties and assignments.[56] The professional bureaucratisation of innovation within German and American science-based oligopolistic industries became commonplace from the 1870s. The research laboratories of large industrial firms were directed to develop inventions that could be patented.[57] In Joseph Schumpeter's words, 'technological progress (was) increasingly becoming the business of teams of trained specialists who turn out what is required and make it work in predictable ways'.[58]

This tendency was particularly apparent in firms created around the consolidation of patent rights, including the Bell Company, Babcock & Wilcox and the various Edison firms, just to name a few.[59] In these companies, research laboratories worked primarily to ensure that patent rights were protected. Sometimes, as Leonard Reich has shown in the cases of General Electric, Bell and AT&T, these enterprise laboratories became a means of identifying and purchasing patent rights.[60] As Kristine Bruland's study on Babcock & Wilcox reveals, multinational corporations frequently resorted to the recourse of aggressive litigation in the management of patents and trademarks in different countries.[61]

The patent trials on telephone technology in Britain and the United States from 1870 to 1900 are a telling example. Bell, United Telephone and AT&T developed an economic monopoly in the early electrical industries through an intense and aggressive patent litigation strategy.[62] These companies controlled the telephone market by securing legal rights over the relevant technology in court. The market structure of this industry thus largely grew out of litigation and the advice of lawyers. Expert witnesses and scientific consultants were, likewise, pivotal in resolving patent disputes involving key late nineteenth-century electrical inventions (such as wireless telegraphy) and chemical inventions (celluloid, aniline red and the incandescent lamp).[63]

High-technology industrial companies used patents as a prime instrument for international competition in foreign markets.[64] These firms followed a common pattern. Their primary strategy was to block imitation, creating barriers to domestic companies. Persistent patenting, along with other mechanisms, served to preserve the potential market positions of industrial firms in the international arena. These companies tended to register patents in as many national jurisdictions as possible before disseminating the content of their inventions. Once their patents were protected abroad, industrial firms could set up factories and subsidiaries in multiple countries. More frequently, these companies proceeded to export their technology or license their patent rights in a given country.[65]

By the beginning of the twentieth century a number of companies had established some distinct unit or department for the management of patent rights. Large British, French, American and German companies began to employ full-time lawyers, engineers and scientists for the management of innovations inside the firm.[66] This tendency reflects a broader transition from individual to corporate inventive activity. In the same way as specialised personnel or marketing departments were created, legal departments (and in some cases centralised patent departments) were introduced with the goal of reducing corporations' transaction- and information-related costs. The hiring of in-house patent agents followed the creation of research laboratories. Good examples of companies with in-house departments for the management of inventive activity are Westinghouse, Schneider, Edison and General Electric.[67] In contrast to what we might expect, the hiring of human resources with patent expertise did not imply a reduction in the number of external intermediaries, given the difficulties that companies faced in coordinating information and securing patents in several national jurisdictions.[68]

In-house corporate agents became an instrument by which large firms could exert their power in the international market for technology and accumulate patents through assignments, pools and litigation. Foreign multinational firms often retained both in-house patent agents and corresponding subagents in countries at the periphery of industrial development. These local intermediaries assisted multinational corporations in patenting, diffusing and commercialising their property rights globally. The various companies set up by Thomas A. Edison in the 1880s and 1890s serve as a good example of the ways in which agents assisted multinational firms. Edison's firms sent, in 1888, powers of attorney to agents in twenty-four countries (including several Latin American countries such as Peru, Mexico, Brazil and Argentina). The New York patent firm Dyer and Seeley, which represented Edison, coordinated transnational patenting with agents in the various countries. The differences among the foreign agents' fees were huge. Interestingly, the fees in Mexico, Portugal, Brazil and India were the most expensive among the experts working for Edison.[69]

Corporate patent management in the agricultural and mining sectors has been less explored in the historiography. An interesting case study is provided by European and American engineering firms that specialised in the mass production of machinery. These companies supplied patented equipment to sugar plantations in Asia and the Americas. Good examples would be Duncan Stewart (Glasgow), McOnie (Glasgow) and Fives-Lille (France), all of which provided plantations with the most advanced technologies for the construction of large-scale central sugar factories throughout the Caribbean, the Indian Ocean and Southeast Asia during the last third of the nineteenth century.[70] These companies built their international expansion on global networks of patent counsels, chemists and machinists.

Similarly, the Cassel Company, founded in 1884 in Glasgow, developed technological improvements to be used in distant peripheries. This firm exploited patents for new mining techniques, among them the Macarthur-Forrest process of gold extraction using cyanide, patented in 1887 by the Glasgow chemist John Stewart Macarthur and physicians Robert and William Forrest and assigned in 1888 to the Cassel Company. This company set up subsidiaries in New Zealand, South Africa, Mexico and Australia during the 1890s to exploit and license their patented mining techniques, thereby bringing patent rights to these four countries as well as Britain and the United States.[71] Apart from exploiting its patents, the Cassel company commercialised its techniques in other major mining centres, including India, Chile and Russia. The Cassel techniques were a commercial success, despite the fact that the company charged royalties that were considered exorbitant by some governments. Lawyers and mining engineers were key to the company's management of its inventions far afield from its headquarters in Scotland. The Cassel Company's patent attorneys, meanwhile, were instrumental in resolving questions of ownership pertaining to originality and technical knowledge, given the large number of cases of patent infringements and lawsuits this company faced in countries such as South Africa and New Zealand, where some of its patents were deemed as invalid by local governments and mining communities.

Conclusion

This chapter has shed light on the rise of specialised patent agents and lawyers over the course of the second industrialisation. The objective has been to broaden the study of the innovative community by analysing different forms of expertise. Historical accounts of patent institutions have almost invariably concentrated on inventors and entrepreneurs while overlooking the activities of intermediate experts. Inventors and entrepreneurs may have changed world history through their diffusion of innovative technologies, but they have done so hand in hand with other actors, from lawyers to consultant engineers.

This discussion of the various forms of patent expertise has gone beyond national histories of patenting, instead reconstructing the internationalisation of professional expertise on intellectual property issues. The pressures of a globalising economy and multilateral agreements stimulated the internationalisation of the patent profession during the 1880s and 1890s. During these two decades, national patent institutions grew more interdependent while continuing to maintain their diversity. In this context, the variations in the evolution of patent expertise among different countries depended on the level of industrialisation and the overall institutional environment. The globalisation of patenting likewise became associated with the development of engineering and legal expertise throughout various countries. This expertise was embedded in specific national technological and legal cultures.

Patent agents emerged in the international arena in order to make possible the transnational transfer of inventions and rights that had been created, and would otherwise have remained institutionally embedded, in specific national or local sites. From this perspective, patent experts were necessary mediators in knowledge transfer. They served as channels in the realm of international patenting, communicating knowledge and property rights globally. Networks of agents reduced the risks and uncertainty that foreign inventors encountered in international technology transfer. Indeed, the circulation of technologies and rights depended on overlapping long-term networks. In these networks, patent experts linked inventors, firms, patent offices, markets and technical consultancies.

The expansion of international networks of intermediary agents reflected the changing imperatives for intellectual property management during the years of the second industrial revolution. The crucial role of patent agents in the transnational transmission of patent rights immediately raises, then, the question of whether these experts were a constraint for international technology transfer. Unfortunately, the existing historiography on the history of patents does not go very far in answering this question. It seems that patent expertise was supporting and reinforcing an asymmetric international patent system that exposed the imbalances of world industrial capitalism. Skilled and elite expert agents remained concentrated in a small group of industrial countries with advanced technological capabilities. That said, this is not just a history of core and peripheral agents, but a history of the global interdependence that existed among experts and of the long-term constraints to technological development that encouraged this interdependence.

Studies often present patent agents as driving actors in the growth of efficient markets for technology. From this perspective, the presence of expert agents removed constraints in international patenting. However, during the late nineteenth century agents in many countries were accused of carrying out rent-seeking activities and maintained privileged relationships with officials and commissioners. A question can thus be raised as to whether intellectual property institutions at the close of the nineteenth century were actually open to a broad segment of the population or were instead primarily accessible only to powerful social classes and corporations with a large amount of capital.

Networks of patent experts as mechanisms of international knowledge transfer posed significant limitations. Given that agents were a tacit requirement for entry into foreign markets, patenting was effectively restricted across national boundaries. While agents increased the security of patentees in transnational operations, they also limited registration to those who could afford agents' fees for moving patent rights across these boundaries. In latecomers, the chain of intermediaries necessary for foreign patentees to register their patents drove costs much higher. Even when transaction and information costs were relatively reduced by agents during the last decades of the nineteenth century, the majority of foreign inventors in advanced industrial countries remained reluctant to extend their rights to peripheral countries. The transactions costs of international patenting remained high.

The rapid expansion of networks of patent experts during the late nineteenth century reflected the need for large companies to exercise international control over valuable inventions. The scholarship has emphasised the competitive advantages created by control over intellectual property rights and the resort to experts to maximise corporate profits. The outcomes of corporate patent management deserve further historical research. Indeed, while the historiography deals forcefully with successful business models, it tends to ignore the contentious consequences of patent management, such as lower degrees of knowledge diffusion, limited competition and predatory strategies. Similarly, alternative and sometimes highly successful strategies, such as secrecy or the eschewing of patents altogether, are overlooked. The most important point remains that agents' powers, and their many services to multinational corporations, had enduring consequences on the structure of knowledge property worldwide.

Notes

1. Shapin, S.: 'Nibbling at the Teats of Science: Edinburgh and the Diffusion of Science in the 1830s', in Inkster, I. and Morrell, J. (eds.): *Metropolis and Province: Science in British Culture, 1780–1850*, London: Hutchinson, 1983, p. 151.
2. See, for instance, the contributions to the two following special issues: Galvez-Behar, G. and Nishimura, S.(eds): 'Le management de la propriété industrielle', *Entreprise et Histoire,* No.82 (2016); Inkster, I. (ed.): 'Patent Agency in History: Intellectual Property and Technological Change', *History of Technology*, Vol. 31 (2012).
3. See Kranakis, E.: 'Patents and Power: European Patent-System Integration in the Context of Globalization', *Technology and Culture,* Vol. 48, No. 4 (2007), pp. 689–728; Khan, Z.B.: 'Selling ideas: An international perspective on patenting and markets for technological innovations, 1790–1930', *Business History Review,* Vol. 87, No. 1 (2013), pp. 39–68.
4. Guagnini, A.: 'Patent Agents in Britain at the turn of the 20th Century', *History of Technology,* Vol. 31 (2012), p. 159.
5. Abbott, A.: *The System of Professions: An Essay on the Division of Expert Labor*, Chicago: University of Chicago Press, 1988.
6. Inkster, I.: 'Patent Agency: Problems and Perspectives', *History of Technology*, Vol. 31 (2012), p. 91.
7. Swanson, K.: 'The Emergence of the Professional Patent Practitioner', *Technology and Culture*, Vol. 50, No. 3 (2009), pp. 519–548; van Zyl Smit, D.: 'Professional Patent Agents and the Development of the

English Patent System', *International Journal of the Sociology of Law*, Vol. 13 (1985), pp. 79–105; Gálvez-Behar, G.: 'Des Médiateurs au Coeur du Système d'Innovation: Les Agents de Brevets en France (1870–1914)' in Corcy, M., Douyère-Demeulenaere, C. and Hilaire-Pérez, L. (eds.): *Les archives de l'invention. Ecrits, objets et images de l'activité inventive*, Toulouse: Université Toulouse-Le Mirail, 2006, pp. 437–447.

8. See the articles on patent agents and patent management in Spain, Sweden, Japan and Germany in the special issues mentioned in Endnote 2. For Japan see as well Nicholas T. and Shimizu, H.: 'Intermediary Functions and the Market for Innovation in Meiji and Taisho Japan', *Business History Review*, Vol. 87, No.1 (2013), pp. 121–149. For Australia see Hack, B.: *A History of the Patent Profession in Colonial Australia*, Melbourne: Clement Hack & Co., 1984.
9. On the writing of patent specifications see, for example, Myers, G.: 'From Discovery to Invention: The Writing and Rewriting of Two Patents', *Social Studies of Science*, Vol. 25, No. 1 (1995), pp. 57–105.
10. Biagioli, M.: 'Patent Republic: Representing Inventions, Constructing Rights and Authors`, *Social Research,* Vol. 73, No. 4 (2006), pp. 1129–1172.
11. Newton, A. V.: 'On the Patent Agent and his Profession', *Transactions of the Institute of Patent Agents*, Vol. I (1882–3), pp. 158–169.
12. Guagnini, A.: 'Patent Agents, Legal Advisers and Guglielmo Marconi's Breakthrough in Wireless Telegraphy', *History of Technology*, Vol. 24 (2002), pp. 171–201.
13. Bowker, G.: 'What's in a Patent?', in Bijker, Wiebe E. and Law, John (eds.): *Shaping Technology Building Society. Studies in Sociotechnical Change*, Cambridge: MIT Press, 1992, pp. 53–75.
14. For patent disputes and their relationship to patent pool agreements see Usselman, S. W.: 'Patents Purloined: Railroads, Inventors, and the Diffusion of Innovation in 19th-Century America', *Technology and Culture,* Vol. 32, No. 4 (1991), pp. 1047–1075.
15. Good examples of scholarly research that acknowledges the various legal actors participating in patenting dynamics are Cambrosio, A., Peter Keating, P. and Mackenzie, M.: 'Scientific Practice in the Courtroom: The Construction of Sociotechnical Identities in a Biotechnology Patent Dispute»' *Social Problems,* No. 37 (1990), pp. 275–93; and Lucier, P.: 'Court and Controversy: Patenting Science in the Nineteenth Century', *The British Journal for the History of Science,* Vol. 29, No. 2 (1996), pp. 139–154.

16. Bijker, W. E.: 'The Social Construction of Bakelite: Toward a Theory of Invention' in Bijker, W., et al.: *The Social Construction of Technological Systems: New Directions in the Sociology and History of Technology*, Cambridge: MIT Press, 1987, pp. 164–173.
17. MacLeod, C.: 'The Paradoxes of Patenting: Invention and Its Diffusion in 18th- and 19th-Century Britain, France, and North America', *Technology and Culture,* Vol. 32, No. 4 (1991), pp. 885–910.
18. Lamoreaux, N. R. and Sokoloff, K. L.: 'Intermediaries in the US Market for Technology, 1870–1920' in Engerman S. L. et al. (eds.): *Finance, Intermediaries, and Economic Development*, Cambridge: Cambridge University Press, 2003, pp., 209–46; Cooper, C.: *Shaping Invention: Thomas Blanchard's Machinery and Patent Management in Nineteenth-Century America*, New York: Columbia University Press, 1991.
19. Pretel, D. and Sáiz, P.: 'Patent Agents in the European Periphery: Spain, 1826–1902', *History of Technology* Vol 31, 2012, pp. 97–114.
20. For the French case, in 1881, 89 percent of "cabinets" of Patent Agents were established in Paris. Gálvez-Behar, *Des Médiateurs*; In Britain in 1893, 53 percent of agencies were located in London. Inkster, *Patent Agency.*
21. Khan, Z.: *The Democratization of Invention: Patents and Copyrights in American Economic Development, 1790–1920*, Cambridge: Cambridge University Press, 2005.
22. Pretel and Sáiz, *Patent Agents.*
23. Other relevant mechanics' and trade journals related with the patent business included *The Artisan*, *The Repertory of Patent Inventors* and *Mechanics' Magazine* in Britain, *The American Artisan*, *American Inventor* and the *Patent Right Gazette* in the United States and *Le Journal des Inventeurs* and *Moniteur des Inventions* in France.
24. The contemporary agents' literature on these topics is vast. These are just some pointers: Munn & Co., *Hints to Inventors*, New York: Munn & Co., 1867; Johnson, J. and Johnson, J. H.: *The Patentee's Manual,* London: Longmans, Green, and Co., 1890; Thompson, W.P.: *The Patent Road to Fortune*, London: Stevens & Sons, 1884; Thirion, C.: *Législations Française et Étrangères sur les Brevets d'Invention: Tableau Synoptique*, Paris: Dupont, 1878. Carpmael, A.: *Patent Laws of the World*, London: W. Clowes, 1889; Edwards E. and Edwards, A. E.: *How to Take Out Patents in England and Abroad*, London: Edwards and Co., 1905.
25. Armengaud, J. E.: *The Practical Draughtsman's. Book of Industrial Design*, London: Longman, Brown, Green and Longmans, 1853. The original in

French is *Cours de Dessin Linéaire Appliqué au Dessin des Machines*, Paris: Z. Mathias, 1840.

26. Dutton, H. I.: *The Patent System and Inventive Activity during the Industrial Revolution 1750–1852*, Manchester: Manchester University Press, 1984, pp. 43–51; Gálvez-Behar, *Des Médiateurs*; Usselman, *Regulating Railroad Innovation*, pp. 149–51.
27. Machlup, F. and Penrose, E.: 'The Patent Controversy in the Nineteenth Century', *The Journal of Economic History*, Vol.10, No.1 (1950), pp. 1–29; Plasseraud, Y. and Savignon, F.: *Paris 1883: Genèse du droit de brevets,* Paris: Litec, 1983, pp. 102–6; Christine MacLeod, C.: 'Concepts of Invention and the Patent Controversy in Victorian Britain' in Fox, R. (ed.): *Technological Change: Methods and Themes in the History of Technology*, Amsterdam: Harwood Academic, 1996, pp. 137–153.
28. Sir Williams Thomson to Dr. Bryce, 'Discussion on Patents', *Glasgow Philosophical Society* (14 December 1869), cited in Smith C. and Wise, M. N.: *Energy and Empire: A Biographical Study of Lord Kelvin*, Cambridge: Cambridge University Press, 1989, p. 708.
29. 'Inaugural Meeting', *Transactions of the Institute of Patent Agents*, Vol. I, 1882–3, p. 46. For a contemporary account of international patenting see also Imray, O.: 'On Foreign Patents', *Transactions of the Institute of Patent Agents,* Vol. II, 1883–4.
30. May, C. and Shell, S.: *Intellectual Property Rights: A Critical History*, London: Lynne Rienner Publishers, 2006, pp. 115–117.
31. The British patent barrister Thomas Webster, a delegate at this conference, wrote about this meeting. Webster, T.: *Congrès International des Brevets d'Invention tenu à l'Exposition Universelle de Vienne en 1873*, Paris: Marchal, Billard et Cie, 1877.
32. Ricketson, S.: *The Paris Convention for the Protection of Industrial Property: A Commentary,* Oxford: Oxford University Press, 2015, pp. 36–39.
33. Cited in Seckelmann, M.: 'The Indebtedness to the Inventive Genius: Global Expositions and the Development of an International Patent Protection', in Barth, V. (ed.): *Identity and Universality / Identité et universalité*, Paris: Bureau International des Expositions, 2002, p. 132.
34. *Congrès international de la propriété industrielle tenu à Paris du 5 au 17 septembre 1878*, Paris: Imprimerie Nationale, 1879.
35. Penrose, E.: *The Economics of the International Patent System*, Baltimore: Johns Hopkins Press, 1951, pp. 48–55; Plasseraud and Savignon, *Paris 1883*, pp. 155–174; Gálvez-Behar, G.: *La République des Inventeurs:*

Propriété et Organisation De l'Innovation en France, 1791–1922, Rennes: Presses Universitaires de Rennes, 2008, pp. 153–177.

36. See http://www.wipo.int/treaties/en/ip/paris/.
37. Penrose, *The Economics of the International Patent System*, Chapter 4; Plasseraud and Savignon, *Paris 1883*, pp. 205–9.
38. Ricketson, *The Paris Convention*, p. 69.
39. Stearns, P. N.: *Globalization in World History*, London and New York: Routledge, 2010, pp. 107–8.
40. Ricketson, *The Paris Convention*, pp. 75–6. See also the article by Max Georgii, founding member of this association, Georgii, M.: 'International Association for the Protection of Industrial Property', *The Inventive Age*, No.3 (March 1898), pp. 42–3.
41. See the *Annuaires de L'Association Internationale pour la Protection de la Proprieté Industrielle* published from 1897.
42. *Annuaire de L'Association Internationale pour la Protection de la Proprieté Industrielle, 1902, Congrès de Turin*, Paris, 1903, p. 31.
43. Callon, M.: 'Techno-economic networks and irreversibility', *The Sociological Review*, No. 38 (1990), pp. 132–161; Biagioli, 'Patent Republic'.
44. For a sociological analysis of the idea of 'interactional' and 'contributory' expertise and the difference between experts and specialist see Collins, H. and Evans R.: *Rethinking Expertise*, Chicago and London: University of Chicago Press, 2007, pp. 23–38 and 77–90.
45. Magee, G. B. and Thompson, A. S.: *Empire and Globalisation: Networks of People, Goods and Capital in the British World, c.1850–1914*, Cambridge: Cambridge University Press, 2010, pp. 143–5.
46. 'Professional Co-operation', *Journal of the Society of Patent Agents*, Vol. II, No. 13 (January 1901), p. 1.
47. For a detailed study of the CIPA in the late nineteenth century see Guagnini, 'Patent Agents in Britain'.
48. Paper read at the Eighth Annual Meeting of the Institute, Howgrave Graham, H.: *On the Progress and Work of the Institute of Patent Agents*, London: Spottiswoode & Co., 1890.
49. Printed by the Calcutta Central Press Co. in 1892.
50. *International Directory of Patent Agents*, London: William Reeves, 1893, 1897 and 1901.
51. According to Ian Inkster, in the *International Directory of Patent Agents* for the year 1893, 2202 agencies were listed: 45% of them in the USA, 27% in France and 13% in Britain. Inkster, I.: 'Engineers as patentees

and the cultures of invention 1830–1914 and beyond: The evidence from the patent data', *Quaderns D'Historia de L'Enginyeria*, Vol. VI (2004), pp. 25–50.

52. 'The Desirability of an International System of Procedure for Protection of Invention', *Journal of the Society of Patent Agents*, No. 44, 45 and 46 (1902–1903), p. 116.
53. There is a large literature on corporations and patent management; see, for instance, Andersen, B.: *Technological Change and the Evolution of Corporate Innovation: The Structure of Patenting, 1890–1990*, Cheltenham: Edward Elgar, 2001; Noble, D.: *America by Design*, New York: Oxford University Press, 1979, pp. 84–109; Wilkins, M.: 'The Role of Private Business in the International Diffusion of Technology', *The Journal of Economic History* 34, No. 1 (1974), pp. 166–188.
54. For Chandler's thesis see Chandler, A.: *Scale and Scope: The Dynamics of Industrial Capitalism*, Cambridge, Mass: Belknap Press, 1990; Chandler, A.: *The Visible Hand: The Managerial Revolution in American Business*, Cambridge, MA: Harvard University Press, 1977.
55. May and Sell, *Intellectual Property Rights*, pp. 122 y 131.
56. According to Leonard Reich, industrial laboratories were 'set apart from production facilities, staffed by people trained in science and advanced engineering who work toward deeper understanding of corporate-related science and technology, and who are organised and administered to keep them somewhat insulated from immediate demands yet responsive to long-term company needs'. Reich, L.S.: *The Making of American Industrial Research: Science and Business at GE and Bell, 1876–1926*, Cambridge: Cambridge University Press, 1985, p. 3.
57. Fox and Guagnini, *Laboratories, Workshops, and Sites*, pp. 158 and 166–7.
58. Schumpeter, J. A.: *Capitalism, Socialism, and Democracy*, 6th ed., London: Routledge, 2003, p. 132.
59. Smith, G. D.: *The Anatomy of a Business Strategy: Bell, Western Electric and the Origins of the American Telephone Industry*, Baltimore: Johns Hopkins University Press, 1985.
60. Reich, *The Making of American Industrial Research*, p. 3.
61. Bruland, K.: 'The Management of Intellectual Property at Home and Abroad: Babcock & Wilcox, 1850–1910', *History of Technology*, Vol. 24 (2002), pp. 151–170.

62. For a recent study on the corporate monopoly in the telephone sector constructed by patent litigation see Beauchamp, B.: *Invented by Law: Alexander Graham Bell and the Patent that Changed America*, Cambridge: Harvard University Press, 2015.
63. Gooday, G. and Arapostathis, S.,: *Patently Contestable: Electrical Technologies and Inventor Identities on Trial in Britain,* Cambridge: MIT Press, 2013; Bijker, 'The Social Construction of Bakelite'; Van den Belt, H.: 'Action at a Distance: A.W. Hofmann and the French Patent Disputes about Aniline Red (1860–63), or How a Scientist May Influence Legal Decisions without Appearing in Court', in Smith, R. and Wynne, B. (eds.): *Expert Evidence: Interpreting Science in the Law*, London: Routledge, 1988, pp. 185–209.
64. Inkster, I.: *Science and Technology in History: An Approach to Industrial Development*, Basingstoke: Macmillan, 1991, p. 113; Fox and Guagnini have also identified 'protective patenting' strategies among large German chemical firms in the late nineteenth century. Fox and Guagnini, *Laboratories, Workshops, and Sites*, p. 158.
65. According to Edith Penrose, in the last decades of the nineteenth century 'patents were used to protect international markets', Penrose, *The Economics of the International*, p. 89. The relationship between the effectiveness of the patent protection in a country and the commercialisation of technology is well developed in Arora, A.: 'Trading Knowledge: An Exploration of Patent Protection and Other Developments of Market Transactions in Technology and R&D', in Lamoreaux, N. and Sokoloff, K. (eds): *Financing Innovation in the United States, 1870 to the Present*, Cambridge, Mass: MIT Press, 2007, pp. 365–403. According to Arora, "patent protection and commercialisation are strategic complements" (p. 371).
66. Noble, *America by Design*, pp. 84–109.
67. Nishimura, S.: 'The Rise of the Patent Department: An Example of the Institutionalization of Knowledge Workers in the United States', *Entreprises et histoire*, Vol. 82, No. 1 (2016), pp. 47–63. An early example of corporate professional management of industrial property rights is the French firm Schneider and Cie and its patents in metallurgy. D'Angio, A.: 'The Industrial and Financial Use of Patent by Schneider & Cie in the 19th Century (1836–1883)', in Merger, M. (ed.): *Transferts de Technologies en Méditerranée*, Paris: Presses de l'Université Paris-Sorbonne, 2006, pp. 345–358.

68. Galambos, L.: 'The Role of Professionals in the Chandler Paradigm', in Lazonik, W. and Teece, D. (eds.): *Management Innovation: Essays in the Spirit of Alfred Chandler*, New York: Oxford University Press, 2012, pp. 125–146.
69. D8846ACD; TAEM 124:80; and D8846, Document Files Series 1888: D-88-46-Patents (Alfred Ord Tate to Richard Nott Dyer, 08/08/1888).
70. Pretel, D. and Fernandez-de-Pinedo, N.: 'Circuits of Knowledge: Foreign Technology and Transnational Expertise in Nineteenth-Century Cuba' in Leonard, A. and Pretel, D. (eds.), *The Caribbean and the Atlantic World Economy: Circuits of Trade, Money and Knowledge, 1650–1914*, Basingstoke: Palgrave Macmillan, 2015, pp. 263–289.
71. Todd, J.: *Colonial Technology: Science and the Transfer of Innovation to Australia*, Cambridge: Cambridge University Press, 1995; Beatty, E.: *Technology and the Search for Progress in Modern Mexico*, Oakland: University of California Press, 2015, pp. 134–153.

7

Networks of American Experts in the Caribbean: The Harvard Botanic Station in Cuba (1898–1930)

Leida Fernandez-Prieto

Introduction

In 1899, Edwin F. Atkins, a Boston-born businessman with properties and sugar concerns in Cienfuegos, Cuba, founded a botanical and agronomic station on his Soledad del Muerto sugar estate. Harvard University provided the station with its scientific design and staff of experts, and by the early twentieth century the Harvard Botanic Station for Tropical

This work was made possible by my stay at the Harvard University as Wilbur Marvin Visiting Scholar at the David Rockefeller Center for Latin American Studies in 2015–2016 and Project HAR2015-66152-R (MINECO). I would particularly like to thank Brian D. Farrell, Jorge Dominguez, Alejandro de la Fuente, June C. Erlick, Marysa Navarro, Jonathan Losos, Ned Strong, Erin Goodman, Edwin Ortiz, María T. Vázquez, Thenesoya V. Martin de la Nuez, James N. Levit, Liz Mineo and Marial Iglesias. I must also thank Harvard library and archives staff for their invaluable help, especially Lynn M. Shirey, as well as staff of the Massachusetts Historical Society and Jardin Botánico de Cienfuegos. I would also thank Deborah Fitzgerald, who made excellent suggestions and comments on my research. I must thank Chet Atkins, who introduced me to some unfamiliar passages of the Atkins family.

L. Fernandez-Prieto (✉)
Instituto de Historia, CSIC-Madrid, Madrid, Spain
e-mail: leida.fernandez@cchs.csic.es

© The Author(s) 2018

D. Pretel, L. Camprubí (eds.), *Technology and Globalisation*, Palgrave Studies in Economic History, https://doi.org/10.1007/978-3-319-75450-5_7

Research and Sugar Cane Investigation was the premier tropical research centre in the western hemisphere. After the 1959 Cuban Revolution, the Harvard Botanic Station became the Cienfuegos Botanical Garden, and still functions today within the Cuban Science and Environment Ministry. The Soledad estate however, now named Pepito Tey, closed down as the sugar sector fell deep into crisis.

The history of the Botanic Station is a compelling case study in examining the role and activities of US scientists, botanists and agronomists in the context of economic practices globally, regionally in the Hispanic Caribbean and more locally and specifically, on the island of Cuba between 1898 and 1930. The period is often qualified as the second economic and environmental conquest of Latin America and the Caribbean. In its time, the Station supplied a training space for US experts in tropical science in a setting of scientific, imperial, nationalist and capitalist tensions. It became a tropical laboratory, a place of research destination and a source of samples for scientists coming mostly from Harvard, but also Europeans, Latin Americans and Cubans. Above all, it provided a venue for innovative summer classes on the tropics intended for Harvard biology students.

This chapter examines the Harvard Botanic Station's contribution to producing agricultural and botanical knowledge on the tropics, in contexts of hegemonic imperial expansion and subsequent decolonisation and economic globalisation. It shows how networks of knowledge and US experts worked in the Caribbean in the early twentieth century in tune with global processes of imperial expansion and the development of capitalism. The Harvard Botanic Station for Tropical Research and Sugar Cane Investigation was created in Cuba in 1899, against the background of the Spanish–Cuban–American war and rising US imperialism. I argue that knowledge creation is like a two-sided coin. The scientists involved in fields of knowledge such as agriculture and horticulture wanted to create pure science, but also had to accept the commercial use of their knowledge. Field sciences are a paradigmatic example in examining the porous frontier between laboratory and field practices, and in highlighting participation both by expert and amateur knowledge.[1] I underline the ambivalent nature of the process of producing knowledge on the tropical world, with complex dynamics involving power, science and capitalism, and multiple actors and source of knowledge and money.

Among others, the historian Stuart McCook has examined the role of the United States in creating global networks of botanists and a network of agronomic stations in the Greater Caribbean.[2] He describes the history of the Harvard Botanic Station in his analysis of the role of science and the new field of economic botany in the context of US imperialism and the modernisation of commercial agriculture in Latin America and the Caribbean in the early twentieth century. More recently, Megan Raby has pointed out transnational links between Harvard, the Botanic Station and the work of Thomas Barbour in Cuba and Panama, to show the Caribbean's importance in the construction of tropical botany, and interrelations of national, academic and business interests that became tighter in the twentieth century.[3] Barbour was a naturalist, herpetologist and director of the Museum of Comparative Zoology at Harvard who headed the Garden between 1920 and 1940. The Botanic Station has also been of interest to studies on global transfers of the biological science of sugarcane hybridisation.[4]

I have taken inspiration from histories of field sciences and commodities and the Actor–Network Theory (ANT),[5] to focus on the mediating role of the main agents at the Harvard Botanic Station (Edwin F. Atkins and his wife Katherine W., George L. Goodale, Oakes Ames and Thomas Barbour, Harvard University staff at Cambridge, Massachusetts, Garden staff, scientists and agronomists, the Cuban government and others) in negotiations and interactions between various scientific, political, commercial and knowledge networks to resolve problems of the global and local economies through tropical agricultural science. These networks followed particular rules and codes of practice that allow me to explore what I term the 'negotiation zone', and reveal the ambivalent and porous dimensions of the tropical scientific knowledge produced after 1898. Atkins was a businessman and botanists Goodale, Ames and Barbour were fired by scientific ambition. Yet their collaboration allowed the Botanic Station to flourish and become one of the world's richest tropical gardens. Atkins provided land and resources, and financed the work of the Harvard scientists who developed a cane variety resistant to disease. The Harvard botanists turned the Garden into a tropical laboratory wherein they planted trees from across the world.

Negotiating US Tropical Research on a Cuban Sugar Plantation

Sugar was the predominant crop in Cuba, which had become the world's leading cane producer in the nineteenth century. This feat forced the Creole elites to adopt a strategy of joining transatlantic commercial, industrial and slave trading flows. The Creole 'art of sugar production' exploited Cuba's natural productive conditions, but also relied on the consolidation of a regional economy of slave-run plantations fed by global and regional transfers of slave labour, on specialist expertise, technologies imported from industrial centres such as the United States and Europe, and the introduction of productive cane varieties.[6]

From 1838, Elisha Atkins connected Boston with Atlantic sugar and slave circuits.[7] His son, Edwin F. Atkins, arrived at Cienfuegos, Cuba, aged sixteen years, to take charge of regional sugar businesses while learning Spanish and local know-how from Hispano-Cuban traders and estate owners. This was in a context Dale Tomich terms a second period of slavery and a time of technological revolution.[8] In 1883, Edwin F. Atkins bought the Soledad estate from the Sarria family, who were from the Trinidad region, and eventually made it one of the most important production centres of its time.[9] Atkins actively took part in processes to reorganise the Cuban sugar industry's productive capacities. In the late nineteenth century, competition from cane's international rival, sugar beet, led estate owners to introduce crucial organisational changes to sugar cultivation and the industry. On the industrial level, plantations became sugar processing and production mills (*Centrales*), and with the abolition of slavery in 1886, the slave workforce gave way to owner or tenant contractors *(Colonos)*. At the cultivation level, the environmental crisis resulting from pressure exerted on western Cuba's ecosystem caused a degradation of the cane and decline in productivity. Estate owners were using more and more fertilisers, trying out new cultivation systems and introducing new varieties such as the *Crystalline* cane, which came to dominate the countryside as it adapted so well to so-called exhausted terrains.[10] Cuba produced in 1894 more than a million tonnes of sugar, 91.71% of which was exported to the United States.[11] In 1895, Cubans began the last phase of their war of independence against Spain, which led to US intervention.

Spain lost the last colonies of its once vast American empire in 1898, having to cede tropical territories to the United States, including the islands of Cuba, Puerto Rico, the Philippines and Guam. The tropics became a military occupation zone and zone of US economic, political, scientific and cultural influence. The Caribbean witnessed then the expansion of US capital with the arrival of transnational farming and sugar companies, the best known of which was the United Fruit Company. Many Americans also settled in the region to work in the recently refounded farming colonies. The United States intervened militarily in Cuba in 1898–1902, and in 1906–1909. The Cuban Republic was itself born in 1902 under US tutelage, manifest in legal terms in the Platt Amendment to the new Cuban constitution.

US imperial expansion automatically converted its overseas territories into reconnaissance points for the environmental and economic conquest of the tropics' resources. The year 1898 saw an explosion of related scientific production with two aspects. On the one hand there was knowledge produced by universities, botanical gardens and experimental stations and on the other its connection to the large food and farming companies expanding across the empire. Agricultural knowledge thus had practical utility in this context and knowledge creation was, potentially, profitable. In the case of Cuba, private sugar companies made up for the absence of an experimental station owned by the government and devoted exclusively to the scientific modernisation of sugar cultivation and production. On their lands, they trained and employed experts who made a contribution to the sum of scientific knowledge on tropical sugar.[12] At the same time, occupation authorities and Cuban governments forged alliances with foreign sugar companies. In 1899, for example, the Chaparra Sugar Company was founded; its chief administrator was Mario García Menocal, a former general of Cuban independence and president of Cuba in the years 1913–1921. People were involved in different aspects of knowledge creation on the tropical world. These were scientific endeavours, certainly, but remained linked to occupation plans and the Cuban government's farming policies. That gave knowledge creation after 1898 an ambivalent quality.

Science applied to medicine and agriculture remained in the newly acquired territories, a perfect instrument of administration and domination in the informal colonialism of the United States after 1898.[13] Cuba turned into a centre of knowledge on tropical medicine to eradicate yellow fever, while the island's sanitisation was a symbol of what Marial Iglesias terms American-style modernity.[14] The United States was also involved in producing agricultural knowledge to control Cuba's natural resources. Literature relating to US hegemonic expansionism established the idyllic narrative of a tropical landscape to be explored, but also civilised in scientific and economic terms.[15] The first interventionist US government inventoried Cuba's natural and productive resources through an agricultural and population census organised in 1899. Research on the island's plant life and agriculture attracted the attention of several US institutions, both public and private, which undertook taxonomic studies and collected botanical species of commercial, medical and scientific value to the United States. American scientists, botanists and agronomists engaged in internal debates to understand the tropics and the dynamics of global and US domestic markets, and how economic botany could help modernise US farming. Many of these ideas were exported to the technical and scientific institutions being founded in recently won territories. The New York Botanical Garden (NYBG), for example, undertook several exploratory trips financed by wealthy patrons in New York, which produced botanical and scientific inventories of the Caribbean.[16] A participant of one of the expeditions was Franklin S. Earle, a collaborator of the United States Department of Agriculture (USDA) and former micologist at the New York garden. The USDA was creating a network of agricultural stations in the tropics then and Earle became, on USDA and NYBG recommendation, the expert the Cuban government appointed to lead and create a scientific design for the Experimental Agronomical Station.[17]

Cienfuegos was the epicentre of the sugar industry in Cuba when the Americans arrived. The region was the physical and symbolic frontier between the old sugar industry based on slaves and plantations in western Cuba and the business's expansion into eastern Cuba. Atkins's lands suffered from the degeneration of the western sector's naturally productive conditions and especially the decline of more than 40% in cane yields.

The debates of the local scientific and agronomical community centred around sugar degeneration, but more so on the erosion of soils due to intensive and prolonged cultivation. Many estate owners used fertilisers to restore fertility, but Atkins emphasised natural factors in the plant's decline. He echoed the botanical revolution in sugar happening in Barbados and Java following the discovery that cane was fertile, and that varieties could be crossed to obtain hybrids that were more productive and resistant to disease.[18] There were no such investigations in Cuba nor was there an agronomic station working on improving sugar cultivation. Atkins tasked the manager of the Soledad estate, the Briton L.H. Hughes, with experimenting with certain cane varieties that had failed to flourish in Cuba's climate and were susceptible to disease.[19] Their failure had convinced Atkins of the need for domestic sugar cane breeding, and he was a pioneer in realising that such experiments must be done on site to obtain varieties suited to the Cuban environment. In 1899, during the first US occupation of Cuba, Atkins had enough political influence with the government to ask James Wilson, the US Secretary of Agriculture, for permission to found an experimental sugar station in Cuba.[20] Wilson rejected the idea as he backed diversifying Cuban agriculture to meet the needs of the US home market, which for Cuba would also mean breaking with sugar monoculture.

Atkins sought out another valid interlocutor, in a closer and more familiar ambit, which brought Harvard University into his Boston business circles, but also effectively tied it to the legacy of slavery and the Caribbean plantation economy. Harvard was the world centre of botany in the late nineteenth century, and its inclusion would bestow scientific authority on Atkins's enterprise and a place for him among the American benefactors of applied research at Harvard. In 1899, he met George Lincoln Goodale, head of the Harvard Botanical Museum and Professor of Economic Botany, seeking the cooperation of botanists to improve sugar cane yields and disease resistance through hybridisation. Goodale accepted the offer, which linked Harvard, initially informally, with the sugar industry.[21] Connecting this international centre of knowledge and research with a mill built on a former slave plantation is certainly controversial, but the participation of Harvard botanists in Atkins's sugar business did mean the introduction of hybridisation techniques to Cuba. That helped modernise tropical sugar science (Fig. 7.1).

Fig. 7.1 George L. Goodale in his laboratory. (Source: Harvard University Herbaria and Botany Libraries)

The application of economic botany to farming problems and the search for "useful" plants became, from the mid-nineteenth century, the fields of research with the widest scope and impact around the world. Selection of varieties through cross-breeding intended to improve yields and resistance to disease, and the role of horticulture, were the focus of debates among international scientists and agronomists who would often invoke Charles Darwin's theories, among others. The production of knowledge among them to revitalise the agricultural economies of tropical colonies highlighted the role of botanical gardens and museums and of agronomic stations as centres for research and the global distribution of plants.[22] Harvard botanists defended the idea of applying research in the new economic botany to the tropics. At a meeting of the American Association for the Advancement of Science held in Washington, Goodale promoted the contribution of botanical research to obtaining new sources of wealth for the United States, and defended the need for America to have its own international research centres such as the Royal Kew Botanical Gardens for British colonies or the Buitenzorg Garden in Java.[23] Likewise, the tropics and the great diversity of island species fit for study in a botanical garden were an attractive part of Atkins's offer. In this way, Harvard was inserted into imperial networks of botanical research through Cuba's sugar industry and, specifically, an enclave founded on slavery.

Atkins paid an initial deposit of US $2500 ($2000 for a travelling fellowship in economic botany and $500 to create a sugarcane bibliography). [24] The entomologist Edwin Mead Wilcox, a student of economic botany specialising in the East and recent Harvard alumnus, was the first to benefit from the travelling fellowship in economic botany. Goodale wanted Wilcox to run the station in Cuba after returning from a trip to Java where he would collect information on obtaining hybrids and a set of samples as primary testing material. Wilcox accepted, but then declined the position. Goodale then turned to Oakes Ames, another recent Harvard graduate appointed Assistant Director of the Harvard University Botanic Garden. Ames was the only one with ample experience of hybridisation, though only of orchids.

On 4 January 1900, Atkins, Goodale and Ames met at the Soledad sugar plant, 'that colossal institution which grinds grass and makes money', in Cuba.[25] This meeting reflected some early tensions between the businessman and scientists. Atkins wanted a cane hybrid but the botanists did not know how to attain his ideal, and this prompted doubts among the botanists who saw in Cuba a wider ranging tropical laboratory. What then was the common ground between sugar production and tropical research, or science and business? Ames defined the comfort zone as 'created interests' between Atkins and the botanists based on introducing plants of commercial and decorative value, diversifying Atkins's agricultural businesses and expanding the agronomic station into a botanical station useful to agriculture. I believe it is only by seeing the revolution of varieties as the paradigm of the new field of economic botany that one can explain the foundation of the Harvard station in Cuba.[26] I would say that scientific plant-breeding or hybridisation, besides being a novelty in Cuba, was the preeminent investigative link between sugar production and botanical research as it was essentially a botanical practice. The botanists sensed that the garden would become a leading place of tropical research, at a time when the tropics were unknown territory for American investigators and when the US government was keen to learn about existing natural resources as part of a process of conquering the tropics. This was the setting for the creation of the Harvard Botanical Station for Tropical Research and Sugar Cane Investigations, the first successful zone of negotiations between research projects in tropical botany and sugar business, between the scientist and a businessman, and a place to train experts at a time of US imperial expansion.

Harvard became a precursor among university institutions with outposts to study tropical agriculture and biology within global and local sugar economy circuits originally tied to slavery.[27] Edwin F. Atkins entered into Harvard academic circles as a member of the Botanic Garden Committee and representing the active patronage of field research, while Ames justified the decision to Harvard staff by presenting him as a Boston businessman who had created a small, experimental garden in Cuba, supplying its land, local workforce and enough funds to begin experiments on eight of the most important species cultivated to improve existing sugar varieties in Cuba.[28] The professors of botany would in turn take part in the garden's scientific design while experts would pursue research there. The garden would be a part of the Botanic Gardens of Cambridge, Massachusetts.

The Garden's early activities and investigations reveal the roles of many agents and knowledge sources in producing agricultural and tropical botanical science, amid the tensions of scientific, production, nationalist and imperial interests. The choice of experts reflected for example differences between the Harvard botanists and Atkins's staff at the sugar plant, or between expert and amateur knowledge. Ames chose the horticulturist Robert Grey who could create hybrid varieties and had worked in America, but was not a scientist. L.H. Hughes, the plant manager previously in charge of research, obstructed the initial experiments by Hugo Bolnhoff, a gardener at the Botanic Gardens at Cambridge, and Grey. Yet it was Grey who found the right spot for creating the experimental garden and obtained the ideal hybrid cane Atkins wanted. The place was the Colonia Limones, a position suited to tropical botanical research thanks to its natural woodland habitats near the Escambray mountain range, and for being over areas already deforested to cultivate sugar.[29]

The Harvard Botanic Station shows the collaboration of networks of imperial gardens and experimental stations throughout the tropics, the interactions of botanists inside transatlantic knowledge circuits in both practical and formal terms, and exchanges between commercial and scientific groups at different levels (global, regional and local). These global networks of botanists exchanged professional knowledge and information applied to the economy, research models, plans and material, or practical projects. For example the material for sugar hybridisation experiments came from British colonies in the Greater Caribbean, Java, Mexico and elsewhere. Plants with a commercial and ornamental value were introduced initially from Florida and later, from several parts of the world.

With the Station's scientific design, Goodale and Ames followed the successful imperial model of botanical gardens in British and Dutch tropical colonies. The two academics were in the global community of practices and knowledge networks of botanists created by Kew Gardens. Ames travelled to Cuba with John C. Willis, a reputed British botanist and head of the Royal Botanical Garden of Peradeniya in Ceylon (now Sri Lanka), who would help the Harvard Botanic Station earn international renown.[30] As expert witness, Willis suggested introducing economic plants, though his considerable ignorance of cross-breeding and sugar production irked Ames, as Willis was unable to advise the garden about certain initial problems.

Goodale and Ames focused on Barbados and Java, which were leaders in sugar hybrid research and experimentation. Goodale contacted the British botanist and horticulturist Daniel Morris, Commissioner at the Imperial Department of Agriculture for the West Indies, to pay for the dispatch of material from Cuba.[31] Morris sent hybrids from Barbados and Demerara, which were supervised in Cuba by the manager of the Soledad estate, L.H. Hughes. The results were not good. Correspondence between Goodale and Morris reflected global exchanges of knowledge apparently typical of agronomists, botanists and scientists, and Goodale informed Morris of trials undertaken in Cuba on pollinating sugarcane. Morris pointed out that J.P. d'Albuquerque, a chemist heading the botanic station in Barbados, had done similar experiments consisting of cross-fertilising two varieties of flowers to produce 3000 seedlings. Paradoxically, Goodale and Ames never invited local cross-breeding experts such as John R. Bovell, the father of the botanical revolution in Barbados, to the Garden.

Goodale also wrote to the botanist Melchior Treub, director of the Buitenzorg Botanical Garden in Java. Their correspondence, however, sheds light on tensions among scientists, agronomists and producers over sharing information. In July 1902, Treub replied that cultivators in Java were reluctant to send planted cane material to Cuba, fearing Cuba could soon steal some of their share of the world sugar market.[32] But the scientists decided to send a set of eleven of the best varieties cultivated in Java, and experiments in Cuba with those varieties did yield good results. Treub also travelled to Harvard to exchange ideas with Goodale. The exchange of seedlings between Java and Cuba illustrated how pathogens could travel the world, and scientists warned Goodale that sending seeds could spread mosaic disease.[33] There were precedents to the collaboration between Cuba and Java. *Crystalline* cane, the main commercial variety on the island, had come from Java, while Cuba had given Java an intensive cultivation system created by the chemist Álvaro Reynoso. The search for material connected networks of Harvard botanists throughout the Americas. For example, Goodale sent C.G. Pringle, a US botanist and collector, to the garden with six sugar varieties Pringle had gathered on research trips to Mexico.[34]

The Harvard Botanic Station shows the value of the tropical countryside as a space for tropical knowledge-building on site, and to train the US expert to solve economic problems. The botanists' unfamiliarity with sugar research was initially resolved with information on orchid hybridisation available to Ames and Grey. In fact Ames felt he was not so much an expert as a work coach for Grey and Bolnhoff.[35] He complained that the humidity in Cuba did not favour work on sugar cane, as humid flowers and anthers did not release pollen. Such problems were solved with a greenhouse built in 1902 for sugar experiments.[36] The panicles were covered with gauze fabrics and flowers were pollinated manually with needles. The abundant sugarcane pollen known locally as *Black Fiji* offered the best results for pollination when crossed with *Crystalline* cane. During 1904–1905, Grey obtained 810 seedlings of twenty-one different hybrids that were named Harvard and numbered.

In 1919, Atkins added more land and offered funding worth $100,000 to officially seal the ties with Harvard University. The donation's scientific objective caused frictions between Ames and Harvard staff, and between the tropical research project and the sugar business. The corporation wanted to use the fund for research generally, not just on tropical botany and sugar. Ames opposed this. He was the one who had sealed the ties between Atkins's garden and the Harvard Botanical Department through the Atkins Fund for Tropical Research in Economic Botany. In 1920, Atkins made it clear to the president of Harvard that the fund was to finance the work of the Harvard station in Cuba. Its research included all studies to benefit tropical agriculture and horticulture but especially sugarcane and forage cultivation. The Atkins Institute for Tropical Research formalised the Garden's ties with Harvard.

International recognition of the Harvard Botanic Station was consolidated under Thomas Barbour between 1920 and 1940. His figure and those of other American scientists, reflect the ambivalent dimension of tropical knowledge generation, as explored in Paul Sutter's research on the role of scientists as 'agents of imperialism' in the early twentieth century.[37] Barbour first visited the Garden in 1908, and documents show him working in the First World War as a secret agent reporting on Cuba for the United States.[38] Barbour expanded the links within the Botanic Station between tropical medicine and agriculture, and basic research.

This interest led to the Botanic Station contributing to publications by the Institute for Tropical Biology and Medicine, which Barbour and the physician Richard P. Strong founded at Harvard.[39] In the case of Cuba, all publications concerned research on sugar. Barbour also became a close friend of the Atkins family. In 1924, Atkins and especially his wife Katherine W. Atkins (Catalina to the Cubans), backed Barbour by creating the Harvard Biological Laboratory and Harvard House, two spaces intended to consolidate research and provide lodgings for scientists at the Garden. Barbour also inaugurated summer courses for Harvard biology students. In contrast with Goodale and Ames, Barbour kept in touch with Latin American naturalists and, above all, with scientists and Cubans such as the naturalist Carlos de la Torre, with whom he forged a lasting friendship, the agronomist Julián Acuña and botanist Juan Tomás Roig of the Cuban Experimental Station.[40] Barbour began forestation studies for the Santa Clara sugar region at the Garden, planting native trees in the area known as Seboruco.

Atkins died in 1926. By then, the Harvard Botanic Station had become a reference in the realm of economic botany, and every year an increasing number of American and international scientists visited its collections. In fact the institution became a representative space for the community of American tropical experts, as shown in the meeting held at the garden between Thomas Barbour, Robert M. Grey, Wilson Popenoe and Franklin S. Earle, to review its work in the context of the Cuban sugar crisis. This again revealed tensions between sugar production and tropical research, or business and science.[41] Each defended his ideas on the Botanic Station's contribution to the economic development of the Caribbean and Cuba. Wilson Popenoe, a botanist and agronomist, worked with the USDA, was a director of the United Fruit Company and the first head of the Panamerican School of Tropical Agriculture in Honduras. Popenoe supported the garden's importance in tropical research and especially to the Caribbean, for being a "great open-air laboratory" for studying and disseminating useful plants, improved through selection and hybridisation and adapted to the Caribbean's needs. Thomas Barbour saw the garden as the jewel in Harvard's crown, with benefits for both Cuba and the United States. Earle had become an expert in tropical sugar science for US sugar companies in the Caribbean after resigning as head of the

Experimental Station in Cuba.[42] Earle had contacted Atkins through the Soledad estate manager L.H. Hughes to work with them on sugar experiments.[43] Atkins rejected Earle's idea as he was already working with Harvard, but remained open to collaboration. Earle was the best qualified to evaluate the importance of Robert Grey's investigations with the *Harvard 12029* hybrid, and exclaimed, 'Why, man, you don't know what you've done. That can be worth millions to Cuba.'[44] If true, the question is, why did this hybrid not come to dominate the Cuban countryside, as Atkins and Grey hoped?

Networks of US Experts in Cuba: Diseases, Knowledge and Seedlings

The history of tropical sugar became after the late eighteenth century one of the introduction and distribution globally of sugar varieties. These introductions and exchanges of varieties and hybrids were related to problems of a deteriorating productive environment, such as the emergence of pests and disease, and the quest for higher yielding sugar plants. Latin American governments and liberal local élites hastened the collapse of agricultural ecosystems by boosting the production of tropical commodities for world markets (causing deforestation, exhausted soils, increased infestations and disease, falling yields and so on). In the late nineteenth and first half of the twentieth centuries, these environmental and economic problems were the focus of efforts to create in situ new codes of practice and scientific know-how in farming and botanics, such as developing higher yielding and more resistant cane hybrids.

The work of the Harvard Botanic Station in Cuba illuminates other zones of resistance to the circulation of sugar knowledge and hybrids at different levels, beyond the global aspect. I focus here on the conduct of Edwin F. Atkins, Robert H. Grey, the Harvard botanists and other US and Cuban experts in studying the causes of sugarcane mosaic disease (SMV, sugarcane mosaic virus), which affected the global sugar industry and mainly Cuban production in the early twentieth century.[45] Mosaic disease control and eradication provided Atkins, Grey and the Harvard botanists with an opportunity to popularise and scientifically verify the tropical knowledge produced in Cuba on sugar and the hybrids created there. US cross-breeding experiments pursued locally at the Soledad estate were the centre of important theoretical debates on the mosaic disease's aetiology, its origin and point of dispersion inside Cuba after the arrival of seedlings from Java, itself a process marked by political, economic and scientific tensions. In 1915, at a time when Cuban sugar production peaked thanks to the high prices and rising demands during the First World War, L.D. Marsen, a pathologist of the Experimental Station of the Hawaiian Sugar Planters' Association, sounded the alarm on the mosaic disease's presence at the Soledad estate, though it appeared then as isolated and relatively non-contagious.[46] In fact Ames and Edward Murray East, a Harvard professor and one of the more prominent US plant geneticists of the early twentieth century, did not notice the disease when visiting the Botanic Station in 1918 to provide Grey with modern perspectives on plant breeding for sugar hybrids with greater industrial yields.[47] Yet that year, the USDA had sent Franklin S. Earle to Puerto Rico to study the causes of the mosaic disease that threatened to destroy its industry. Earle recommended spreading the *Uba* cane variety for its resistance to mosaic.[48]

The presence of infection in the plantations of the Soledad mill meant that the Harvard botanists had to validate Atkins's sugar business and Grey's research, which provoked differences with the Cuban government and USDA. In 1919, Harvard's cross-breeding experiments at Soledad were back in the public eye when mosaic became of serious concern to the Cubans. E.W. Brandes, an entomologist and pathologist heading the USDA's Sugar Plant Investigations, and John R. Johnston, a Harvard botanical collector and then chief pathologist at the Cuba Experimental Agronomic Station, were asked by the government to isolate and study the disease. Brandes identified three possible points of dispersion for the infestation following various hybrid imports. The most virulent point was on the lands of the Experimental Station, which had seeds from Java, the US state of Louisiana and Tucumán in Argentina. The second infectious point was the Mercedita mill in Matanzas, with imports from Java and Argentina, and the last focal point, the Soledad estate in Cienfuegos, with hybrids from Java.[49] Brandes said the disease had existed in latent form at the Atkins mill for some twenty years and that practically all the seedlings Harvard had obtained were infected. He claimed the infestation had not developed virulently because Hughes, the Soledad factor manager, discarded seedlings in the fields both for selectivity and to ensure the absence of *Aphis maiclis*, the disease vector.

Correspondence between Ames and Johnston sheds light on the ambivalent aspects of the networks of biologists working with private and official institutions in Cuba. As a former Harvard staff member, Johnston wrote to Ames to exchange views on the possible gravity of mosaic disease at the Harvard Botanic Station in Soledad.[50] Johnston understood that eliminating all infected seedlings meant throwing away the work of many years at the Botanic Station, even if replicating the disease was clearly a very serious problem. Thus he trusted Grey to find varieties that were highly resistant or immune to the disease. As a Cuban civil servant, however, Johnston reminded Ames that they were obliged to assist them at the Quarantine Inspection Service and respect the Cuban government's orders banning cane importation and use of any infested canes as seed.

Ames's reply was, as a scientist and the Garden director, equally ambiguous because of his interest in the mosaic pandemic and its spread in Cuba, though he doubted its presence in Soledad. Ames believed the disease remained largely unknown, since its only symptoms were for now the appearance of a nuanced green colour on leaves and the plant's degradation. Ames had seen these signs in the plantations fifteen years previously, and attributed them to soil conditions or seasonal influences rather than the mosaic identified by the Cuban government's two US scientists. Yet he admitted to Johnston that working together was a great deal for both, as he saw in him an 'impartial investigator' seeking ways to benefit the Cuban economy and because it was always a pleasure learning from the interests of a 'Harvard man'. Atkins was more critical of Johnston who, as a Cuban government pathologist, was accusing his sugar estate of spreading the disease when writing in the Experimental Station bulletins. In 1920, during the so-called Dance of the Millions in Cuba, Brandes, Johnston, Grey and Hughes convened at the Botanic Station to discuss Grey's theory on what had caused the epidemic and how to eradicate it.[51] Grey maintained that the disease had come from erroneous cultural practices and could be controlled and eradicated through scientific cultivation. Brandes and

Johnston observed that experiments showed that the infected plant died within three years. Hughes favoured checking Grey's theory. A while later, seedlings obtained by Grey were tested by Harold L. Lyon, the US pathologist and botanist heading the Hawaii Agronomic Station, who saw no evidence of disease.

Ames took the opportunity of the disease's outbreak in the Soledad lands to counsel Atkins to reassure himself with written opinion from an economic plant pathologist. The advice reflected tensions between amateur and expert knowledge but also links between US and British sugar circuits in the Greater Caribbean, as Ames recommended Noël Deer, a prestigious British technologist, expert and sugar historian.[52] Deer was hired both by the Cuban government and by transnational corporations to modernise Cuban sugar production. He represented the typical expert who was trained and who earned repute by working in private sugar companies and by travelling throughout the tropics for sugar.[53] Ames was confident of Grey's abilities as an intelligent horticulturist and specialist of plant breeding, but Grey was no scientist and that could hamper studying the disease mosaic. Nevertheless, Ames continued to back Grey's theory before the opinion of the pathologists Brandes and Johnston, given the lack of evidence on the disease being more biological than physiological. Deerr also backed Grey's scientific work as being as legitimate as the work in any prestigious laboratory, but warned that as this was a private enterprise, it would probably be the only party to benefit from the knowledge created here, suggesting the Cuban government should keep an eye on these studies. He was in other words highlighting the identity of the owners of locally produced knowledge.

The tensions between Ames and Atkins show the porous borders between the public and private spheres, between professional and amateur knowledge, and business and academia, when it came to certifying Harvard's scientific work in Cuba. In contrast with private sugar companies that hired experts from various countries and with differing training or technical backgrounds (chemists, engineers and so on), Atkins wanted Harvard University botanists to give the station its scientific authority. In exchange he would win prestige, power, credibility and recognition from this global centre of scientific knowledge par excellence. Yet Atkins was the one financing investigations, which gave him rights to the publication of their results and to the backing of Robert Grey's work even if he was not a professional scientist. Atkins insisted the reports be kept confidential, probably to safeguard his business's comparative advantage over other plantation owners, but he also sensed that studying the mosaic disease was an opportunity to validate the Agronomic Station's work, which he financed, and Grey's work within the scientific world. Ames, however, maintained that a pathologist must run the station for Harvard to recognise the Garden as a scientific enterprise. Differences between Ames and other Harvard professors also revealed doubts about the Botanic Station. For example, Roland Thaxter, a micologist, plant pathologist and entomologist, questioned its scientific value and the pertinence of Grey's knowledge.[54] Ames referred to East, who believed Grey was the best plant breeder who had worked with them. Ames was also using the economic reasoning of a businessman who demanded practical results to consider the Botanic Station's scientific worth, as he had been working with Atkins for twenty-five years and insisted on both the investigative lines of economic botany at the garden: the agricultural side and the scientific aspect that examined particular problems.

In 1924, the Cuban Agriculture Ministry ordered another general inspection and the propagation of cane varieties immune to mosaic, especially *Uba* variety and different types of *POJ* (Proestation Oost Java) hybrids.[55] This provoked another instance of friction between Atkins and the Experimental Agronomic Station, which had become an official centre of research and hybrid dissemination. Atkins complained to its director, Gonzalo Martínez Fortún, about the Soledad plantations being cited as the chief focus of mosaic disease in Cuba, as they had records of its presence some forty years previously. Likewise, he and Harvard had studied the disease without finding any harm to tonnage or reduction of sugar production. He stated that Harvard botanists had inspected his fields and only found two canes with mosaic. For Atkins, the *Uba* variety was not suited to commercial purposes, but he hoped to obtain certain valuable hybrids with *Crystalline* cane that seemed immune to mosaic. Fortún asked him to send in the data from experiments with the *Uba* variety and the report by the two US pathologists.

Atkins was referring to research by Harvard botanists Edward M. East and William H. Weston Jr, an Assistant Professor of Botany, who had to corroborate Grey's experiments and declare on the gravity of the disease in the Soledad plantations. The two professors confirmed the theses defended by Brandes and other scientists, agronomists and botanists, that mosaic was viral and its vector was the *Aphis maiclis*.[56] East and Weston explained that the disease was not as virulent on Atkins's plantations owing to the equilibrium *Crystalline* cane and mosaic had attained after a long period of coexistence. They recommended Hughes's method of selecting seeds in the countryside and endorsed Grey's plant breeding capabilities as suited to the Harvard–Atkins combination. Some authors maintain that the participation of US botanists in studying the mosaic epidemic exemplified a new ecological thinking characteristic of biology in the early twentieth century, taking into consideration as it did for the first time the life of a plant and making a comprehensive analysis of all environmental and agricultural factors favouring or restricting the disease's propagation.[57] East and Weston represented this ideal and included other fields in their study of the mosaic (genetics, pathology, phytopathology, entomology and so on). They saw a need for more studies on agents of transmission and agricultural and environmental factors (such as climate, cultivation practices or the role of other plants). In 1926, George Salt and J.G. Myers, two economic entomologists, joined Harvard's sugar investigations at the Botanic Station.[58]

The study of diseases and improvement of varieties through scientific breeding of sugarcane involved multiple socio-economic and political agents (landowners, transnational corporations, the Cuban government and other US institutions, both private and public) interested in modernising tropical agriculture. The transnational sugar companies likewise illustrate the link between science and economy, as creators and disseminators of global modernisation programmes, in using experts to head their agro-industrial research and by founding chemical laboratories and agricultural stations at production centres. For example, Willian Cornellius van Horne, president of the Cuba Company, hired Noël Deer and Paul Karutz, a German chemist, to create an experimental garden at the Jobabo estate. Its aim was to study different economic plants as part of plans to make Cuba the United States's granary.[59] Landowners gathered in turn around the Cuba Sugar Club to defend the sector's interests in the early twentieth century. In 1924, the US National Research

Council founded the Tropical Plant Research Foundation to undertake agricultural and biological research at temporary and permanent stations in the tropics, in response to the high demand in the United States and world markets for tropical products.[60] The foundation also represented a meeting of business and science with its staff of five scientists and four businessmen, such as representatives of the United Fruit Company and the Cuba Railway Company. The Foundation and the Cuba Sugar Club established an experimental sugar station in the Baragua estate in eastern Cuba, linking up with other US experts employed by sugar corporations. The aforementioned Franklin S. Earle was the author of the foundation's structural design and in charge of studying varieties at the Baragua mill.

The scientific investigation of the problems of tropical sugar cultivation provided a meeting point between the Foundation and Harvard botanists, which reflected other areas of negotiation between networks of US experts working in Cuba in the early twentieth century. In 1925, Thomas Barbour and W. A. Orton of the USDA's Bureau of Plant Industry and director of the Tropical Plant Research Foundation, signed a memorandum of collaboration between the two institutions.[61] This was to authorise studies by James A. Faris, a pathologist of the Foundation, and James B. Weir, a pathologist of the USDA Bureau of Plant Industry, to use the Harvard Botanic Station laboratory without interfering in Harvard's investigations. Like the Harvard botanists, the Foundation's pathologists and entomologists understood that studying the disease mosaic encompassed agricultural and environmental factors (such as parasites, soils, conditions or adaptability of varieties). The Foundation understood the Soledad estate was an excellent field of studies for the prevalence of the disease at Cienfuegos, which prompted the logical idea of the Harvard Botanic Station becoming a centre to complement the research carried out in Baragua.

The exchange of information and hybrids between Harvard and the Tropical Plant Research Foundation raises questions about cooperation between networks of US scientists in the tropics, and between Harvard and the Cuban government. Robert Grey wrote to Thomas Barbour on Harvard's request for hybrids from different places, including the Cuba Experimental Station. Grey suggested they write to D.L. Van Dine, the local director of the Baragua station, following his decision to collaborate only with the Baragua station in studying varieties, to avoid confusion.[62] At the same time, Grey complained that the Foundation, and in this case Earle, was giving priority to trials with the *POJ* types, to which Barbour replied that he was not surprised, suggesting tensions between the two laboratories. The Foundation thanked Grey for the varieties sent, and confirmed that they only had a comparative field of experimentation, with priority given to *POJ* varieties for their resistance to mosaic. Even so, the Foundation wanted to keep Harvard's seedlings for the purpose of comprehensive studies when possible, and as reproductive stocks for breeding adapted to Cuban conditions. These studies showed that the *Harvard 12029*, bred from the Barbados Sport Blanco and *Crystalline* varieties, yielded the greatest quantities but only on terrains rich in organic material. Private correspondence between Grey and Barbour sheds light on other tensions among Harvard botanists themselves and especially between Barbour and Edward M. East, whom Barbour did not wish to see at Soledad.

The Baragua station, however, became a bridge between scientists hired by Cuba and the Harvard Botanic Station. For example, S.C. Brunner, an American pathologist who worked at the Cuban Agronomic Experimental Station, and the Italian entomologist F. Sylvestri worked with the Baragua station and later with the Harvard Botanic Station as part of their studies on sugar infestations and hybrids.[63] In 1932, the Foundation ended the investigations and debated whether to hand over its research and collection of seedlings to the Cuban Agronomic Station or the Harvard Botanic Station, which finally became the beneficiary.[64] The star project of improving sugar cane through hybridisation declined at the Botanic Station as in the wake of the Cuban sugar crisis of 1929 the *POJ 2878* hybrid from Java gradually imposed itself. By 1940, the *Harvard 12029* hybrid had disappeared from Cuban plantations.[65]

Conclusion: Networks in Tropical Botany and Agricultural Research in Cuba

The history of the Harvard Botanic Station illustrates the participation of American scientists as mediating agents in economic practices globally, regionally in the post-slavery Hispanic Caribbean and locally in Cuba.

They acted in a period of US expansion in the tropics and regional decolonisation. The Botanic Station's long history confirms that a balance was reached between the respective aims of sugar research and tropical biology in the new field of economic botany, and between science and business, which was characteristic of scientific knowledge production in the early twentieth century. While the Harvard Botanic Station was an American institution designed to develop the needs of US cultivators, it had to deal with economic and environmental problems in Cuba such as deforestation, the mosaic disease and declining outputs in areas saturated by industrial-scale sugar production, such as Cienfuegos, a former nucleus of the slave-plantation economy. Likewise we can say that Harvard recreated a tropical world on the biggest 'sugar island', which meant breaking the sugar monoculture in one of the principal tropical production enclaves in the region and the world. The study of mosaic disease itself brought into Cuba modern fields of study such as tropical genetics, pathology and entomology.

The Botanic Station placed Edwin F. Atkins and Harvard at the productive epicentre of world tropical science, making them exponents of the geopolitics of knowledge, while also associating them with slavery and the plantation economy. This highlighted the complexity of building knowledge in settings of imperial and hegemonic expansion and subsequent decolonisation, but also of economic globalisation. The unfolding of political, scientific and economic controversies of their time played out by the chief actors of the Harvard Botanic Station sheds light on the ambivalent and porous aspects of tropical scientific knowledge. It also reveals certain failed negotiation zones in the circulation of knowledge involving power, science and money within global and local sugar economy circuits. Atkins and Harvard were pioneers in obtaining hybrids adapted to Cuba's productive and environmental conditions, but could not displace *Crystalline* cane and other seedlings, which prompted questions such as: to whom did the knowledge belong, and which knowledge flowed with the financial means, economic and trading profits and global knowledge circuits?

The Experimental Station and transnational sugar companies reflected Cuba's attraction as a source of knowledge generation for networks of American, European, Latin American and Cuban scientists concerned with tropical agriculture and especially sugar cultivation. The Harvard Botanic Station, however, gave priority to participation by networks of scientists from Boston, Harvard and Great Britain.

Notes

1. Kuklick, Henrika and Kohler, Robert E.: 'Science in the Field', *Osiris*, No.11 (1996), pp. 1–16. Kohler, Robert E: *Landscapes and Labscapes: Exploring the Lab-Field Border in Biology*, Chicago, IL: University of Chicago Press, 2002.
2. McCook, Stuart: *States of Nature. Science, Agriculture, and Environment in the Spanish Caribbean, 1760–1940*, Austin: University of Texas Press, 2002. McCook, Stuart: 'The Neo-Columbian Exchange: The Second Conquest of the Greater Caribbean, 1720–1930', *Latin America Research Review*, No.46S (2011), pp. 11–31.
3. Raby, Megan: *American Tropics: The Caribbean Roots of Biodiversity Science*, Chapel Hill, NC: University of North Carolina Press, 2017.
4. McCook, *States of Nature*; Raby, *American Tropics*. Fernandez-Prieto, Leida: 'Islands of Knowledge: Science and Agriculture in the History of Latin America and Caribbean', *Isis*, Vol. 104, No. 4 (2013), pp. 788–97; Curry-Machado, Jonathan: 'Cuba, sugarcane, and the reluctant embedding of scientific method Agete's *La caña de azúcar en Cuba*', in van Schendel, Willem (ed.): *Embedding Agricultural Commodities. Using Historical Evidences, 1840s–1940s*, London and New York: Routledge Taylor & Francis Group, 2016, pp.119–45.
5. Latour, Bruno: *Science in Action. How to Follow Scientist and Engineers through Society*, Cambridge, MA: Harvard University Press, 1987; Vetter, Jeremy: *Knowing Global Environments: New Historical Perspectives on the Field Sciences*, New Brunswick: Rutgers University Press, 2011; Curry-Machado, Jonathan: *The Global and Local History of Commodities of Empire*, London: Palgrave Macmillan, 2013.
6. Moreno Fraginals, Manuel: *El Ingenio*, three volumes, La Habana: Editorial de Ciencias Sociales, 1978; Curry-Machado, Jonathan: *Cuban Sugar Industry: Transnational Networks and Engineering Migrants in Mid-Nineteenth-Century Cuba*, London: Palgrave Macmillan, 2011; Leonard, A. B. and Pretel, David: *The Caribbean and the Atlantic World Economy. Circuits of Trade, Money and Knowledge 1650–1914*, London: Palgrave

Macmillan, 2015; Rood, Daniel: *The Reinvention of Atlantic Slavery: Technology, Labor, Race, and Capitalism in the Greater Caribbean*, Oxford: Oxford University Press, 2017.

7. Allen, Benjamin: *A Story of the Growth of E. Atkins & Co. and the Sugar Industry in Cuba*, New York, NY: E. Atkins, 1925.
8. Atkins, Edwin F.: *Sixty years in Cuba: Reminiscences*, Cambridge: Riverside Press, 1926; Tomich, Dale: *Through the Prism of Slavery: Labor, Capital, and World Economy*, Lantham: Bowman & Littlefield Publishers, Inc., 2004.
9. Pite, Rebekah E.: 'The Force of Food Life on the Atkins Family Sugar Plantation in Cienfuegos, Cuba, 1884–1900', *The Massachusetts Historical Review*, No. 5 (2003), pp. 59–94; Scott, Rebecca: 'A Cuban Connection: Edwin F. Atkins, Charles Francis Adams, Jr., and the Former Slaves of Soledad Plantation,' *The Massachusetts Historical Review*, No. 9 (2007), pp. 7–34.
10. Iglesias, Fe: *Del Ingenio al Central,* La Habana: Editorial de Ciencias Sociales, 1999; Funes Monzote, Reinaldo: *From Rainforest to Cane Field in Cuba: An Environmental History since 1492*, trans. Alex Martin, Chapel Hill, NC: University of North Carolina Press, 2008; Fernández Prieto, Leida: *Cuba Agrícola: Mito y Tradición, 1878–1920*, Madrid: CSIC, 2005.
11. Sanger, Joseph P., Gannett, Henry and Willcox, Walter F.: *Informe sobre el Censo de Cuba, 1899*, Washington: Imprenta del Gobierno, 1900, p. 533.
12. Fernández Prieto, Leida: 'Saberes híbridos: las Sugar Companys y la moderna plantación azucarera en Cuba', *Asclepio*, Vol. 67, No.1 (2015).
13. Anderson, Warwick: *Colonial Pathologies. American Tropical Medicine. Race, and Hygiene in the Philippines*, Durham, NC: Duke University Press, 2006; Espinosa, Mariola: *Epidemic Invasions: Yellow Fever and the Limits of Cuban Independence, 1878–1930*, Chicago, IL: The University of Chicago Press, 2009; Neill, Deborah J.: *Networks in Tropical Medicine: Internationalism, Colonialism, and the Rise of a Medical Specialty, 1898–1930*, Stanford CA.: Stanford University Press, 2012; Adas, Michael: *Dominance by Design: Technological Imperatives and America's Civilizing Mission*, Cambridge, MA: Belknap Press, 2006.
14. Iglesias, Marial: *A Cultural History of Cuba during the U.S. Occupation, 1898–1902*, Trans. by Russ Davidson. Chapel Hill, NC: University of North Carolina Press, 2011.

15. Pérez, Louis A.: *Cuba in the American Imagination: Metaphor and the Imperial Ethos*, Chapel Hill, NC: University of North Carolina Press, 2008.
16. Mickulas, Peter Philip: *Britton's Botanical Empire: The New York Botanical Garden and American Botany, 1888–1929. Memoirs of the New York Botanical Garden*, Bronx: New York Botanical Garden, 2007.
17. Martínez Viera, Rafael: *Estación Experimental Agronómica de Santiago de las Vegas. 100 años de historia al servicio de la agricultura cubana (1994–2004)*, La Habana: INIFAT, 2004, pp. 21–5; Fernández Prieto, Leida: 'Making Tropical Agriculture: Science, Knowledge and Practice in Cuba, 1881–1906', *Studies in the History of Biology*, Vol. 6, No. 1, (2014), pp. 5–25.
18. Galloway, J. H.: 'Botany in the Service of Empire: The Barbados Cane-Breeding Program and the Revival of the Caribbean Sugar Industry, 1880s–1930', *Annals of the Association of American Geographers*, Vol. 86 (1996), pp. 682–706.
19. Atkins, *Sixty years in Cuba*, pp. 331–32.
20. Cahan, Marion D: 'The Harvard Garden in Cuba-A Brief History', *The Harvard Tropical Garden*, Vol.1 (1991), pp. 22–32; Harris, Christopher: 'Edwin F. Atkins and the Evolution of United States Cuba Policy, 1894–1902,' *The New England Quarterly*, Vol. 78, No. 2 (2005), pp. 202–31.
21. Hazen, Dan: 'Cienfuegos Botanical Garden: Harvard's Legacy, Cuba's Challenge', *ReVista-Harvard Review of Latin America*, (fall 1998), pp. 6–8.
22. Brockway, Lucile H.: *Science and Colonial Expansion: The Role of the British Royal Botanic Garden*, New York: Academic, 1979. Drayton, Richard: *Nature's Government: Science, Imperial Britain, and the "Improvement" of the World*, New Haven, Conn.: Yale University Press, 2000; Schiebinger, Londa: *Plants and Empire. Colonial Bioprospecting in the Colonial World*, Cambridge, MA: Harvard University Press, 2007.
23. Goodale, George L: 'Horticultural Experiments and Botanical Investigations at the Harvard Station in Cuba', *American Journal of Science*, Vol.13, No. 76 (1902), pp. 325–26.
24. Harvard University: *Annual Reports of the President and Treasurer of Harvard College 1898–1899*, Cambridge, MA: Harvard University, 1900, pp. 242–43.
25. HUA UAV231 S1 Ames Oakes num.7, Cuban Letters to Blanche Ames (wife), Records of the Atkins Garden and Research Laboratory, 1898–1946, 01.15.1903.

26. Fitzgerald, Deborah: *The Business of Breeding: Hybrid Corn in Illinois, 1890–1940*, Ithaca, NY: Cornell University Press, 1990.
27. The Atkins Garden was not the only project involving Harvard University in Cuba. In 1900, two Harvard alumni, Alexis E. Frye, the Superintendent of Schools for Cuba appointed by the US government, and Ernest L. Conant, a lawyer working in Havana, proposed organising summer courses in Harvard for Cuban teachers. Harvard hosted 1450 Cuban teachers for training in the US teaching system whose principles would then be spread across the island. Atkins and his wife Katharine contributed by financing a part of the project and inviting teachers from the Trinidad and Cienfuegos regions to dinner. Harvard University: *Annual Reports of the President and Treasurer of Harvard College 1899–1900*, Cambridge, MA: Harvard University, 1901, pp. 344–54. Iglesias, *A Cultural History*, pp. 65–86.
28. Harvard University: *Annual Reports of the President and Treasurer of Harvard College 1900–1901*, Cambridge, MA: Harvard University, 1902, pp. 227–28.
29. Atkins, *Sixty years in Cuba*, pp. 331–32. Grey, Robert Melrose: *Report of the Harvard Botanical Gardens, Soledad Estate, Cienfuegos. Cuba (Atkins Foundation) 1900–1926*, Cambridge, MA: Harvard University Press, 1927.
30. Harvard University: *Annual Reports of the President and Treasurer of Harvard College 1902–1903*, Cambridge, MA: Harvard University, 1904, p. 229.
31. HUA UAV231 M, Morris a Goodale folder, Harvard Archives Botanical Garden 1898–1940, 02.07.1901.
32. HUA UAV231 S5 B, Records of the Atkins Garden and Research Laboratory, 1898–1946, 07.23.1902.
33. HUA UAV231 S2 Grey, Robert M. folder, Records of the Atkins Garden and Research Laboratory, 1898–1946, 06.21.1926.
34. Harvard University: *Annual Reports of the President and Treasurer of Harvard College 1901–1902*, Cambridge, MA: Harvard University, 1903, p. 224.
35. HUA UAV231 S1 Ames Oakes folder num.8, Cuban Letters to Blanche Ames (wife), 01.16.1902. Grey, *Report of the Harvard Botanical Gardens.*
36. HUA UAV231 S1 Ames Oakes folder num. 8, 12, Cuban Letters to Blanche Ames (wife), 01.16.1902, 01.17.1902.
37. Shuter, Sutter, Paul: 'Nature's Agents or Agents of Empire? Entomological Workers and Environmental Change During the Construction of the Panama Canal', *Isis*, Vol. 98, No. 4 (2007), pp. 724–54.

38. Atkins, *Sixty years in Cuba*, p. 332. Barbour, Thomas and Robinson, Helene: 'Forty years of Soledad', *The Scientific Monthly*, Vol. 51, No. 2 (1940), pp. 140–6; Barbour, Thomas: *A Naturalist in Cuba*, Boston, MA: Little, Brown and Co., 1945. Raby, *American Tropics*, pp. 99–104.
39. Raby, *American Tropics*, pp. 129–31.
40. Martínez, Modesto: 'El Jardín Botánico de la Universidad de Harvard (Fundación Atkins) en Soledad, Cienfuegos, Cuba', *Revista de Agricultura, Comercio y Trabajo*, No. 10 (1931), pp. 9–13; Acuña, Julián: 'Jardín Botánico de Harvard, en el central Soledad', *Revista de Agricultura, Comercio y Trabajo*, Vol. 14, No. 11 (1933), pp. 51–9.
41. Popenoe, Wilson: 'The Harvard Botanic Garden at Soledad', *Harvard Graduated Magazine*, Vol. 37, No. 3 (1929), pp. 281–85.
42. Fernandez-Prieto, *Making Tropical Agriculture*, pp. 5–25.
43. MHS 4 f. 6, Soledad Sugar Co. Letterbook. Edwin F. Atkins, 12.28.1908–12.06.1910.
44. Popenoe, *The Harvard Botanic Garden*, pp. 283–85. HUA UAV231 Grey, Robert M. folder, 01.16.1928.
45. Galloway, *Botany in the Service of Empire*, pp. 686–702; Garcia Muñiz, Humberto: 'Interregional Transfer of Biological Technology in the Caribbean: The Impact of Barbados' John R. Bovell's Cane Research in the Puerto Rican Sugar Industry, 1888–1920s', *Revista Mexicana del Caribe*, Vol. 2, No.3 (1997), pp. 6–42; McCook, *States of Nature*, pp. 98–102.
46. Agete Piñeiro, Fernando: *La caña de azúcar en Cuba*, I, La Habana: Ministerio, 1946, p. 22.
47. HUA UAV 231 S1 Ames Oakes folder, Harvard Archives Botanical Garden 1898–1940, 01.07.1918.
48. Earle, Franklyn S: *Sugar Cane and its Culture*, New York, London: J. Wiley & Sons, INC, 1928.
49. Brandes, E. W: *The Mosaic Disease of Sugar Cane and Other Grasses*, Washington: US Department of Agriculture, 1919.
50. HUA UAV231 S1 Ames Oakes folder, Records of the Atkins Garden and Research Laboratory, 1898–1946, 12.01.1919, 12.09.1919.
51. HUA UAV231 S1 Ames Oakes folder, Records of the Atkins Garden and Research Laboratory, 1898–1946, 04.13.1920, 04.13.1920.
52. HUA UAV231 S1 Ames Oakes folder, Records of the Atkins Garden and Research Laboratory, 1898–1946, 12.13.1919, 01.25.1920.
53. Zanetti Lecuona, Oscar; García Muñíz, Humberto and Venegas Delgado, Hernán: 'Noël Deerr en la Guayana Británica, Cuba y Puerto Rico

(1897–1921). Memorandum para la historia del azúcar en el Caribe', *Revista Mexicana del Caribe*, Vol. VI, No. 11 (2001), pp. 57–154.

54. HUA UAV231 S1 Ames Oakes folder, Records of the Atkins Garden and Research Laboratory, 1898–1946, 12.31.1919.
55. AEEA 782 f. 44, expediente relativo a varios individuos sobre caña Uba, 1924.
56. East, Edward M: 'The Harvard Botanical Garden', *Science*, Vol. 59, No. 1533 (1924), pp. 433–34. East, Edward M. and William Henry Weston: *A Report on the Sugar Cane Mosaic Situation in February, 1924, at Soledad, Cuba*, Cambridge, MA: Harvard University Press, 1925.
57. McCook, *States of Nature*, p. 9. Raby, *American Tropics.*
58. HUA UAV231 S4 reports George Salt, Records of the Atkins Garden and Research Laboratory, 1898–1946, 1925. Salt, George W. and Myers, John G.: *Report on Sugar-Cane Borers at Soledad, Cuba*, Cambridge, MA: Harvard University Press, 1926.
59. Fernández Prieto, *Saberes híbridos*, *Asclepio*, Vol. 67, No.1 (2015).
60. The Tropical Plant Research Foundation, *Science*, Vol. 59, No. 1538 (1924), p. X. See McCook, *States of Nature* and Raby, *American Tropics.*
61. HUA UAV231, Orton, W.A: Memorandum to the Harvard Committee in Charge of the Botanic Garden and laboratory at Soledad, Records of the Atkins Garden and Research Laboratory, 1898–1946, 03.02.1925.
62. HUA UAV231 Grey, Robert M folder, Records of the Atkins Garden and Research Laboratory, 1898–1946, 05.01.1925, 01.16.1928, 03.20.1930, 01.07.1930, 02.04.1930; Barbour, Thomas: 'Cane Breeding at the Harvard Cuban Station', *Harvard Alumni Bulletin*, Vol. 30, No. 31 (1928), pp. 925–26.
63. AEEA 348 f. 141, expediente relativo al ofrecimiento de una colección de "seedlings" de caña por el Sr. Vice Pte del Club Azucarero de Cuba al Sr. L Scaramuza para la EEA, 11.22.1932.
64. AEEA 348 f.141, 11.22.1932.
65. Agete Piñeiro, *La caña de azúcar en Cuba*, pp. 93–94.

8

Breaking Global Standards: The Anti-metric Crusade of American Engineers

Hector Vera

Introduction

During the Gilded Age and the Progressive Era in the United States, groups of scientists and exporters pushed for legislation that would make the decimal metric system of weights and measures the exclusive system of measurement in the country. As is well known, this—and further—efforts to metricate America failed repeatedly (today the United States in one of the six countries in the world that have not adopted the metric system).

One of the key reasons to explain the failure of the US federal government to secure metric adoption in this period was the forceful opposition mounted by mechanical engineers. This group of experts participated in mass media, scientific publications, and political debates against the

I would like to thank Cristóbal Henestrosa for his help in preparing the images used in this chapter.

H. Vera (✉)
Universidad Nacional Autónoma de México, Mexico City, Mexico
e-mail: hectorvera@unam.mx

© The Author(s) 2018

D. Pretel, L. Camprubí (eds.), *Technology and Globalisation*, Palgrave Studies in Economic History, https://doi.org/10.1007/978-3-319-75450-5_8

convenience of adopting the metric system. Their well-executed campaigns made the compulsory introduction of the metric system a highly contested political issue. Mechanical engineers joined forces with associations of manufacturers that gave their movement economic and political strength. The main objectives of this coalition were to allow for the continuous use of the English customary system of measurement and to prevent government intervention in technical and economic matters.

The engineers' tenacious opposition prevented the pro-metric front from presenting a unanimous support from experts in favour of the metric system. Engineers were crucial to contest the widespread idea (pushed partly by exact scientists, such as physicists and mathematicians) that the metric system was the only scientific solution to standardise weights and measures, both nationally and globally. Engineers claimed a particular expertise—that of men who actually constructed and made things—that disputed the monopoly of scientists to define what was the most rational solution to the problem of measurement standardisation. Even though there was not a clear winner in the discursive debate on speculative versus practical knowledge, engineers and manufacturers achieved their ultimate goals by undermining the scientific authority of metric proponents and by intimidating politicians not to pass an unpopular law.

This chapter follows key organisations and individual actors in this context. On the one hand are the American Society of Mechanical Engineers (ASME) and the National Association of Manufacturers (NAM), which represented the major organised opposition to metrication in the United States. On the other hand are Frederick A. Halsey (a mechanical engineer from New York) and Samuel S. Dale (a textile industrialist from Boston), who founded in 1916 the American Institute of Weights and Measures, an association that for two decades led the fight against metric adoption.

Measurement is one of the key cognitive operations in economic exchange. When different products and commodities are sold or bartered it is necessary to know how much of a product is being bought or exchanged. Commodities are sold using measures of weight, length, and volume. For these operations to be carried out smoothly, it is crucial to have standardised units of weights and measures. There is an important relationship between the standardisation of measures and economic transactions, or more specifically to the issue of transaction costs (the costs 'of specifying and enforcing the contracts that underline all exchange'),[1] the problem of asymmetric information (that is 'buyers and sellers [exchanging] goods on the basis of different amounts of information about costly-to-measure attributes of goods or services'),[2] and measurement.[3]

The reigning system of measurement that was able to standardise almost the whole globe is the metric system. Engineers have been, in almost every national case, instrumental in metric adoption. Their basic forms of operation and their world view are akin to the purpose of standardisation and simplification of practices and codes. In the words of Bruce Sinclair, 'engineers usually found purposeful non-standard usage offensive to their sense of order'.[4]

Even in the United States, some of the most active metric proponents used to emulate engineering procedures to justify their own activities. For example, Melvin Dewey, who was named president of the National Efficiency Society as a result of his efforts to promote metrication, library classifications, and spelling reform, described the Society as part of 'a nation-wide movement, not for promoting efficiency in the field where engineers have already done so much, but rather for extending the principles of efficient engineering to all other relations of life'.[5] Paradoxically, in the particular case of the United States, engineers were instrumental in impeding the adoption of the metric system.

Engineers and the Establishment of Metrological Standards in the United States

In January 1898, during the Third Annual Convention of the National Association of Manufacturers (NAM), industrialist Albert Herbert read the report of the Committee on Language, Weights and Measures,[6] a lengthy and spirited document that asked for Congress to make the metric system of weights and measures the only legal system in the United States, starting in 1901. The report maintained that the metric system was an instrument to foster universality and compare it to other globally accepted codes: 'a very important part of the language of commerce is what is properly called the "language of quantity", or the terms and divisions employed in determining weight and measure. [...] The Arabic numerals are a labor-saving, trade-extending tool, and as a language of commerce it is today the only language that is absolutely universal. These signs when written or printed are understood in all commerce everywhere on Earth.'

Herbert's report finalised portraying an enthusiastic future of progress in which the metric system would be the norm of the day and old ideas—as old systems of measurement—would be thrown away:

> We live in an age of marvelous progress. We have now machines which annihilate space and enable us to see and hear and talk over the distance of half a continent, which reproduces the speech of yesterday in its exact tones, and transmit to us with the actual photograph of the speaker if it is desired. Man is entering the twentieth century, equipped with inventions which will compel him to overcome his present ultra conservatism. There is clearly no obligation to retain in our later and higher civilization any more of the past than is entirely suited to our altered condition. There is equally the necessity to find new tools for our new and greater needs—eliminating all waste and discarding the old framework of past customs as we discard the skins and blankets of the patriarch for the modern clothes of the citizen.[7]

Attached to the report was a letter by Andrew Carnegie—member of NAM's Language, Weights and Measures committee—who could not attend the session. Carnegie's letter said, 'that [the metric] system is one of the steps forward which the Anglo-Saxon race is bound to take sooner or later. Our present system inherited from Britain is unworthy of an intelligent nation of today. The advantage we possess over Britain in our decimal dollar system as compared with their pounds, shillings, and pence, would be fully equaled by the adoption of a metric system of weights and measures.'[8]

Despite some objections to the idea of industrialists binding themselves to a legal chain to the metric system and risking being forced to remake 'every tool in the hands of American manufacturers', the report was accepted by the convention and referred to the Committee on Resolutions.[9]

On the government's side, in 1901 the National Bureau of Standards was created, and his director, Samuel Stratton, was an open and unapologetic supporter of the idea of the United States going metric. To this end he wrote articles, participated in congressional hearings, and collaborated with pro-metric organisations.[10]

That same year the National Board of Trade—an organisation interested, among other things, in securing uniform trade practices—passed a resolution backing a pro-metric bill before Congress, arguing that 'The use of weights and measures is universal in all civilization, and uniformity in such matters is important in economizing time and in facility for adjustment of calculations in every branch and line of industry,' adding that 'the decimal plan, which is the basis of the metric system, has become the accepted and established form of notation throughout the entire civilized world'.[11]

Early in 1902, in hearings on the subject of the metric system before the House Committee on Coinage, Weights, and Measures, the inventor Elihu Thomson, of General Electric, was one of many supporters of the metric system.[12] And in a supplemental hearing the British physicist William Thomson (better known as Lord Kelvin), arguably the most celebrated scientist of the time, went to the United Stated Congress to plead American lawmakers to pass favourable metric legislation under discussion—this after he had made several public interventions in England urging Parliament to take action in favour of the metric system.[13]

Besides these groups and individuals—scientists, inventors, public officials, legislators, and traders—educators had for a long while been in favour of metric adoption. Among them was Melvil Dewey (founder in 1876 of the first pro-metric organisation in the United States, the American Metric Bureau), who championed the metric system as part of his efforts to improve public education (along with spelling reform and library efficiency, the field in which he became notorious for his decimal classification system), and who received generous donations from Andrew Carnegie to pursue these goals.[14]

In general terms, this loose coalition portrayed a vision of universality in which humanity, in a modern world, would share common codes and languages. In this world, the metric system was compared to Hindu-Arabic numerals, the alphabet, a universal calendar, musical notation, and international auxiliary languages such as Esperanto (and sometimes even with a single universal language and a single currency). The more utilitarian interest of helping the international trade of the United States was married with that universalistic world view. To convey these ideas they frequently depict the metric system as a 'universal language'.

At this point the pro-metric movement had considerable momentum in its favour and few people showed opposition to it. In newspapers, journals, and congressional hearings metric advocates visibly outnumbered their counterparts.

But in 1902 the tide started to turn. In the meeting of the ASME some members again raised alarm about a government-mandated change in the measurement system. Frederick Halsey (1856–1935) and Samuel Dale (1859–*c.* 1935) presented strong opinions warning about the dangers of adopting the metric system; and they become the voice of the anti-metric movement in America for decades to come.[15]

Halsey in particular was a critical actor owing to his reputation, gained in part for his inventions (such as the 'slugger' rock drill) and for the design of a premium plan of paying for labour in the 1890s, also known as the Halsey Premium Plan (a widely recognised method, contemporary of the work by Frederick Taylor—another mechanical engineer—on industrial efficiency).[16] Significantly, Halsey started his anti-metric activities at the heart of the powerful ASME, with additional support from members of other organisations, such as the Society for the Promotion of Engineering Education. Halsey dedicated the rest of his life to the cause of opposing metrication in America, wrote profusely about the topic (both in technical journals and the popular press), and became the single most visible anti-metric figure in American history.[17]

A lesser-known figure, Dale was editor of the magazines *Textile World* and *Textile World-Record,* and from 1815 owner of *Textiles*. Like Halsey, Dale spent an astonishing amount of time, energy, and resources during almost half of his life to the defence of customary weights and measures. He produced a deluge of articles, pamphlets, and letters, and was a constant presence in conferences, meetings, and public discussions (besides amassing a large collection of books on historical metrology that he later donated to Columbia University alongside his enormous correspondence).

In a private letter to Charles H. Harding, President of the Bank of North America, Dale articulated his social and economic vision with a simplicity and cohesiveness that was not always present in his pamphlets, and that in many ways summarises the position of many other metric opponents:

> the defense of our weights and measures has become part of that greater problem involved in the defense of all of our established institutions. That time has come for all of us to stand by the United States of America. That is a duty we owe to those who come after us in return for what others did for us before we saw the light of day. [...] Protection for our form of government, our language, law, weights and measures and our right to consume what we produce and produce what we consume. That protection is our first duty. We shall win if everyone does his or her duty.[18]

Following the ideas of the British sociologist Herbert Spencer (who in 1896 sent to all members of the House of Commons, to some members of the House of Lords, and to all representatives in the United States Congress copies of his booklet *Against the Metric System*),[19] Halsey and Dale elaborated their main arguments questioning the alleged advantages brought by metrication and the assumed technical superiority of metric over customary units. Certainly, Halsey and Dale employed nationalistic cries to attack the metric system (as was characteristic of metric opponents during the nineteenth century), but that was not at the core of their arguments. These were educated and intelligent 'practical' men who possessed the intellectual resources to challenge scientists' claims on their own terrain.

Spencer began what he called a 'rational opposition' to the metric system and Halsey and Dale followed him in that path. They articulated their ideas in what became the single most influential book in the history of the metric system in America, *The Metric Fallacy*.[20] The book revolved around the idea that based on practical and economic grounds the transition to the metric system was undesirable. More specifically, Halsey argued that changing a system of weights and measures was enormously difficult and the transition would never be fully completed; that the change of system represented the destruction of the existing mechanical standards; that foreign commerce does not require the adoption of a new system in manufacture; that for industrial processes the metric system was not better suited than the English system; and that England and the United States have 'the simplest and the most uniform system of weights and measures of any country in the world'.[21]

Finally, in the *Metric Fallacy* Halsey also questioned the mere possibility of implementing a universal system: 'The experience of a century has shown that the idea of a universal system of weights and measures is an "iridescent dream". We must make up our minds to get along with diverse systems of weights and measures in the world as we do with divers languages and systems of currency.'[22]

This opposition to the metric system, based on economic and technical arguments, was not restricted to mechanical engineers, and soon enough other groups and associations expressed similar beliefs. The National Machine Tool Builders Association passed, in 1902, a resolution protesting at the prospect of metrication in America, arguing that 'the adoption of the metric system would entail an enormous first cost of new equipment to conform to the new standards and a constant increased cost in the maintenance of a double standard for repairs and renewals, and a consequent increased cost of the product to the consumer'.[23]

That same year the NAM, forgetting its fervent endorsement of the metric system from four years previously, adopted a mildly negative resolution that asked for a stop to any immediate change in the country's weights and measures:

> It appears to this association, first, that the compulsory adoption of the metric system would probably affect the manufacturers interests of this country as follows: One-third who are exporters to European countries and dependencies would be benefited; one third who do business in this country and all other countries would neither be benefited nor greatly injured; one third who do business in this country and in England and dependencies would be seriously injured. For all this the expense and incontinence would be very great. In view of these conditions and of the further fact that the metric system is already legalized for the use of those who find it profitable, this association recommends that no further action can be taken on this matter at this time.[24]

By 1904 the members of the National Association of Manufacturers completed the U-turn, from metric enthusiasts to hostile adversaries of metrication, and asked no other than Frederick Halsey to represent them in a Congressional hearing, in which he delivered a forceful testimony against any change in current metrological legislation.[25]

Overall, for the anti-metric movement at large, Halsey and Dale articulated the ideas to oppose metrication, and the National Association of Manufacturers provided the political muscle to stop metric initiatives in Congress. For example, correspondence from 1906 between Dale and the Secretary of the National Association of Manufacturers, Marshall Cushing, shows how the Association used its political influences to oust key pro-metric members of the Committee on Coinage, Weights and Measures in the House of Representatives, starting with its chairman, James Southard (who lost his nomination in the Republican Party). Lucius Littauer and Solomon Dresser, other active metric proponents in the committee, abandoned Congress as well. Referring to these events, Dale wrote to Cushing, 'I think I see in these events the fine Italian hand of Marshall Cushing.' To which Cushing responded: 'It is, as I can say to you only in the strictest confidence, of course, that I am closely in touch with the [Republican] party managers, and I think that we have reason to fear nothing in the future if political pressure can be made to enter into the situation at all. In other words, we are entitled to feel good and yet to keep our ammunition dry.'[26] Cushing was also in close contact with Halsey in organising the opposition against the metric movement; it looks as if he was instrumental in the anti-metric campaign, suggesting where, when, and against whom actions should be taken.

It is relatively easy to understand why manufacturers opposed mandatory metrication, as it was they who faced the financial burden of retooling and acquiring new equipment if metric legislation was passed.[27] Why mechanical engineers joined manufacturers in this battle is not as evident. Their position against the metric system can in part be explained by a long-standing tradition among 'practical men' in America. As Leo Marx notes, since the 1840s there was in the United States a vision in which inventors were depicted as intellectual heroes whose ingenuity and creativity was on an equal footing with high culture, and 'mechanical arts' were as worthy as 'fine arts'. These celebrations of the machine and utility had a defensive tone. Practical subjects and the creations of 'men from the workshop' were presented in opposition to classical education and theoretical science.[28]

Furthermore, even though this was a dispute among different kinds of experts who embodied divergent forms of specialised knowledge, there were some old elements of anti-intellectualism present in the criticisms made by Halsey and Dale against the 'metricists'. These included attitudes such as a disdain for pure science, a certain aversion to university professors, and a conception of 'pure' intellectuals as pretentious and snobbish.[29] Mechanical engineers in general emphasised these traits in their debates regarding metrication. They underlined that theirs was a 'plain, practical profession', and the opinion on weights and measures expressed by 'men who calculated eclipses' was of little importance for them.[30]

As Monte Calvert pointed out, there was a confrontation, among the members of the ASME, between 'school culture' and 'shop culture', the former represented by engineering educators and the latter by individual shops and firms. Regarding the establishment of standards, school culture was inclined to the idea that they should be set by a centralised body, be that the government or an engineering body (as the ASME itself). Shop culture, on the other hand, stood up for the idea that the logic of the open market should reign on these matters, any party could develop their own standards, and the best of them would eventually prevail over the others. Among the ASME members, the perspective of the shop culture won easily over the school culture.[31]

In the hands of Dale, that conjunction of free market and metrological laissez-faire was mixed with a touch of social Darwinism. Describing what he called the 'natural selection' of measures, he argued that

> The simplicity of textile calculations in American mills is the more remarkable because our standards of weight and measure on which they are based have apparently been adopted by chance. English speaking people have been left to use those standards they found most convenient. And as a result we find that textile operations are based on a very few convenient standards, which have been in use from the first English settlements on the Western Hemisphere and from time immemorial in the old world. The people have adopted the most convenient standards for their work, and the government has confined itself to keeping those standards uniform.[32]

The opposition by engineers and manufacturers against compulsory metric legislation should also be framed in the larger context of the adamant defence by American industries of their perceived right to set technical standards by themselves (without government intromission), and the reluctance by the federal government to intervene decisively into the matter, in one direction or the other—something it almost never did, not even in cases where national coordination seemed highly desirable, such as standard time zones and the daylight savings time.

In the long run, the opposition by the NAM and the mechanical engineers to the metric legislations proposed during the first two decades of the twentieth century proved to be crucial in the ultimate failure by the United States to adopt the metric system, as it effectively halted the momentum in favour of metrication and successfully made the topic a contentious political issue, dissuading many politicians and public servants from showing any kind of support.

Then, midway through the first decade of the twentieth century, a landscape of forces battling this metric war was set, and this remained fairly stable for the next four decades or so. Favouring the metric system were exporters, scientists, and educators; opposing it were manufacturers and mechanical engineers. Congress—in general and in legislative committees—was divided and prone to inaction. The federal government was divided as well, with some (particularly in the National Bureau of Standards) supporting metric legislation, but with other key actors legitimately interested in accomplishing standardisation (e.g. the Secretaries of Commerce and members of the military) but rather reluctant to support any compulsory measure. The general public was indifferent and rather uninformed about what the metric system was, what it was useful for, and what the implications of adopting it were.

As the experiences of other countries have shown, the mandatory adoption of the metric system is a daunting and prolonged enterprise even when governments are unified, determined, and resourceful. But conditions in the United States (where the government itself was torn apart on the issue) at this critical juncture only produced a long-lasting deadlock and failure to act.

During the 1910s and 1920s metric proponents made several new tries to advance numerous pieces of legislation, but every time their opposition was strong, obstinate, and, ultimately, victorious. Eventually these camps coalesced into formal associations. Halsey and Dale—who had received financial backing from Brown & Sharpe Manufacturing Co.—organised in 1916 the American Institute of Weights and Measures, aimed to stop the metric advance and to 'perfect' English measures.

On the other hand, that same year during the annual meeting of the American Association for the Advancement of Science, a group of scientists and educators held the first meeting of the American Metric Association—with Italian educator Maria Montessori as a special speaker for the occasion. The first president of the Association was the renowned mineralogist and vice-president of Tiffany and Co. George Kunz (who fruitfully mixed his knowledge on metrology and precious stones to invent the 'metric carat', that helped to standardise measurement in that field); Samuel Stratton (director of the NBS), was a member of the executive committee.

In 1919 Albert Herbert (the same who twenty years earlier had championed the metric system with Andrew Carnegie in the Association of Manufacturers) hired Aubrey Drury, a professional advertiser, to operate the World Trade Club of San Francisco (later the All-American Standards Council) that promoted full national metrication—with special emphasis on facilitating commerce with countries from Latin America and the Pacific and to make American companies more competitive in those markets, vis-à-vis European industrial powers such as Germany. Herbert also paid for a full-time representative in Washington to lobby his cause.

Even though he could not claim any expertise on metrology, Aubrey Drury became a relentless worker for the metric cause. His activity from 1920 to 1950 in several pro-metric civic organisations made him one of the key actors in the history of the metric movement in the United States (he served as director of the World Trade Club and its successor the All-American Standards Council; he also was vice-president of the Metric Association).

These groups wrestled in all imaginable settings (newspapers, popular magazines, technical journals, political campaigns, Congressional hearings, and, as we shall see, international conferences) and on all conceivable topics. One of their most animated subjects of controversy was the extent of metric diffusion in the world. Both camps conducted investigations to determine if the metric system was actually a worldwide metrological convention or just a nominal legal requirement. They synthesised their results in world maps that portrayed the metrological state of the world from their respective standpoints. These were widely distributed in press briefings and other propagandistic materials.

The anti-metric map, made by Dale (Fig. 8.1), emphasised the 'commanding position of the English system'. It was divided into five groups of countries. The United States, Canada, Great Britain, Ireland, Australia, New Zealand, South Africa, Kenya, and Nigeria were all described as countries where English weights and measures were 'established and fundamental'. For Latin America, Dale considered that Mexico and Brazil were nations where the metric system was mixed with English and local units; the rest (Central America, Colombia, Peru, Chile, Argentina, etc.) were marked as 'local and English prevail and are closely identical. Metric also used.' Russia, India, and Thailand were marked as 'English basis for linear measurements'. Finally, the only countries where 'metric prevail with mixture of old and English' existed in continental Europe, plus a couple of countries in Africa and Southeast Asia.

On the other hand, the pro-metric map, drawn by Drury (Fig. 8.2), was intended to show the 'well-nigh worldwide use of metric units'. It was arranged in four categories. The United States and Canada were countries that used the metric system 'in many important and practical fields', such as science, medicine, coinage, education, trade, and industry. South Africa, India, Australia, and New Zealand were a group of countries which 'have officially petitioned the central British government to adopt the world metric units thruout Britannia'. Great Britain and Ireland were described as 'countries in transition, now making a declaration of independence against the obsolete German jumble of weights and measures'. The rest of the world was depicted as countries that 'now use meter-liter-kilogram more or less exclusively'.

Both maps were a mixture of facts and wishful thinking. Dale's map was correct in emphasising that the traditional local measures were widely used in countries where the metric system had officially been adopted; but it underestimated the momentum that the metric system had already gained worldwide. Drury's chart correctly pointed out that metric units were clearly conquering the globe, but greatly misjudged how entrenched the customary measures were rooted in countries such as the United States and England.

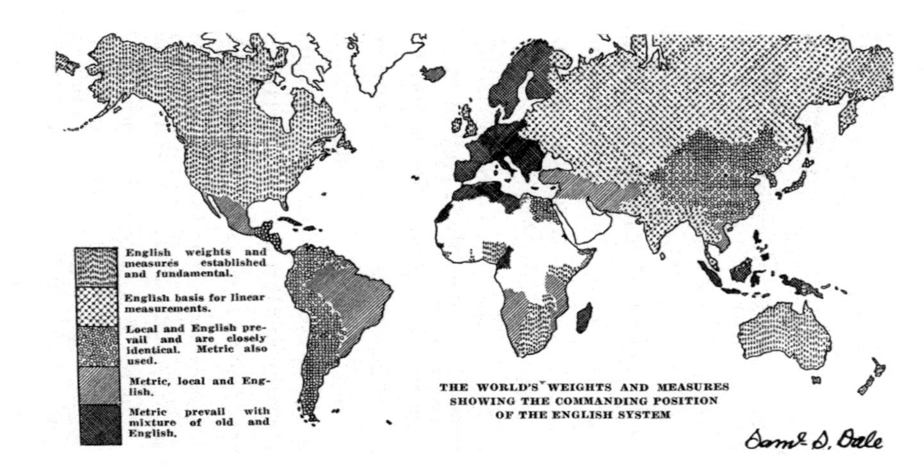

Fig. 8.1 Samuel S. Dale's world map showing 'the commanding position of the English system'. (Source: *The Metric Fallacy*, p. 125)

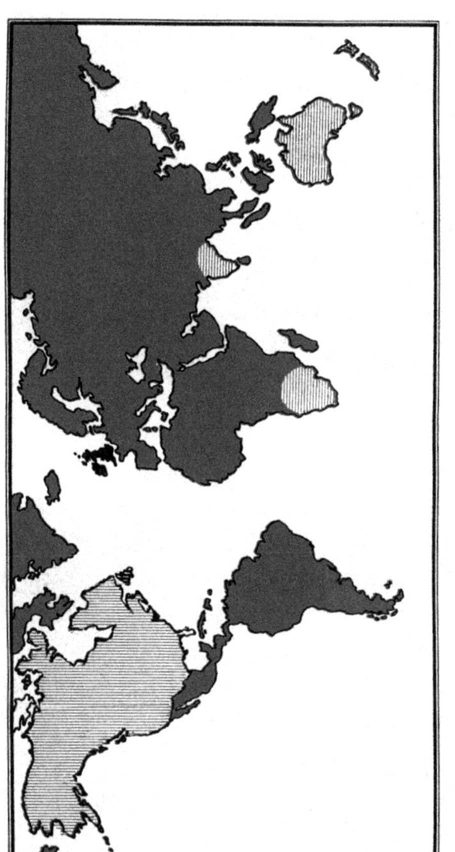

Fig. 8.2 Aubrey Drury's world map showing the 'worldwide use of metric units'. (Source: *World Metric Standardization*, pp. 2–3)

Battling for Pan-American Uniformity

The clash between pro- and anti-metric associations stepped quickly to the international arena, primarily around the question of the economic integration of the Americas. A long battle was fought with and about Latin America around the issues of what system of measurement was better suited to bring unity to the 'western hemisphere', what measures were actually used in the Spanish- and Portuguese-speaking countries, and whether the metric system was a suitable instrument to bring those countries together.

Halsey, Dale, Kunz, Drury, and the rest of the American disputants brought their fight south of the Rio Grande River, pretty much reiterating the same arguments but revamped and adjusted to the continental circumstances. In the numerous pan-American conferences dedicated to science, commerce, and standardisation during the 1910s and 1920s, the members of the Metric Association, the World Trade Club, and the American Institute of Weights and Measures made numerous appearances to present papers and to lobby representatives from other countries.[33]

This confrontation was critical because there was a genuine interest in the United States to have a more sizeable influence in Latin America. Since the majority of the Latin American countries had adopted the metric system during the second half of the nineteenth century, this was a perfect opportunity for the metric camp to score some points by showing how crucial the international arena was for the establishment of a system of weights and measures. In this they had the support of Americans working in the Pan-American Union itself, such as William Wells—chief statistician of the Union—who insisted that the world was gradually becoming one big market where the sale of manufactured products was becoming increasingly more important than the sale of gross materials, and this required a common system of measures to produce standardised goods whose dimensions were the same in the manufacturer's country as in the consumer's. That system was the metric system.[34]

In this context the interest in pan-Americanism was a door to pass metric legislation in the United States, as Halsey quickly recognised.[35] Faithful to his arguments, Halsey questioned whether the metric system was actually used by the people in Latin America and suggested that several systems of measurement coexisted there. To back up this opinion he sent questionnaires to selected people in South American countries asking them about what units of measurement were used for international and domestic commerce in their place of residence. With that information he prepared a report that allegedly demonstrated the failure to introduce the metre, litre, and kilogram as exclusive units of measurement.[36] Another study, this time by the metric proponents, had to be prepared to counter this suggestion.[37]

In 1919, during the Second Pan-American Commercial Conference, the president of the American Metric Association presented a paper on 'The Metric System as a Factor in Pan American Unity', in which he advocated for the metric system and for the introduction of a common continental currency. Kunz maintained there that 'nothing can better pave the way for a good understanding among the American nations than uniformity of currency and of weights and measures, for this will obviate many causes of misunderstanding and dispute, and will aid powerfully in developing trade among these nations'.[38]

On that same occasion Halsey advocated for a plan to standardise the units of customary English measures with the Spanish colonial measures (with the former as standard to the latter, of course) and to use that system in all the Americas as a real pan-American system of measurement (this plan was based on some similarities between some units in those systems). In an interesting historical and genealogical move, Halsey proclaimed: 'Let us unify the weights and measures of the two Americas and the British Empire on the basis of the system which came to us from the mother of us all—the Roman Empire.'[39]

This movement served Halsey and Dale as a counterpunch to the internationalistic pretensions of the metric camp, and theoretically speaking it sounded plausible. The three centuries of Spanish dominance in the New World created something close to a Hispanic-American system of measurement, with virtually all countries from Chile to Mexico sharing a considerable number of units (vara, carga, fanega, libra, onza, cuartillo among them)—even if this was an imperfect 'system' from the standpoint of uniformity and standardisation, owing to the high variability in the magnitudes of those units between and within countries.

Halsey, Dale, and their anti-metric associates used the existence of this apparent Ibero-American unity, along with its similarities with the English system, to recommend an alternative pan-American system of measures—which basically meant the adoption in Latin America of the English units and calling them by the Spanish names: pound–*libra*, ounce–*onza*, yard–*vara* and so forth. But this plan was, for all practical matters, intended solely to influence the metrication debate in the United States. The actual situation in Latin America was never a serious concern for them: their objective was exclusively stopping the penetration of the metric system in the United States. Halsey and Dale never engaged in significant conversations with Latin American experts and governments—they were usually more enthusiastic to receive attention from other English-speaking countries than from people in Latin America.

Moreover, unfortunately for Halsey and Dale, their pan-American plan arrived too late. By the 1920s Latin American countries had already invested a lot of time and money to advance the introduction of the metric system and to seriously consider starting a new plan of standardisation. New because Dale's plan was not a simple 'get back to where you once belonged', as it may look. It implied rather a unification of all units of measures according to physical standards that besides the names were only randomly used. In the end it represented wasting all the efforts invested in the metric system and starting again from scratch. Not surprisingly the plan was not taken seriously by any government or interest group in the continent. What Latin American representatives sought was *metric* continental unification, and they were much more interested in listening to the plans advanced by Drury and company.[40]

Thus, despite Dale and Halsey's candour, Latin Americans did not receive well the idea of a non-metric basis for pan-American metrological unity. For example, in response to the paper 'Uniformity or Confusion in Pan America?' presented by Samuel Dale in the First Pan-American Standardization Conference in Lima, Peru (where he reinstated the argument that the metric system only introduced confusion in commercial transactions and that it would be better to standardise the different pounds and inches in the continent), a member of the Sociedad Nacional Agraria of Peru replied saying that Latin American countries were well on their way to metric uniformity and that it would be better for the United States to go metric.[41]

The same message came again and again from Latin America, with key actors from different countries voicing their disappointment over the United States not making good on its pledge, made in 1890, to join the other American countries in the use of the metric system—something that hurt their common business interests. For example, in 1927 the Secretary of the Ministry of Industry of Colombia wrote to Aubrey Drury saying:

> The United States are proud of the large amount of out-of-date machinery that they discard to be replaced with more efficient equipment as soon as a practical innovation appears—without considerations of expense. A scientific and uniform system of weights and measures is a vital equipment for domestic and international commerce. Abandoning the antiquated and incoherent system inherited from colonial times is worth as much as throwing away dated machinery. Adopting the French metric system would be a convenient and lucrative achievement, worth of such a great people; it would create a bridge to the other American republics and enhance their commerce with them.[42]

Regarding the case of Mexico, merchants, bankers, and diplomats all received information that contradicted Halsey's and Dale's claims on the prevalence of customary measures and metrological confusion. At least regarding foreign commerce, Mexico had become a metric nation. For example, this is what the US vice-consul at Ciudad Porfirio Díaz (today Piedras Negras), on the Mexican-American border, reported:

> The metric system is the legal standard for weights and measures in the Republic of Mexico. Our English weights and measures are stumbling-blocks to the Latin countries whose trade we seek. The persistent use by our merchants in their catalogues, circulars, etc., of our old English weights and measures causes us to lose considerable trade with Mexico and other Latin-American countries, as they are unintelligible to many foreign buyers. It would be well for our merchants and manufacturers desiring to extend their trade to Mexico and other countries using the metric system to base descriptions of their goods and estimate prices upon this system.[43]

And this is what John Lind, personal representative of President Woodrow Wilson in Mexico, commented on the same topic:

> We must get into line with the commercial world in the matter of weights and measures. I asked an intelligent German merchant in Vera Cruz one day to explain to me how it had come about that Germany had absorbed so much of the trade that at one time went to England. He reached into a drawer, pulled out an invoice from England and said, 'Do you see those denominations or yards, feet and inches, gallons and pints, two kinds of ounces, grains and pennyweights, the whole summed up in pounds, shillings and pence? Well,' he continued, 'a Mexican, even if he can read a little English, needs an interpreter and an accountant to put this into the language of civilization.' [...] We should be in accord not only with Latin America, but with all the rest of civilization, except England. The American manufacturer and merchant must learn to understand that a foreign market is always a 'buyers' market.[44]

In 1926 a more emphatic call of attention to the necessity of using the metric system in Mexico came from within, when the Mexican Confederation of Chambers of Commerce, following an initiative by the Chamber of Commerce of the border state of Chihuahua, made a call for the 'urgent necessity to use the metric system'. They expressed their disappointment with the United States' and England's refusal to adopt the metric system, noting that the adoption 'would help them tremendously in their domestic dealings, it would bring immensely positive results to improve their relationship with clients all over the world, and they would also collaborate with the establishment of universality in weights and measures'.[45] The document recalled that the United States had participated in the 1890 agreement for pan-American metrication. Finally, the Confederation asked for its American and British counterparts to take steps towards metrication.

Overall, the reluctance by the United States to adopt the metric system generated shock and dissatisfaction in Mexico and other Latin American countries, where the United States was seen by many as an advanced nation and their hope-to-be future; its unwillingness to implement the metric system was considered as a weird anomaly—it was certainly something that did not match with their image of a prosperous nation. In the end, as we know, all exhortations—within and outside America—to persuade the government and major economic actors to embrace the metric system were not enough to prompt any substantial action.

The American government recognised the importance of the metric system to enhance the commercial relations of the country, but that was insufficient to convince them of the necessity to pursue a drastic metrication policy. Illustrative of this is a letter by Lyman J. Briggs (director NBS) to J.T. Johnson (president of the Metric Association) in 1943:

> There is no question but that we should cooperate to the fullest extent with the Latin American countries and should promote trade with them. But there is a serious question as to whether it is either necessary or desirable for this country to adopt the metric system in order to use it in foreign trade. A knowledge and understanding of the metric system and a willingness to use it properly are essential. No one proposes that we should change our language to Spanish or Portuguese in order to promote trade relations with Latin America because in the matter of language it is recognized that knowledge and understanding are the essential items. Perhaps the analogy is not complete but the idea deserves consideration. If the metric system is a sufficient step forward, then its use in our trade with our neighbors to the South may give it the impulse for use among ourselves.[46]

That impulse never came about, and for the rest of the twentieth century the obstacles in the road to metrication in the United States proved insurmountable—as they had been in the past.

Final Comments

In the long run, the opposition orchestrated by engineers and manufacturers to the metric legislations during the first decades of the twentieth century proved to be crucial in the ultimate failure by the United States to adopt the metric system at a crucial juncture. Contrary to the predictions of many of its detractors, the metric system became a truly global language of measurement, officially as the sole legal system of measurement in countries that concentrate 95% of the world population. Today there are only six countries that have not committed to metrication: Liberia, Samoa, Micronesia, Marshall Islands, Palau, and the United States.

What have been the consequences of not adopting the metric system for the United States economy? It is difficult to quantify. An estimate made at the beginning of the present century said that it represented more than $2 billion of lost opportunities for non-metric American industries.[47] To this we have to add all the extra expenditure that metric American industries have by doubling some production processes (such as slight changes in what are basically the same car model, creating one for the United States and another with metric specifications for Europe). Is this a validation of the claims of pro-metric groups about the necessity for the United States to go metric? Just partially.

Frederick Halsey said at the dawn of the twentieth century that foreign commerce does not require the adoption of a new system of measurement in manufacturing. Probably he was right—at least in the case of the United States. American industries have been a global force for more than a century now. Working from that position of power they were in the position to wait for the world to adopt their standards, instead of the other way around. But this was a luxury that only a very powerful country could afford (as England did in the nineteenth century).

Will a greatly globalised economy and a decline in American hegemony change this situation? Some think so.[48] I am sceptical, though. The change to the metric system requires forceful intervention by the state—the absence of which has ultimately crippled all tries to metricate the United States—but it does not look as if the American state now has the strength to muster such an effort. A radical change in the economic position of the United States in the world or a substantial transformation in the structure of the federal state appear to be necessary for the metric system to have a realistic chance to set foot in this territory.

Archival References

RBML – Rare Book & Manuscript Library, Columbia University.
SIBL – Science, Industry, and Business Library; New York Public Library.

Notes

1. North, Douglass C.: 'Transaction Costs in History', *The Journal of European Economic History*, Vol. 14 (1985), p. 558.
2. North, Douglass C.: 'Review of *Measures and Men,* by Witold Kula', *The Journal of Economic History*, Vol. 47 (1987), p. 594.
3. For an overview of the importance of standardised weights and measures for economic life, see Vera, Héctor: 'Medición y vida económica. Medidas panamericanas y la lucha por un "lenguaje universal para el comercio" ', *Estudios Sociológicos,* Vol. 32, No. 95 (2014), pp. 232–39.
4. Sinclair, Bruce: A Centennial History of the American Society of Mechanical Engineers, 1880–1980, Toronto: University of Toronto Press, 1980, p. 47.
5. Melvil, Dewey: 'Efficiency Society', in *The Encyclopedia Americana*, New York: The Encyclopedia Americana Corporation, 1918, IX, pp. 719–20.
6. The members of the Committee on Language, Weights and Measures were Albert Herbert (chairman of the committee), president of Hub Gore Makers; Charles A. Schieren, of C. A. Schieren & Co. and ex-mayor of Brooklyn; Charles H. Harding, of Erben Harding & Co.; Henry Fairbanks, vice-president of E. and T. Fairbanks Scale Co.; Theodore C. Search, Manufacturer, president and founder of American National Association of Manufacturers; and Andrew Carnegie.
7. A typed copy of the report can be seen in RBML Melvil Dewey Papers, box 67. A somewhat modified version of the text was reproduced in Drury, Aubrey (ed.), *World Metric Standardization: An Urgent Issue*, San Francisco, World Metric Standardization Council, 1922, pp. 38–44.
8. RBML Melvil Dewey Papers, box 67.
9. 'Manufacturers Meeting: Opening of the Third Annual Convention of the Association in Masonic Temple,' *New York Times* (26 January 1898).
10. Cox, Edward Franklin: 'A History of the Metric System of Weights and Measures: With Emphasis on Campaigns for its Adoption in Great Britain and in the United States prior to 1914', Unpublished PhD Dissertation, Indiana University, 1956, pp. 584–7.
11. National Board of Trade: *Proceedings of the Thirty-First Annual Meeting of the National Board of Trade*, Philadelphia: John R. McFetridge & Sons, 1901, pp. 55, 73–4, 299–300. That same resolution was supported again in 1902, see National Board of Trade: *Proceedings of the Thirty-Second*

Annual Meeting of the National Board of Trade, Philadelphia: John R. McFetridge & Sons, 1902, pp. 137.
12. Hearings before the Committee, Feb. 6-March 3, 1902, on Bill H. R. 2054, Washington: Government Printing Office, 1902, pp. 1–5.
13. Supplemental Hearing on the Subject of the Metric System of Weights and Measures: Hearings before the United States House Committee on Coinage, Weights, and Measures, Fifty-Seventh Congress, First Session, on Apr. 24, 1902, Washington: Government Printing Office, 1902, pp. 1–11.
14. See Vera, Hector: 'Melvil Dewey, 'Metric Apostle', *Metric Today*, Vol. 45, 2010, pp. 1, 4–6.
15. As an illustration of Dale's position on metrication: Dale, Samuel S.: *The Foreign Attack on Our Weights and Measures*, Boston, 1926.
16. Halsey, F. A.: 'The Premium Plan of Paying for Labor', *Transactions of the American Society of Mechanical Engineers*, Vol. 12 (1890), pp. 755–80.
17. For a couple of examples of Halsey's non-specialised anti-metric articles, see 'Disputes Metric Success', *New York Times,* 23 August 1925; 'Continuing the Metric War', *New York Times* (5 June 1927).
18. Samuel S. Dale to Charles H. Harding, 26 March 1919 (RBML Samuel Dale Papers, vol. 7).
19. Spencer, Herbert: 'Against the Metric System', in *Various Fragments,* New York: D. Appleton and Company, 1914, pp. 130–56. On Spencer's opposition to the metric system, see Vera, Hector: *The Social Life of Measures: Metrication in the United States and Mexico, 1789–2004*, Unpublished PhD Dissertation, New School for Social Research, 2012, pp. 341–59.
20. The first edition of Halsey's *The Metric Fallacy* was published in a single volume with Dale's *The Metric Failure in the Textile Industry*. The second edition of the book, this time without Dale's text, was published as Halsey, Frederick A.: *The Metric Fallacy: An Investigation of the Claims Made for the Metric System and Especially of the Claim that Its Adoption Is Necessary in the Interest of Export Trade*, New York: The American Institute of Weights and Measures, 1920.
21. Halsey Frederick A. and Dale, Samuel S.: *The Metric Fallacy* and *The Metric Failure in the Textile Industry*, New York: D. Van Nostrand, 1904, pp. 16–17; see also Cox, 'A History of the Metric System', pp. 606–17.
22. Halsey *The Metric Fallacy*, 127. As part of their reaction against the introduction of the metric system, Halsey and others developed a not

fully articulated but evident defence of diversity as positive value, not only in the somewhat limited field of technical standards, but in broader cultural aspects; they critiqued, for example, what they called 'over standardization' or applauded the defence of traditional national languages. On these issues, see 'Over-Standardization', *Bulletin of the American Institute of Weights and Measures* (1 October 1923), p. 3; and in that same publication 'Elusiveness of World Uniformity,' 1 July 1924, p. 14.

23. Resolution and Protest of the National Machine Tool Builders Association, 1902.
24. Quoted in 'Manufacturers and the Metric System', *New York Times* (25 April 1902). See also Steigerwalt, Albert K.: *The National Association of Manufacturers, 1895–1914*, Ann Arbor: University of Michigan, 1964, pp. 93–4.
25. The Metric System: Hearings before the Committee on Coinage, Weights, and Measures on H. R. 93 (58th Congress, 1st Session); H. R. 2054 (58th Congress, 2nd Sessions), and H. R. 8988 (59th Congress, 1st Sessions), Washington: Government Printing Office, 1906, pp. 1–19.
26. Samuel S. Dale to Marshall Cushing, 6 October 1906; Marshall Cushing to Samuel S. Dale, 8 October 1906 (RBML Samuel Dale Papers, box 4).
27. For an analysis of the distribution of costs of going metric (for a different national case, though) see Faith, Roger, McCormick, Robert, and Tollison, Robert: 'Economics and Metrology: Give 'em an Inch and They'll Take a Kilometre', *International Review of Law and Economics*, Vol. 1, 1981, pp. 207–21.
28. Marx, Leo: The Machine in the Garden: Technology and the Pastoral Ideal in America, New York: Oxford University Press, 2000, pp. 198–209.
29. Hofstadter, Richard: *Anti-Intellectualism in American Life*, New York: Vintage, 1963, pp. 9–19.
30. Calvert, Monte A.: The Mechanical Engineer in America, 1830–1910: Professional Cultures in Conflict, Baltimore: Johns Hopkins Press, 1967, pp. 179–86.
31. Calvert, The Mechanical Engineer in America, 179–86.
32. Dale, Samuel S.: 'A Talk on Textile Arithmetic', Reprint from *Textile World Record*, June 1908, SIBL 3-VBDP+ (Dale, S. S. Weights and measures).
33. See, for example, the papers presented by Drury, the Metric Association, Dale, and W. R. Ingalss in one of the Pan American standardization

meetings: Inter American High Commission: *Report of the Second Pan American Standardization Conference*, Washington: Government Printing Office, 1927, pp. 75–87.

34. Wells, William: 'The Metric System from the Pan-American Standpoint', *The Scientific Monthly*, Vol. 4 (1917), pp. 196–202.
35. 'To Star New Fight on Metric System: Manufacturers See Insidious Move to Make It a Feature in Pan-Americanism', *New York Times,* 30 March 1916.
36. Halsey, Frederick A.: *The Weights and Measures of Latin America*, New York: American Society of Mechanical Engineers, 1918.
37. 'The Weights and Measures of Latin America', *Decimal Educator*, Vol. 2 (1920) pp. 178–86, 218–19 and 259.
38. Kunz, George: 'The Metric System as a Factor in Pan American Unity', in Barret, John (ed.), Report of the Second Pan American Commercial Conference: Pan American Commerce; Past, Present, Future from the Pan American Viewpoint, Washington: Pan American Union, 1919, pp. 270. See also George Kunz, 'The International Language of Weights and Measures,' *The Scientific Monthly* 4 (1917): 215–19. In a letter to Melvil Dewey Kunz drew a succinct explanation on his view on measurement and international languages: 'Has it ever occurred to you that musicians in every part of the world can read and play the music written by anybody in the civilized world? Why not the meter-liter-gram?' (George Kunz to Melvil Dewey, 18 December 1924, RBML Melvil Dewey Papers, box 66).
39. Halsey, Frederick A.: 'Pan Americanism in Weights and Measures', in *Report of the Second Pan American Commercial Conference*, p. 274.
40. On Drury's plan for continental metrological unification, see *One Standard for All America*, San Francisco: American Standards Council, 1927.
41. Dale, Samuel S.: 'Uniformity or Confusion in Pan America?' and a letter by A. F. Del Solar to the First Pan American Standardization Conference, 19 December 1924. RBML Melvil Dewey Papers, box 67.
42. From Ministry of Industry, Colombia to Aubrey Drury, 5 November 1927 (Aubrey Drury Papers, box 3).
43. Quoted in Hearings Before the Committee, Feb. 6-March 3, 1902, on Bill H. R. 2054 (Washington: Government Printing Office, 1902), p. 163.
44. Lind, J.: *The Mexican People* (Minneapolis: The Bellman, n.d.), 29–30.

45. Confederación de Cámaras de Comercio de los Estados Unidos Mexicanos, 'Urgente necesidad de hacer universal el uso del sistema métrico decimal. Estudio para la 9na. Asamblea General de Cámaras de Comercio de la República. Septiembre de 1926', RBML Aubrey Drury Papers, box 3.
46. Lyman J. Briggs to J. T. Johnson, 9 April 1943 (a copy of the letter was sent to the Vice President of the United States, Henry A. Wallace). RBML Aubrey Drury Paper, box 3.
47. Jourdan, Louis: *La grande métrication*, Nice: France Europe éditions, 2002, p. 166. Jourdan mentions the *Wall Street Journal* as the source of that figure, but unfortunately does not quotes an article or date in particular.
48. See Zakaria, Fareed: 'The Rise of the Rest', *Newsweek* (12 May 2008).

9

Statistics as Service to Democracy: Experimental Design and the Dutiful American Scientist

Tiago Saraiva and Amy E. Slaton

Introduction

This chapter recounts the twentieth-century use of statistical methods to make science serve American ideals of democracy on a global scale. It details how methods elaborated during the New Deal era at an American land-grant college, now Iowa State University, and at the United States Department of Agriculture (USDA) became after the Second World War a hallmark of American development initiatives through United Nations (UN) agencies and Harry S. Truman's Point Four program.[1] Deployments of randomization and analysis of variance not only established a new standard for the design of experiments and their interpretation but also widely signaled just how scientific an experimental statistical study might be.[2] Iowa, although overlooked by historians of statistics, was one of the main entrance points in the USA of these techniques first developed in England by Ronald Aylmer Fisher, and that promised a new relation

T. Saraiva • A. E. Slaton (✉)
Department of History, Drexel University, Philadelphia, PA, USA
e-mail: tsaraiva@drexel.edu; slatonae@drexel.edu

© The Author(s) 2018

D. Pretel, L. Camprubí (eds.), *Technology and Globalisation*, Palgrave Studies in Economic History, https://doi.org/10.1007/978-3-319-75450-5_9

between theory and practice in which the scientific description of a phenomenon suggested ways of interfering with it.[3] Statistical methods determined specifically how to tinker with different variables in order to reach a desired outcome (whether in the case of food supply, workforce preparation or the sustenance of developing economies), granting them a leading role in the design of science-based public policies. While statistics enlarged the political imagination and the realm of scientific intervention in the world, it also imposed constraints on what was considered a legitimate object for democratic discussion and on who, and on what condition, was involved in such discussions.[4]

These statistical practices moved steadily outward from American university departments and federal agencies through the 1940s and 1950s as increasing numbers of settings, both domestic and global, appeared to the statistical experts and their patrons to be in need of systematic economic intervention. The racial unrest and poverty of the American south, and developing economies in South Asia, Africa and South America, all drew the statisticians' scrutiny. Their inclinations towards service, expressing Protestant sensibilities that had sustained both the land-grant colleges and many "Social Gospel" projects in the United States since the late nineteenth century, included sharing their techniques with local personnel.[5] Education and training of future statistical practitioners and the establishment of sustainable statistical agencies around the world were central to the Americans, including through their work in the nascent UN following the Second World War. But however generous in intent, this energetic dissemination of statistical methods represented a project of *constrained* democracy. The Iowans' work of outreach demarcated identities of many kinds: expert and not; American and not; white, male and Protestant and not. In this chapter we follow a number of these demarcations to suggest how scientific knowledge initiated in American institutions carried knowledge abroad with the service-driven aim of improving human welfare, but with social hierarchies and differentiations nonetheless integral to that mission.

Through the middle decades of the twentieth century statistical thought leaders trained at Iowa State worked through political offices and federal agencies, international academic organizations and the UN to expand globally the role of statistics in scientific research in general and in economic planning in particular. These attributions configured the statisticians' subjects—the agricultural and industrial communities, the groups of workers or students, under study—*and* statistical actors, including both the experts and those in whom they hoped to instill their experimental techniques. There existed, in other words, a continuity for American statisticians between the populations and traits awaiting study and their sense of themselves, and all others, as members of given human and cultural types, as knowable elements of knowable societies. White, American, Protestant, identifiably gendered and with knowable (heterosexual) sexualities, the credentialed statistical experts moved around the globe to deliver what can be seen as a privileged empirical perspective. This is an understudied historical pattern that suggests how the evidently generous and discriminatory could so seamlessly combine in US social sciences of the post-colonial era. To explore this pattern we will focus on the statistical practices of Henry A. Wallace and Gertrude M. Cox, who carried these social visions outward from Iowa with tremendous influence from the 1930s onward.

Both are well-known characters among different historiographies. While Wallace is an obligatory reference in the recent revisionist literature on the New Deal and international development, Cox is known among historians of science as an exemplary case of overlooked women in science.[6] Although from very different angles, both of those historiographic traditions approach our characters similarly as historical actors to be reclaimed from the oblivion of main narratives of American history. This chapter is not, however, a matter of shedding light on "hidden figures," as recuperative accounts of understudied scientists often attempt to do. That light is of course welcome: paying attention to Wallace and Cox points at paths not taken, at promises of inclusiveness of American democracy that could have been fulfilled but have not been. Wallace is a clear alternative to depictions of American science dominated by the military industrial complex. He was a progressive politician and scientist committed to visions of science serving community empowerment, people's welfare and world peace. Instead of the hegemonic American Century he strived for a globally egalitarian "century of the common man."[7] As for Cox, she seems to confirm not only the many difficulties women had to face to assert their position as respected knowledge-makers in American academia but also the particularity of the epistemic niches women have historically occupied.[8] Recognition of women's roles in science is taken in most of this literature as a necessary condition for the future success of women and other minorities in science, technology, engineering and mathematics.

However, this chapter, while building on this literature, aims at complicating the notions of democracy and gender at work in the worlds of Wallace and Cox, integrating their life experiences and epistemic commitments as applied statisticians. The ordering of people by race, gender, sexuality, nationality and class was both an epistemic and ontological effort for the Iowans. We see their work as part of the globalizing reach for American science at mid-century, a kind of Western triumphalism that conflated prosperity and very particular kinds of modernity; here, statistical modernity made up of universalizing namings and orderings. Alistair Bonnett makes the point that for many Eurocentric thinkers through the twentieth century it was not only military effort and direct domination that could bring a sense of white mastery. Working in an administrative or scientific capacity, one might also experience a sense of conquest as the rest of the world "'opens up' and begins to see things 'our way' and acts accordingly."[9] Transported from America to so-called developing nations after the Second World War, the use of statistics for agricultural, economic, health and educational projects constituted an expert intervention that was to supply both means of uplift, via empirical study and intervention, and epistemic conversion experience.

Engaging with the detail of Wallace and Cox's scientific practices suggests the fruitfulness of understanding the statistics they performed as a form of both politics and service and, indeed, the inseparability of those two kinds of projects among US scientists at mid-century. Wallace imagined how to serve communities of farmers by developing the first major econometric studies in the USA. Only multiple correlation could, he believed, enable his Social Gospel- inspired millennial visions of democracy in a modern world. In Cox's case, her research in experimental design served other scientists' endeavors, helping them conduct their research in a proper scientific manner. The gendered notion of social service she brought with her from her Methodist Episcopalian experience became integral to the new kind of statistics she performed. The gender and racial implications of Cox's positioning complicates the apparently inclusive nature of statistics as understood by Wallace. More than taking gender and democracy as stable characterizations, we explore how statistical practice and the statistician's identity were coproduced. In carrying her educational and training initiatives to international forums, through consulting and as an important member of multiple UN statistical units, Cox enacted highly structured visions of centralized epistemic infrastructure. From their institutional vantage points in academia, federal government and international institutions, with great influence in the training of scientific experts, both Wallace and Cox redefined American scientists as dutiful servants of global democracy.

Wallace's Statistical Gospel

American pig farmers, for whom a hog was no more than the animal form of corn, had traditionally considered they were getting a fair deal when they sold their hogs for a value per hundredweight of the value of ten bushels of corn, or a ratio of ten. When the United States Food Administration, the agency responsible for mobilizing American farm production during the First World War, announced in November 1917 that it would pay the equivalent of thirteen bushels of corn for a hundred pounds of hog flesh, farmers gladly fed more corn to hogs and increased the number of animals brought to market. This encouragement for transforming a larger amount of corn into hog than usual (typically, hogs consumed 40% of the American corn crop) aimed at preventing the tragic shortage of fats available to the ally countries, namely France and the United Kingdom. Herbert Hoover, the Food Administration leader later to become president during the Great Depression, emphatically declared, "Every pound of fat is as sure of service as every bullet, and every hog is of greater value to the winning of this war than a shell."[10]

The fixing of the ratio at the value of thirteen resulted from the calculations undertaken by the young Henry A. Wallace, a statistical enthusiast better known in the annals of history of science for his plant breeding work with hybrid corn and as founder of the Hi-Bred Corn Company.[11] Wallace was the grandson of Henry Wallace, popularly known as Uncle

Henry, the founder of *Wallace's Farmer*, a newspaper mixing agricultural technical news and moral advice very popular among Corn Belt farmers. This Iowan Presbyterian preacher made the journal into a major mouthpiece of the influential Social Gospel movement and its call for the "rejection of the divide between a private ethical realm and a competitive public order."[12] Social Gospellers were invested in bringing to an end the inequalities produced by the unbridled capitalism of the Gilded age patent in the 1890s crisis by applying Christian ethics to social issues abiding by Matthew (6:10): "Thy kingdom come, Thy will be done on earth as it is in heaven." Post-millennialists assembled in organizations such as the American Institute of Christian Sociology, a summer program held at Iowa College, asserted that Christ's thousand-year reign had already arrived, and that their task was to build the Kingdom then and there.[13] Uncle Henry's son, Henry C. Wallace, the father of Henry A. Wallace, would take over the publication of *Wallace's Farmer* from 1916 to 1921 and it was in this role as spokesperson for the interests of Midwest farmers that he integrated a commission appointed by Hoover's Food Administration to determine the right incentive for producing more hogs during the First World War and make American farmers contribute to the victory in Europe of the Allied forces. The commission recommended a thirteen ratio after Henry C. Wallace consulted with his son Henry A. Wallace.

The younger of the Wallaces had been busy producing long series of agriculture prices (from 1858 to 1914) from the data made available by USDA and by the Chicago Board of Trade, which, as he said, registered "prices of corn belt food staples … more promptly and more delicately than anywhere else in the world."[14] Wallace produced tables showing ten-year average values of the ratio based on prices of No. 2 Chicago corn and values of Chicago hog flesh. Using simple and multiple correlation, he investigated how hog prices varied with bank clearings (a proxy for business activity) and hog receipts at Chicago, as well as domestic and foreign demand. To refine his method of ratio calculating he put forward a composite corn value to account for the fact that hogs had been made out of corn at varying values and that the amount of corn consumed also varied in function of the age of the animals. Wallace had produced what prominent statistician Mordecai Ezekiel, later Wallace's main economic adviser, characterized as "the first realistic econometric study ever published."[15] More than that, however, through Wallace's calculations knowledge of technicalities in the making of index numbers had far-reaching political implications.

The problem was this. The Food Administration promised a thirteen ratio calculated from Chicago corn prices and Chicago hog prices. But when the ratio was put into effect, Hoover's agency made use instead of farm corn prices, a value produced from USDA crop reports. This would not have been an issue in most years, since corn values on the farm and Chicago no. 2 corn tended to be similar in value. But for the winter of 1917–1918 the quality of the crop was so poor that only a small amount of the crop was graded at no. 2, thus reducing the actual value of farm corn price. As vehemently denounced by Wallace, instead of a thirteen ratio, the government paid no more than 10.8. He thus spoke of "chicanery and deceit" and of a concerted effort of the British and American administrations to obtain cheap hogs on the back of American farmers, as more than 60% of these farmers raised hogs in this period.[16]

The lessons from this first experience in discussing state intervention in agricultural price formation would stick with Henry A. Wallace. He acknowledged that lawmakers, although familiar with markets and prices, were not aware of the actual conditions of raising hogs on the farm, and it was thus only natural that they were not sensitive to the special conditions of the winter of 1917–1918. In contrast, farmers knew a lot about how to feed hogs, but they were in general ignorant of markets and the making of index numbers and prices. Thus, for Wallace, those claiming to defend farmers' positions had to be acquainted with both the realities of markets and the farm. Only a thorough knowledge of statistics could guarantee the informed participation of agricultural producers in public democratic debate. In other words, only statistics enabled farmers to become full American citizens.

While Henry C. Wallace would become Secretary of Agriculture under presidents Harding and Coolidge from 1921 to 1924, guaranteeing the support of the West and Midwest farm bloc for the Republican Party, Henry A. Wallace would continue in the pages of *Wallace's Farmer* his campaign for the importance of statistics in the design of fair policies in a democracy. His main challenge was preparing Midwest farmers for the war's aftermath: overproduction, which was caused by the recovery of European farming output and the associated decrease in income of American rural communities. He insisted that farmers had to arrange with packers for a price "representing the cost-of-production," instead of relying on the unstable "supply and demand price": "in the future more and more attention has to be paid to production and less to price speculation," thus making the corn/hog ratio the decisive number.[17] A system of establishing hog prices on the basis of hog production costs would not allow for the large profits made at Chicago futures exchange pit, but it would avoid big losses among farmers. The moral economy of the corn/hog ratio should, according to Wallace, be extended to every staple, leaving the rule of supply and demand only for non-basic things such as "theaters, luxuries, newspapers, etc."[18] Instead of the state imposing ratios to the different agents, he maintained that citizens in general, and farmers in particular, had to be educated in statistics in order to be able to make the case for a fair price. Here is his 1920 vision of such a statistical democracy:

> Even in the grade schools and country schools, ratio methods of price judgments should be taught … It is suggested that not only should the ratio method of price judging be taught in high school, but also the practical use of correlation coefficients and lines of regression in determining prices from business conditions and the supply. In college … students should have access to adding machines, calculating machines, rechentaffels, and other modern devices for making calculations easy and accurate. But the most important thing of all just now is adequate research by colleges, by experiment stations, and by governmental departments.[19]

Wallace's vision of democratically deployed statistics, involving technically capacitated citizens empowered by the work of research experts, was both inclusive and stratified.

Statistical Academics

In 1923, after seeing how statistical calculations were made in Washington at the Bureau of Agricultural Economics during his father's tenure as secretary of agriculture, Henry A. Wallace was quick to put his hands on a tabulator from a Des Moines insurance firm. He taught himself how to use it to calculate correlations. He would punch data cards and bring them to the insurance company for tabulation. As an alumnus of Iowa State College (now University), Wallace would also engage with the local faculty members, namely with the professor of statistics, George W. Snedecor, in discussing applications of correlation and regression techniques. Never modest, Wallace himself describes how Snedecor was so impressed with his skills that in the spring of 1924 Snedecor invited him to present a Saturday afternoon seminar for the university faculty and graduate students on multiple correlation and machine calculation. For these lectures Wallace would bring back and forth the borrowed card-handling equipment from the insurance company and use it to exemplify calculation of correlations between land value in Iowa and different variables, such as corn yield, percentage of farm land in small grains or number of brood sows per 1000 acres.[20]

While some results had already been published by Wallace in the pages of the *Wallace's Farmer* that he had directed since 1921, the course resulted in the joint publication in 1925 by Wallace and Snedecor of the bulletin "Correlation and Machine Calculation."[21] This pamphlet was a success for it offered "explicit directions for the use of the usual commercial forms of calculating machines, either key-driven, such as the Comptometer and Burroughs Calculator, or crank driven, such as the Monroe or Marchant, in finding correlation coefficients or related constants."[22] Agricultural economists exploring the variables that affected commodity prices or breeders making experiments to distinguish between hereditary and environmental factors in animal performance or crop yielding did not need to understand the mathematical background of the methods. They just had to follow the algorithm detailed in "Correlation and Machine Calculation."[23] The algorithm was a tool that extended the use of statistics to different scientific fields, which pointed at a division of labor in which mathematical statisticians were the only ones who understood the nature of the methods and the possibilities and conditions of their applications.

After the joint course with Wallace, Snedecor was increasingly looked for by fellow faculty at Iowa State, "especially those doing research with farm crops, genetics, and farm animals [who] were often uneasy about the differences in their results when they repeated an experiment." By the fall of 1927, Snedecor was already able to justify the acquisition of IBM card sorting and tabulating machines for the university and the establishment of a Statistical Service within the Department of Mathematics. During that first year the equipment would be, not surprisingly, used mostly for agriculture research such as inheritance of characteristics in corn, relations among factors affecting farm profits, farm mortgage studies or bank and credit conditions in Iowa.[24] Snedecor thus took advantage of the presence of the multiple researchers in agricultural sciences typical of a Midwestern land-grant college committed to putting science at the service of a constituency largely made of farmers. The other resources readily available to Snedecor were students themselves. In that first year, Snedecor and associates also used the machines of the new Service to calculate the "relation of college grades to high school averages, mental tests, physical condition and other characters," as well as "correlation among mental test grades." Agriculture (breeding and economics) and education would constitute in the upcoming years the two main fields in which statistics would prove its value.

To perform the tedious calculations demanded by statistical methods, Snedecor enrolled undergraduate students of calculus as computers of the Statistical Service. One of these students was Gertrude M. Cox, whom Snedecor enlisted for the work hoping that the "only woman in the class, would have more patience for detail work than men."[25] In 1931, under Snedecor's mentorship, Cox received her MS degree, the first MS in statistics from Iowa State College. Two years later, while working at UC Berkeley toward her doctorate, she was invited to take responsibility at Iowa over the computing service of the now Statistical Laboratory. Here is Cox's recollection of Snedecor's invitation: "Our work is opening up rapidly … Are you interested? Immediately you would have charge of the girls, 140 calculating machines, and all the *stray* jobs that I can rustle for you."[26] Not only was Snedecor considerate of feminine attention to detail, he also assessed women as well adapted to "stray jobs."

The Dutiful Miss Cox

What are the implications of this naturalization of differing aptitudes among men and women, or persons of ascribed intellectual capacities? There are of course intimations of limits on individual opportunity and achievement: not everyone will have the chance to become a statistical theoretician or analyst, or, by extension, achieve administrative stature. Some persons are simply suited to low- or mid-level occupational niches. But perhaps less obvious are the service-related implications of such a world view; these expand our sense of how social contributions functioned for the Iowans as a factor in scientific productivity. In memorializing Snedecor, Cox commended her mentor for being "impatient with people who did not strive to live up to their full potential."[27] What we begin to see are not just the limits on individuals' productive possibilities envisioned by the statisticians, but concurrent *demands for* productivity. Productive activity is cast in the land-grant climate as a matter not only of self-fulfillment but also of sufferance and duty. This would become an ideal of stratified, and linked, competence and responsibility that, as we will see, also shaped UN statistical projects under Cox's influence following the war, as conceptions of statistical science as professionalized labor became embedded in international cooperation efforts under Western auspices. At all times, Cox's own "potential" as white, female, American and Protestant established her position, just as did her protégés' identities, relative to white men of intellectual or governmental influence.

A student of Snedecor's at Iowa State, as a young woman Cox had become a Methodist deaconess and sought a bachelor's degree to provide her with eligibility for the position of headmistress at a Montana orphanage. At Iowa she studied mathematics and instead committed to a career in statistics, becoming a widely praised protégée of Snedecor. Snedecor was certainly familiar with these transitions from religious service to scientific service, since his father was a Presbyterian minister who had been superintendent of the American Bible Society and secretary of the Executive Committee of Colored Evangelization of the Southern Presbyterian Church and his mother the Dean of Women and head of the Bible Department of the University of Alabama, whom her husband described as "our Church's missionary to the Negro par excellence, for she has served the race in a land where such service often brings a social ostracism."[28] "Miss Cox," as generations of correspondents referred to her, went from serving on the staff of Iowa's Statistical Laboratory in 1940 to establishing the program in statistics at North Carolina State College (later University), became the first director of statistical research at

Research Triangle Institute and through the 1950s and 1960s created vast global networks through her work with international societies and government agencies in the USA and abroad.[29] As member of the International Statistical Institute (ISI) from 1948 and later as chairman of its Education Committee, Cox helped to design elaborate coordinated statistical projects among UN member nations. She traveled widely and spent extended stays in Egypt, Thailand and elsewhere integrating existing institutions and new practitioners into educational and government statistical operations. In her work we see features of the US land-grant service ideology, including the clear coproduction of knowledge and identity that configured her predecessors' and colleagues' work. The understanding of statistics as fundamentally social (i.e. deriving from human relations and differences) is in Cox's work as in Wallace's and Snedecor's revealing of how the field enacted a constrained democracy.

Through this lens, some of Cox's encounters with sexism take on a new significance for our understanding of statistics as sociability. Service for the Iowan statisticians comprised both selflessness and individual initiative, both suffering and reward. Adversity is in this social world made normal and desirable, including that perpetrated by men in regard to women, bosses in regard to staff, faculty in regard to students. Clearly through the decades of her career Cox faced in professional settings both overt discrimination and talk of female desirability or lack thereof. Accounts of Cox's life written since the 1980s universally mention that Snedecor failed to recommend her for a job at North Carolina until she prompted him to do so.[30] In Cox's own memorialization, Snedecor "loved to tease" her and, we may recall, to spend time with the "prettiest girls" at any gathering. In that same account, we learn that Snedecor's son James remembers meeting Wallace's family; he doesn't "remember much about the boys, but the daughter was impressive." On at least one occasion recalled by Cox, Snedecor responded to a staff shortage by complaining to Cox that "All smart women get married." Cox herself never married, and the derisive and heteronormative features of all these exchanges are hard to ignore.[31]

From the vantage point of 2017 we are readily offended by these objectifying and demeaning episodes. But what does Cox's experience indicate about statistics as a professional site, concertedly trying to expand on the global stage? Cox is not best understood as working in defiance of the identity-steeped world of American science; rather she functioned within in and because of it. She is teased, intermittently disdained and condescended to by her mentor, and described as working "behind the scenes" even by an organization in which she was instrumental.[32] But she is also a necessary part of the establishment in which she is positioned. She is a woman working among men, and one who is necessarily both tolerant and dutiful. It is not inconsequential that she also fondly quotes Snedecor as having told her, "'Nobody can hurt your feelings unless you let them'."[33] Withstanding insult was a skill that both apparently thought Cox would do well to acquire.

How did notions of tolerance, dutifulness and service relate to Cox's practices as a statistician? Immediately after joining the Statistical Laboratory at Iowa State, Cox systematically visited the different laboratories and experimental fields of the college to inquire how experiments were being conducted and how statistics might be of service. The following year, in 1934, she started to teach "Design of Experiments," which followed Snedecor's introductory course on "Statistical Methods," a sequence of classes required for graduate students in agriculture, most of whom were doing research into animal and plant breeding. Snedecor's course would be first published as a book in 1937 and would become a bestseller in the field, selling more than 130,000 copies in the USA and reprinted in Portuguese, Spanish, French and Japanese.[34] As in his previous publication with Wallace, the success of the book was ascribed to its usefulness to the researcher, who was now able to employ statistical methods without having to go through the painful mathematical work behind the methods. As Snedecor insisted, the only mathematics needed to follow the argument was after all arithmetic.

Cox's course was also made into a book, *Experimental Designs*, coauthored with W.G. Cochran, but published only in 1950.[35] While Snedecor's book was explicitly based on Ronald A. Fisher's *Statistical Methods for Research Workers* (1925), Cox was inspired by Fisher's *The Design of Experiments* (1935). In the 1931 edition of "Correlation and Machine Calculation," Wallace and Snedecor had already acknowledged how they had "adapted some of the elegant methods devised by Dr. R. A. Fisher for testing the significance of the various correlation statistics,"[36] thanking Fisher's British publishers for their gracious consent for use of material drawn from "Statistical Methods for Research Works." Historians of science have long recognized the overwhelming influence of Fisher in the field of statistics, not to say anything about his famous contributions to the evolutionary synthesis that finally brought together Darwinians and Mendelians.[37] Nevertheless, it has been overlooked that Fisher's methods, first developed at Rothamsted Experimental Station in England, achieved their influence on the way American scientists interpreted their data or performed their experiments mainly through the joint work of Snedecor and Cox at Iowa State. In the summer of 1931 Snedecor invited Fisher to Ames for a three-week course on methods and on the general theory of natural selection; this was attended by Iowa State faculty from the departments of mathematics, genetics, agronomy and animal husbandry. Members of the Iowa Statistical Laboratory would study with Fisher in London, and in 1936 Fisher would return once more to Iowa to teach about his new book on experimental design.[38]

The indifference of historians to the activities and impact of the Iowan context is probably owing in part to the lack of pedigree granted to agriculture in the history of scientific ideas. Scientific work on agriculture is traditionally perceived as pertaining to the history of application and not as a relevant context for the emergence of groundbreaking theories. Importantly for our argument, the statistical practices developed by Fisher suggested a new way to think about the relation between theory and application, and one, as Snedecor and Cox quickly understood, particularly appropriate for a land-grant college such as Iowa State. The statistical canon Galton–Pearson–Fisher can be simplified in the history of statistical ideas as the evolution from regression to correlation and into analysis of variance, a path from identifying what variables were associated with one another (Pearson's correlation coefficients) to evaluating the significance of such association (Fisher's tests of significance).[39] But in order to be applied, Fisher's analysis of variance demanded also that experiments be performed in certain ways, that the statistician worked closely with researchers not only when analyzing data produced in an experiment but also at the time of *designing* the experiment. In order to apply Fisher's statistical analysis of experimental data in the manner explained by Snedecor in his course, one needed to perform experiments according to Fisher's experimental design as explained by Cox in her course. Moreover, Fisherian statistics identified the significance of different variables in explaining a certain phenomenon, thus suggesting a way of tinkering with them to produce a desired outcome. Agriculture statisticians, such as Wallace, not only identified which variables affected hog prices. The model of explanation indicated also how variables should be tinkered with to produce significant results. The intimate relation implied by Fisher's methods between theory and practice, between describing phenomena and the ability to tinker with them, was the perfect fit for the ethos of a land-grant college such as Iowa State whose motto was "Science with Practice."

The service Cox performed through statistics to fellow faculty and to grad students at Iowa was to guide them through the steps in planning experiments that produced relevant and reliable knowledge. She particularly emphasized *randomization* to avoid bias, *replication* to increase accuracy of comparison and estimating magnitude of experimental error, and *experimental controls* such as optimal size and shape of experimental unit or control over external influences to guarantee "every treatment operates under conditions as nearly the same as possible."[40] Her guidance was central to making Iowa State one of the most productive research universities in the country. "Owing to the utility of the new designs, they have been readily adopted by research men in our Agricultural Experiment Station. Nowhere else are so many of these complex designs in use."[41] The deployment of statistical methods was at Iowa State as far from mere counting and sorting of data as the field had yet been. As scientific experimentation, it involved the demarcation of units of analysis and devising instruments of sorting, with all such conceptualizations configured to best enact desired results. With these amalgamated commitments of scientific and technological work, Snedecor, Cox and their colleagues melded empiricism and application, knowledge and impact, means and ends, in service to scientific productivity.[42] Wallace would be the key figure in expanding this service to science into service to democracy.

Experiments in Democracy

The interventions envisioned by the Iowans, with service to the public welfare paramount, could reach an enormous scale. With New Deal policies ambitiously casting entire economic sectors into reformist mindsets, the statisticians helped bring about huge changes to agricultural planning.[43] In 1935 Cox was calculating and publishing index numbers of Iowa farm prices in support of New Deal's Agriculture Adjustment Act, one of the signature policies of Franklin Delano Roosevelt's administration to fight the Great Depression.[44] Cox's work was exemplary for the research cooperation agreement signed in 1938 between the Iowa Statistical Laboratory and the Bureau of Agricultural Economics of USDA. As historiography has rightly emphasized, USDA, under the leadership of none other than Henry A. Wallace from 1933 to 1940, was the source of some of the most innovative and ambitious federal policies of the New Deal era.[45] Wallace's USDA, which gathered the largest concentration of scientists in the entire federal government and probably in the world, has been aptly described as an unprecedented case (in terms of scale) of science at the service of public policy design. What has been less noticed is that the science in question was not there waiting to be used by agency personnel. Wallace considered that American democracy needed not just more science but a different kind of science than previously known. American scientists had certainly demonstrated their ability to discover new medical cures, increase industrial production or breed more productive animals and plants. But what the Great Depression and the crisis of overproduction that Wallace had been announcing since the end of the First World War demanded was that science concentrate less on increasing economic output and more on making the American economy work for the whole of the American population: "The science of production is useless when products can't be distributed."[46] For Wallace, statistics was the solution; not the traditional statistics of counting and amassing data, but the new statistics as practiced at Iowa, to which he had been an early contributor. A statistics that in its descriptions of reality also pointed at ways of how to tinker with it. In the cooperation agreement, the Iowa State Statistical Laboratory promised to supply USDA with "more efficient experiments, improved sampling techniques and appropriate statistical methods."[47]

Let us look at the Agriculture Adjustment Act (AAA) for which Cox was calculating index numbers of farm prices to understand how it embodied those above-mentioned dreams of a statistical democracy as envisioned by the young Wallace. Before the inauguration of FDR's presidency in January 1933 agriculture prices had reached record lows. Mortgage foreclosures, unpaid taxes, bank failures and a rapid decline in land values all expressed the farm distress across the whole country. Urgent measures were necessary to "increase agriculture's share in the national income," not only to help farmers but to save "the agricultural assets supporting the national credit structure."[48] This was the justification for the AAA, a large-scale experiment in public policy launched in May 1933 by Wallace's USDA to increase agricultural commodities prices and save American farmers from bankruptcy, and with that the entire economy of the country. The act was applied to seven basic agricultural commodities: wheat, cotton, corn, hogs, rice, tobacco and milk. Government agencies were now authorized to experiment with administrative controls over both production and marketing, rewarding farmers through benefit payments for participating in public programs.

Wallace had nevertheless insisted in his previous advocacy for the use of statistics in the design of public policies that statistical expertise should not transfer decision-making from the people to the federal state, replacing democracy with a centralized planning system. And in fact the AAA corn-hog program was from the beginning designed with such concerns in mind, taking a representative producer group into active partnership in its development. By mid-June 1933 the Iowa Federation of Farm Organizations had called a meeting of corn and hog producers from which a first state corn-hog committee was formed. By mid-July the Administration promoted a meeting in Des Moines with representatives of the ten Corn Belt states (Iowa, Illinois, Nebraska, Indiana, Missouri, Minnesota, Ohio, South Dakota, Kansas, Wisconsin), forming a National corn-hog committee of twenty-five members. It included farmers, leaders of farmers' associations, agriculture journalists, directors of marketing associations and the president of the National Grange. When members of the Adjustment Administration met this National Committee, with the presence of Wallace himself, a consensus was formed over the need for reducing simultaneously corn acreage and the number of hogs in the market as the only permanent solution to the problem.[49]

The hog reduction campaign started with a series of educational meetings at county level in which local agents assisted by committee men and extension specialists explained with the help of a series of charts the aims and the expected results of the program. Two main pamphlets were produced to be used in these meetings: *The corn-hog problem* and *Analysis of the corn-hog situation*. This was no less than the materialization of Wallace's dream of producing informed farmers in price formation. Community meetings were then organized to explain the contract forms to individual farmers followed by community sign-up meetings. Finally the campaign promoted the organization of permanent "county corn-hog control associations," making the control of corn-hog production a permanent feature of the daily lives of local communities. By March 1, 1934 Iowa had already signed 144,000 contracts, representing about 90% of the corn and hog producers in the state. Cox's calculations of Iowa agriculture prices indexes performed at the Statistical Laboratory were made to guarantee the fairness of such contracts. And what was at stake was not only a final fair price for farmers, but also to reach consensus by involving hundreds of thousands of people in a democratic experience. As John Dewey, the most distinguished public philosopher of the New Deal era, reminded: "social advance depends as much upon the process through which it is secured as upon the result itself."[50]

The delineation of communities relevant to or worthy of participation in the resolution of the corn-hog problem was not unproblematic. The apparently vast number of farmers enlisted to help shape and undertake this price reform effort did not include those without their own property; neither the statisticians nor their collaborators sought the participation of sharecroppers, many of whom were African American, in the reform.[51] This echoes other New Deal efforts in which black Americans were marginal participants at best and disadvantaged bystanders at worst; there is considerable literature on how New Deal cotton policies led to the departure of black Southerners to Northern cities, for example.[52] But we point this out to indicate not a racist proclivity among the statistical reformers, as such. Rather we emphasize both the taxonomic features of their work, formulating differences and communities that reflect those differences. We also stress that the Iowans and followers working with Cox in North Carolina later addressed the sharecroppers' distress. What is significant is that setting the boundaries of democratic collectivities, to determine whose well-being mattered, was a constant project for the statisticians; the taxonomies on which experimental statistics relied were only intermittently stable, as application brought to light the need for redrawing boundaries. These boundary-making efforts went well beyond Iowa and the nation itself.

Global New Deal

Wallace's painstakingly detailed discussions around corn and hogs always had world affairs in the background. In 1918 considerations of the depressing effect of the First World War's aftermath on hog prices guided his statistically informed early public campaigns against Herbert Hoover's actions. At the same time, Wallace's understanding of the global nature of the Great Depression justified his belief that the reduction of American agricultural production through the AAA was no more than a first stage of a more comprehensive and long-lasting response.[53] FDR's secretary of agriculture was unapologetic in pointing out the myopia of US political leaders' vision of international relations as a major cause of the Depression. In order to guarantee domestic economic security, the USA should acknowledge that it had to intervene on a global scale. The Coolidge and Hoover administrations of the interwar period had been oblivious to the role of the US economy as international creditor of European countries that were highly indebted from a war effort waged in large measure through the purchase of American raw materials. America's high import tariffs resulted in the decreased ability of European countries to pay their debts through exports, turning them increasingly to protectionism. The excess production of the American economy which led to the Great Depression was not just caused by increased capacity and productivity, but also by the inexistence of an open international market able to buy American produce. The increased revenue of American farmers achieved through the measures of the AAA was only sustainable in the long run if these farmers could export their hogs, corn, cotton or wheat. According to Wallace, America had to assume its new leading positioning in the world and the responsibilities it entailed if a new global crisis was to be averted.[54]

An early supporter of FDR's "Good Neighbor" foreign policy towards Latin America, Wallace made USDA into a central piece in this effort of replacing previous US interventionism on the continent with international cooperation aimed at forming a single economic zone.[55] Building on the International Conferences of American States, USDA experts identified the potential complementarities of agricultural production of the continent's different regions. While the USA hoped to export its surplus of grains (corn and wheat), fibers (cotton) and livestock (cows and hogs) to the rest of the Americas and thus increase domestic farm income on a more stable basis, the country could also overcome its deficit of rubber, quinine, cocoa, coffee, cassava and nuts through imports from Central and South America, and thus allegedly help to improve the standard of living of peasants across the continent. USDA promoted an American Society of Agricultural Sciences with delegates from Argentina, Brazil, Chile, Colombia, Costa Rica, Cuba, Ecuador, Peru, Dominican Republic, Uruguay, Venezuela and, of course, the United States. Latin American scientists were welcomed at American land-grant colleges and at courses at USDA in Washington. USDA statisticians as distinguished as Mordecai Ezekiel, Wallace's economic adviser responsible for drafting the details of the AAA, traveled south not only to sell US expertise but to learn as well from agricultural reform experiences such as the expropriation and subdivision of large tracts of land by Mexican president Lazaro Cardenas. A few days after the ticket Roosevelt/Wallace won the 1940 US presidential election by a landslide, the vice-president elect drove his own car to represent the United States at Avila's Camacho inauguration in Mexico City. According to the *Washington Post*, Mexico's revolutionary government could not have been more welcoming of Wallace, "America's apostle of social experimentation."[56]

Wallace's pan-Americanism only became more pressing with the Second World War, as he insisted with his co-workers on making "every minute of every day count toward building security around those sacred rights for which the peoples of the hemisphere stand."[57] As a member of

FDR's secret "war cabinet," Vice-President Wallace now chaired the Board of Economic Warfare, the agency responsible for supplying of strategic resources which in 1942 was given complete control for directing the production and procurement of all raw materials from abroad, namely from Latin America. Revisionist historians praising the democratic experimentalism associated with Wallace's USDA during the New Deal years tend to lament how the war effort curtailed such reforming endeavors.[58] Here we would like to insist that the war was actually instrumental in scaling up the experiment from the national to a continental scale, and later to a global scale.

Wallace's most famous speech, "The Century of the Common Man," broadcast in May 1942 to mobilize American producers for the war effort, points at this global ambition, or we might better say at universalism.[59] He depicted the war as a battle between light and darkness, good and evil, God and Satan, democracy and dictatorship, "democracy [being] the only true political expression of Christianity." Stating that the people of the USA "have moved steadily forward in the practice of democracy," he refused nevertheless the notion of an "American Century," preferring instead a more egalitarian "Century of the Common Man" in which America only "suggest[ed] the Freedoms and duties by which the common man must live." The new global order Wallace envisioned, in which "no nation will have the God-given right to exploit other nations" and in which older nations, such as the USA, had the privilege to "help younger nations get started on the path of industrialization," was a millenarian one. Reproducing the Social Gospel rhetoric he had learned in his Iowan youth, he called on Americans to fulfill their duty, concluding that "The people's revolution is on the march, and the devil and all his angels cannot prevail against it. They cannot prevail, for on the side of the people is the Lord."

Social Gospel followers at the turn of the century, such as Wallace's grandfather Henry Wallace, strived for the realization of the Kingdom of God on this earth. As mentioned above, the Gospelers were harsh critics of American capitalism of the Gilded Age and eager to put science at the service of "God's plan for democracy in the New World."[60] Wallace now urged America to face its world role and extended such vision to the entire globe. He placed American capitalist cartels, the Satan of Social

Gospel preachers, side by side with Hitler, the "Supreme Devil," urging modern science to be "released from German slavery" and from cartels "that serve American greed." Only then could science fulfill its sacred democratic duty: "Modern science, when devoted whole-heartedly to the general welfare, has in it potentialities of which we do not yet dream."

Science was thus central for Americans to realize "a just, charitable, and enduring" peace. But again, for Wallace and those with similar expansive social impulses this was not just any science. This was science as the one practiced in American land-grant colleges, one in which the union of mind and hand, theory and practice, served the people in their millennial march towards democracy. As we have seen, Wallace, Snedecor and Cox had made experimental design and statistical analysis obligatory methods for science to serve democracy. After having proven its value in the American experiment with democracy, statistics was to become integral to the millenarian American vision for world peace; a vision in which the USA would not colonize other nations, but instead assist them in achieving the same democratic standards as the USA.

In 1943 Wallace would take an active role in the plans and arrangements for the Hot Springs Conference on Food and Agriculture, which led to the founding of the Food and Agriculture Organization of the UN (FAO).[61] His previous economic advisor at USDA, Mordecai Ezekiel, another statistician who broke his teeth studying corn/hog ratios, served in 1945 as a member of two of FAO's first missions, to Greece and Poland. Two years later, Ezekiel was overseeing the FAO's Economic Analysis Branch, later being promoted to head of the Economics Department. He would only leave FAO in 1961, to assume a position in the United States Agency for International Development (USAID). As Ezekiel would recall, "many of the American agricultural economists who in their younger years served under Wallace in the USDA are finishing out their professional careers in AID, seeking to help the rest of the world make the same kind of progress in agriculture that the United States made in the first two-thirds of this century."[62]

The personal trajectory of Wallace has been used to confirm the failure of his idealistic views in the post-Second World War period. Truman was the choice of the Democratic Party to run as vice-president in FDR's reelection in 1944, and he would defeat Wallace in the 1949 presidential election when the latter ran as the progressive party candidate, winning no more than 2.7% of the popular vote.[63] Visions of world peace such as the one offered by Wallace, in which the Soviet Union was not considered America's enemy, had given place to Truman's Manichaeism that would characterize the Cold War era. This said, it is hard to miss how much Truman owed to Wallace and fellow New Dealers such as Ezekiel when he concretized what America had to offer to the world in opposition to Communism. Truman's famous Point Four of his 1949 inaugural speech promised "a bold new program for making the benefits of our scientific advances and industrial progress available for the improvement and growth of underdeveloped areas." The Point Four initiative was to be carried out through the UN, expanding the latter's program of technical assistance to developing countries, already largely funded by the USA. As recent revisionist historiography of American development aid has emphasized, the New Dealers identified by Ezekiel had found an institutional new home in the UN and in American development agencies defining the nature of their projects. More than insisting on the application of the lessons of the war to development policies, historians have called attention to how the New Deal experience constituted the basis of Cold War American international aid programs.[64] And as Daniel Immerwahr has convincingly demonstrated, while only through this nexus can one understand the insistence of American experts on community development in India and the Philippines in the 1950s, the results of such voluntarist international transfer of American democratic commitments were no less tragic than the high modernist top down solutions (large dams, nuclear energy) famously denounced by James C. Scott and his myriad of disciples.[65]

Statistical Frontiers

As the Second World War drew to a close, any project aimed at bettering the health, food or education status of a population was understood to need statistics to establish the welfare standard for the country in question. A 1949 UN report on "Existing Statistical Deficiencies" complained that of the thirty-two countries in Latin America, the Middle East and

East Asia studied, only eight had supplied national income figures, data on external trade or numbers relating to principal types of livestock. Unsurprisingly, a UN statistical commission was formed to coordinate the collection of demographic and economic data from the different members. What is more interesting for our argument is the insistence in going beyond such collection activity to accomplish the aims of American global assistance. The uplifting of nations from poverty, according to the aid experts' view, certainly needed the systematic application of science in agriculture, medicine and industry. But to make science contribute to the general welfare, to make science serve the people, statistical methods, as Iowans had demonstrated, were considered mandatory. Gertrude Cox's involvement with the Education Committee of the ISI was instrumental in enlarging the statistical scope of the UN from traditional collection of statistical data to a comprehensive change of scientific practices.

First, to characterize Cox's extensive contributions to the internationalization of American statistical practices, we can consider that for Cox as for Snedecor statistics had an almost infinite mobility. Almost like air or water, statistical techniques could seemingly flow unimpeded into any site of human activity. Also like air and water, their presence was seen to be salutatory for those whom the techniques reached, a guarantor of enhanced well-being. In Iowa through the 1920s and 1930s statistical experimentation had moved vigorously beyond the purview of state accounting to become a means of adjusting operations of agriculture, economic development, public health and other sectors. The mobility of statistical thinking included both geographic and disciplinary flows.[66] At North Carolina State from 1940, Cox undertook statistical studies of Southern poverty and the discrepant economic experiences of black and white communities, receiving funding from the Rockefeller Foundation in 1944 to "improve statistical competencies" in the regions she studied, part of her "lifelong interest in the use of statistics to study human relationships."[67] Her post-war work with the ISI, and positions as editor of the Biometric Society journal and as a traveling consultant, all put her in line with others concerned at this point with "worldwide insecurity and want."[68] This global intervention in human welfare was of course the domain of the UN, and Cox fully embraced the new organization's potential for the dissemination of statistical expertise.

The service commitments of the statisticians in some ways fitted into other UN efforts as hegemonic or post-colonial extensions of Euro-American economic structures, but they enacted as well the Social Gospel conceptions of sharing and caring. Professional social workers in this period saw the UN as itself an institution of infinite mobility, taking up problems of "poverty, ignorance and disease" in ways that were previously unknown to the world.[69] The war had at least had this welcome outcome for the Iowans, along with assurances of American geopolitical and economic authority. For Cox, this was a landscape of frontiers for American science, of places awaiting the presence of statistical techniques. Built into the methods and aims of statistics were possibilities for expansion beyond existing communities of science, as her 1957 description of how best to present statistical results indicates:

> ...you will be asked to swear allegiance to logical organization, preciseness and *ease of comprehension* [italics added].[70]

In this extended metaphor on the state of the field, Cox indicates that when done well, statistics would reach endless frontiers; horizons and audiences not yet reached by statistical experimentation and intervention serve as opportunities.

Patti Hunter suggests that Cox was likely instrumental in bringing UN resources to statistical training activities.[71] In 1948 the UN Educational, Scientific and Cultural Organization (UNESCO) gave the ISI a $5000 grant for statistical education at the recommendation of the UN Statistical Commission, in which Cox had an established role. As the ISI sponsored statistical education in numerous "developing" nations, Cox organized the training of new statistical practitioners. She published widely on the topic and coauthored a textbook on experimental design that reached very wide distribution.[72] We find in this literature distinct ideas about the field of statistics: a body of applied knowledge best approached through stratified opportunities for learning and hierarchical labor structures. Again, we find that identity and epistemic features or statistical practice are inseparable. According to Cox and her cohort, the conduct of statistical study, constituted of training, data collection and data analysis, was to be undertaken in a world of differing cultures (UN member nations included the USA along with Britain and its European neighbors, and also many so-called developing countries), but all cultures were seen to have unified scientific needs: needs determined by Western experts. Statistics could both find and resolve instances of poorly understood health, industrial production or economic planning. For the Statistical Commission as for other expert units within the UN, empiricism (here embodied in statistical science) was seen to empower development: countries were now seen to have a relative level of "statistical progress" to be gauged by the American and European statistical experts.[73] Cox's work with the UN and ISI to support improved statistical capacity among member nations conveys the combination of judgment and service inhering in that world view.

Centralization and Collectivity: Global Flows of Statistical Capacity

The role of the UN in securing the power of Western institutions to demarcate "other" cultures as such and to evaluate their contributions to global peace and prosperity is well established; Mazower summarizes the UN as a "product of evolution not revolution" that grew out of existing imperial ideas and institutions challenged but not displaced by the war.[74] The activities of the Statistical Commission in the late 1940s carried outward the Iowans' ideas of prosperous, healthy communities and also their ideas of optimized epistemic conducts represented by good statistical practice. Both embodied constructions of white/non-white, American/non-American, European/non-European binaries, mapped onto nations' perceived ability and willingness to undertake statistical modernization projects.

In descriptions issued by Snedecor, Cox and others in the US statistical mainstream, skills related to and work with statistics were appropriately divided on a ladder of theory, experimental design, analysis and data collection, in order of most intellectually challenging to least. Lines between the sets of activities were proposed in overviews of the field, and then drawn by the content of instructional materials and job descriptions.[75] This stratification is of course in keeping with scientific and industrial labor of all kinds in the nineteenth and twentieth century, and it is clear that the traditional relegation of women to the repetitive work of calculation (as "computers") in places of large-scale applied mathematics, or the near complete exclusion of persons of color from higher technical education altogether, is not unimportant in explaining this distribution. But the correlation of philanthropic or service functions for statistics with such divisions of learning and labor suggests a constrained vision of democratization. On the ground, this meant that Cox's vision of programs in developing "member" nations for training statistical analysts reasonably excluded the preparation of local individuals for theoretical work.[76] Gifted scientists from developing countries considered able to produce such innovative theoretical work would be trained in US institutions, namely in Cox's own North Carolina State University, which thus became together with Iowa State one of the beacons of American statistical hegemony in the Cold War years.[77] Reports of the UN Statistical Commission frequently argued that many countries needed to upgrade their statistical infrastructures, including through new educational and publishing initiatives; yet this upgrade would be best served, the Commission wrote, by the creation of a "field service" designed to send technical experts outward from UN offices rather than through a thorough ground-up preparation of local expertise. All such schemes indicated that people belong in certain productive roles, each person's efforts perhaps vital for the collectivity yet corroborative of innate human differences.

Of course we must attend to the ways in which the Statistical Commission at the same time appeared to respect "the force of local desire" supporting any such international interventions. The standardization of statistical training programs for member nations, according to commission documents, would have to be done slowly so as not to become "obnoxious to member countries" because "fashioned by non-nationals."[78] Commission authors recognized a risk of "too-pedantic professionalization" in their extension efforts. Yet a strong sense of centralized expert authority pervades the UN statistical activities. The perceived needs of member nations lacking their own means to recruit and train specialists in statistics were, after all, dictated by the white, Western, statisticians. This is conveyed in the most detailed features of

UN-prescribed statistical practice: one commission report notes that local differences must be overcome; for example, when "no term for gainfully employed person" exists in a locality, such a term must be created. It is also clear in the most general character of the Statistical Commission, which created the obligation for member nations to supply statistics. The production of data by way of established metrological practices, and publication, not concealment, were necessary for the betterment of all member nations.[79] While purportedly a flat organization with a "from each … to each…" operating premise, the UN supported a system of Western-driven scientific development with a thinly disguised normative role for American and European experts. The self-assigned importance of these experts is palpable. As a 1949 Statistical Commission report put it, "Education is growth from small beginnings and an international organization is in a well disposed position to communicate to one country the fact and the promise of small beginnings made in others."[80]

The Production of Statistical Selves

The dual nature of statistics to their Western proponents as a means of knowing and making the world is evident in these post-war efforts to enact an internationalized science. What comes across in the Statistical Commission's work, and much of Cox's other writing on education and outreach, is not only some limits to the redistributive potential of statistical interventions in places of strong class divisions, poverty or lagging economic growth, but also that in the Iowans' view epistemic capacity is naturally divided among individuals. Taxonomies—be they of gender, class or intellectual and productive capacity—are the basis not just of statistical research but also of statistics as a scientific enterprise. In no sense should we be surprised that this knowledge-making sector bears out gender, racial, heteronormative and other categorizing efforts that historians have previously located in US insurance, communications, computing and engineering spheres.[81] We can detect in Cox's work in the United States and as a representative of North American expertise globally many sociabilities of American social science at mid-century.[82] In a gendered, raced, classed and nationalized social landscape, statistical research took particular forms; the research in turn enacted all of those human differences and others, including optimized social conducts along lines of optimized Protestant, heteronormative and capitalist conducts. We can see these patterns when we step away from our depiction of Cox as a woman achieving occupational success under discriminatory conditions, as she has been painted by Rossiter and others in the literature of the "missing persons" of science: women and minorities previously absent from the historical record when not entirely excluded from professional opportunities.[83] If read as achieving influence and scientific innovation "in spite of" sexism, Cox's own (ascribed and self-ascribed) identity in a binary gender system loses its explanatory power as part of the history of statistics. Similarly, if Cox's tolerance of discrimination, exclusion and even harassment, all experiences recorded in autobiographical and biographical accounts of her that are given explanatory primacy, she appears to have capitulated to social norms that seem to be antithetical to (her own) rigorous statistical work. Sexist behaviors among university and government experts were not antithetical to rigor but constitutive of it, as racialized world views constituted educational and health research or policy through these decades. Through consideration of Cox's work and its resonance with ideologies of service, we want to suggest that the constrained democracy envisioned by American statisticians coproduced identities (gender, talent and others) and legitimate knowledge.

Cox was committed to the internationalization of cutting-edge statistical practice from wartime onward, and approached statistics education and labor as stratified experiences. We are limited in space here, and both emphases can be much more fully explored as to how they express the relational character of identity. Cox's own experiences and the broader operations of statistics as a knowledge-making realm deserve considerable exploration as representing American modes of difference-making. But here we suggest some reasons why this exploration may yield significant questions to ask about the role of empirical labor, such as statistical research, in the production of gender, nationality and other categories of subject.

We do not see Cox as having been a willing victim of sexism or heteronormativity or in any sense as producing her own marginality. What we want to convey instead is the fluidity and dialogic character of identity-based relations in statistics. To see her as either triumphing over adversity *or* capitulating to demands from men for her subordination and sufferance is to miss the deliberate production of difference within the field of statistics. Mohanty suggests that we reread now-familiar forms of discrimination, such as sexism, as more thoroughly relational; that is, as produced not merely by encounters between those with and without power (here, men and women), but as constantly produced as relations of ruling among much wider sets of actors, in Dorothy Smith's terminology.[84] In the case of Cox, her energy and forbearance helped to constitute gender binaries and also what would count as legible generosity and caring in institutions such as universities, the UN or national statistical bodies: statistics as an entirely social enterprise.

If we insist on seeing Cox as a marginal figure fighting for inclusion in a professional realm, we fail to see the integral role that that marginality played in statistical practice, both in consolidating the role of duty and service in this expert practice and in the sorting of individuals into meaningful taxonomic categories, as with the Statistical Commission's characterizations of member nations lacking in statistical facility. In a sense, Cox's challenges echo the historical role of physical disability in defining ability: her embodiment of deficit, as a female, articulated male assets, much in keeping with constitution of imperial (white) masculinity, as Mohanty and others describe that process.[85] But Cox's deficit was in further ways a constructive one, naturalizing her role as dutiful and maternal (female) provider of new knowledge and opportunities to international subjects. As a white American woman she bore a different relationship to Egyptian or Thai subjects than would a non-white or non-American woman. As a statistician she bore a different standing in those settings than would have a social worker or nurse or tourist. Transnational movements of statistical science were made up of all of Cox's identities, and those of all whom she encountered. The so-called developing communities and nations to which she and her colleagues sought to bring statistical capacity were similarly rendered as *potentially* productive and *potentially* of service to other more fully endowed cultures, should they comply with the expectations set out by the well endowed.

Conclusion

Gertrude Cox's experiences, frustrations and achievements show us that American experimental statistics of the pre- and post-war period while dominated by men is not best understood as a nominally masculine field, but rather one in which knowledge was predicated on difference. This is an important distinction because historians of science and of other cultural enterprises who are concerned with discrimination have suggested that individual conducts in gendered settings can be more flexible than cultural constructions of gender.[86] While that formulation grants agency to individual actors and avoids a deterministic accounting for individuals' behavior, it skirts the ontological character of identity in places of knowledge-making. That which Cox did constituted creditable statistical research because she embodied a particular commitment to service; Snedecor embodied another. Notice that any essentialized ideas of gender that our actors may have taken up are not recapitulated by us; instead we try to follow precise articulations of viable selves and valuable knowledge.

In recognizing these powerful functions for social identity at work among the statistical experts, we need not in any way seek an inauthentic basis for their service ideology. Wallace, like Snedecor and Cox, approached his work with intellectual generosity. Put differently: only statistical ideas and findings that led to, or seemed likely to lead to, beneficent application seemed rigorous to these experts. Other sorts of cognitive efforts were either entirely illegible to the Iowans or beneath scientific notice. Among the implications of that evaluative tendency is the permeable boundaries we detect between theory and application, science and technology, mind and hand, in the statisticians' work that we hope to investigate further. If as we suspect the distinctions the Iowans made among tasks of observation, understanding and action were few and fragile, we may also find that scientists of different ideological bent created far sturdier lines between these modes. What implications, we might now ask, did the relative strength of such lines have for enacting democracy through the use of statistical techniques on a global scale? We will not find answers to that question by imputing to Wallace, Cox or other statistical experts identifiable racist, sexist or colonialist impulses. Discriminatory conduct was neither their intention nor their felt experience; just the opposite. Yet this was, again, a constrained democratic project, and we need next to consider the ways in which their reliance on taxonomies, on the systematic ordering of peoples, things and economic mechanisms, brought about not reform or the redistribution of global power but structural stasis.

Notes

1. There is a growing body of historical scholarship linking New Deal policies and American development initiatives after the Second World War. We find the following references particularly useful: Immerwahr, Daniel: *Thinking Small: The United States and the Lure of Community Development*, Cambridge, Mass: Harvard University Press, 2015; Ekbladh, David: *The Great American Mission: Modernization and the Construction of an American World Order*, Princeton, NJ: Princeton University Press, 2010; Garfield, Seth: *In Search of the Amazon: Brazil, the United States and the Nature of a Region,* Durham: Duke University Press, 2013; Olsoon, Tore C.: *Agrarian Crossings: reformers and the Remaking of the US and Mexican Countryside,* Princeton, NJ: Princeton University Press, 2017; Borgwardt, Elizabeth: *A New Deal for the World: America's Vision for Human Rights,* Cambridge, Mass: Harvard University Press, 2007.
2. There is an important body of work in history of science dealing with the historical emergence of these statistical methods. Porter, Theodore M.: 'Statistics and Statistical Methods', in Porter, Theodore M. and Ross, Dorothy (eds.): *The Cambridge History of Science, Vol. 7, The Modern Social Sciences*, Cambridge, UK: Cambridge University Press, 2003; Gigerenzer, Greg et al.: *The Empire of Chance: How Probability Changed Science and Everyday Life*, Cambridge, UK: Cambridge University Press, 1989; Hacking, Ian: 'Telepathy: Origins of Randomization in Experimental Design', *Isis*, 79 (1988), pp. 427–51; Hall, Nancy S.: 'R.A. Fisher and His Advocacy of Randomization', *Journal of the History of Biology*, 40 (2007), pp. 295–325; Dehue, Trudy: 'Establishing the Experimenting Society: The Historical Origin of Social Experimentation according to the Randomized Controlled Design', *American Journal of Psychology*, 114 (2001), pp. 283–302; Parolini, Giuditta: 'The Emergence

of Modern Statistics in Agricultural Science: Analysis of Variance, experimental design and the Reshaping of Research at Rothamsted Experimental Station', *Journal of the History of Biology*, (2015), pp. 301–335.

3. Although historians have overlooked the Iowan context in history of statistics, the practitioners themselves never forget to mention it. See namely Cox, Gertrude M. and Homeyer, Paul: 'Professional and Personal Glimpses of George W. Snedecor', *Biometrics*, 31 (1975), pp. 265–301; David, Herbert A.: 'Statistics in U.S. Universities in 1933 and the Establishment of the Statistical Laboratory at Iowa State', *Statistical Science*, 13 (1998), pp. 66–74.
4. On the relationship between politics and new statistical categories, see Porter, Theodore M.: *Trust in Numbers: The Pursuit of Objectivity in Science and Public Life*, Princeton, NJ: Princeton University Press, 1995; Hacking, Ian: *The Social Construction of What?,* Cambridge, UK: Cambridge University Press, 2001; Tooze, Adam: *Statistics and the German State: 1900–1945: The Making of Modern Economic Knowledge,* Cambridge, UK: Cambridge University Press, 2001.
5. Marcus, Alan I. (ed.): *Service as Mandate: How American Land-Grant Universities Shaped the Modern World,* Tuscaloosa: University of Alabama Press, 2015 [Second Ed.]; Bateman, Bradley W.: 'Clearing the Ground: the Demise of the Social Gospel Movement and the Rise of Neoclassicism in American Economics', *History of Political Economy* 30 (1998), pp. 29–52. Jewett, Andrew: 'The Social Sciences, Philosophy, and the Cultural Turn in the 1930s USDA', *Journal of the History of the Behavioral Sciences*, 49.4 (2013), pp. 396–427.
6. Gilbert, Jess: *Planning democracy: Agrarian intellectuals and the intended New Deal*, New Haven: Yale University Press, 2015; Jewett: 'The Social Sciences'; Olsoon: *Agrarian Crossings*; Kleinman, Mark L.: *A World of Hope, a World of Fear: Henry A Wallace, Reinhold Niebuhr, and American Liberalism*, Columbus: Ohio State University Press, 2000; White, Graham and Maze, John: *Henry A. Wallace: His Search for a New World Order*, Chapel Hill: University of North Carolina Press, 1995. For Gertrude Cox see, Hunter, Patti W.: 'Gertrude Cox in Egypt: A Case Study in Science Patronage and International Statistics Education During the Cold War', *Science in Context*, No. 22 (2009), pp. 47–83; Rossiter, Margaret: *Women Scientists in America: Before Affirmative Action, 1940–1972*, Baltimore: Johns Hopkins University Press, 1995.

7. Wallace, Henry A.: *The Century of the Common Man*, New York: Reynal & Hitchcock, 1943.
8. Rossiter, *Women Scientists.*
9. Bonnett, Alistair: 'Whiteness and the West', in Bressey, Caroline and Dwyer, Claire (eds.): *New Geographies of Race and Racism*, Oxford: Taylor & Francis, 2009, p. 26; Mazower, Mark: *No Enchanted Palace: The End of Empire and the Ideological Origins of the United Nations*, Princeton: Princeton University Press, 2009, pp. 13–17.
10. For the general context of American economic mobilization for World War One see, Adam Tooze: *The Deluge: The Great War and the Remaking of the Global Order, 1916–1931,* New York: Viking, 2015. For American food production and the international context of the first world war see Collingham, Lizzie: *The Taste of War: World War II and the Battle for Food,* New York: Penguin, 2012; Culhater, Nick: 'The Foreign Policy of the Calorie', *The American Historical Review*, No. 112.2 (2007), pp.: 337–364. For Hoover and his food policies: Nash, George H.: *The Life of Herbert Hoover: Master of Emergencies, 1917–1918*, New York: Norton, 1996; Robert D. Cuff: 'The Dilemmas of Voluntarism: Hoover and the Pork-Packing Agreement of 1917–1919', *Agricultural History,* 53.4 (1979), pp. 727–747.
11. Fitzgerald, Deborah: *The Business of Breeding: Hybrid Corn in Illinois, 1890–1940,* Ithaca, NY: Cornell University Press, 1990; Culver, John C. and Hyde, John: *American Dreamer: A Life of Henry A. Wallace*, New York: WW Norton & Company, 2001; Edward, L. and Schapsmeier, Frederick H.: *Henry A. Wallace of Iowa: The Agrarian Years, 1910–1*940, Ames: Iowa State University Press, 1968; Lord, Russell: *The Wallaces of Iowa,* Boston: Houghton Mifflin, 1947.
12. Jewett, 'The Social Sciences', p. 402.
13. Bateman, 'Clearing the Ground'.
14. Wallace, Henry A.: *Agricultural Prices,* Des Moines: Wallace Publishing Company, 1920, p. 7.
15. Ezekiel, Mordecai: 'Henry A. Wallace, Agricultural Economist', *Journal of Farm Economics,* 48.4 (1966), pp. 789–802.
16. Winters, Donald L.: 'The Hoover-Wallace Controversy During World War I', *The Annals of Iowa,* 39 (1969), pp. 586–597.
17. Wallace, *Agricultural Prices,* p. 108.
18. Wallace, *Agricultural Prices,* p. 110. The concept of 'moral economy' refers to E. P. Thompson's famous discussion about the contrast between fair prices and market price in 'The Moral Economy of the

English Crowd in the Eighteenth Century', *Past & Present*, 50 (1971): 76–136. Historians of science have successfully appropriated the concept to account for the conventions governing scientific sociability. Steven Shapin, 'The house of experiment in seventeenth century England', *Isis* 79 (1988): 373–404; Kohler, Robert: *Lords of the Fly: Drosophila Genetics and the Experimental Life*, Chicago: University of Chicago Press, 1994.

19. Wallace, *Agricultural Prices*, p. 109
20. Wallace, H. A. and Snedecor, George W.: *Correlation and Machine Calculation*, Ames: Iowa State College, 1925); Ezekiel, 'Henry A. Wallace, Agricultural Economist'; Culver & Hyde, *American Dreamer*.
21. Wallace and Snedecor, *Correlation and Machine Calculation.*
22. Ibid., p. 5.
23. Cox, Gertrude M. and Homeyer, Paul: 'Professional and Personal Glimpses of George W. Snedecor', *Biometrics* 31 (1975), pp. 265–301.
24. Snedecor, George W.: 'Uses of Punched Card Equipment in Mathematics', *The American Mathematical Monthly*, 35.4 (1928), pp. 161–169.
25. Anderson, Richard L.: *Gertrude Mary Cox, 1900–1978: A Biographical Memoir*, Washington, DC: National Academies Press, 1990, p. 119
26. Cox and Homeyer, 'George W. Snedecor', p. 289.
27. Ibid., p. 265.
28. Ibid., p. 269.
29. Hunter, Patti W.: 'Gertrude Cox in Egypt', *Science in Context* 22 (2009), pp. 47–83; Anderson, 'Gertrude Mary Cox'.
30. Hunter, 'Gertrude Cox in Egypt'.
31. Cox and Homeyer, 'George W. Snedecor', pp. 272 and 280.
32. Billiard, Frances K. Kernohan: 'The Impact of the United Nations on Poverty, Ignorance and Disease', *Social Service Review*, 29.1 (1955), p. 260.
33. Cox and Homeyer, 'George W. Snedecor', p. 266.
34. Snedecor, George W.: *Statistical Methods Applied to Experiments in Agriculture and Biology*, Ames, IA: Collegiate Press, 1937. The trajectory of the book and the identification of its community of readers will certainly offer major clues on the global reach of statistical methods.
35. Cochran, William G. and Cox, Gertrude M.: *Experimental Designs*, New York: John Wiley, 1950.
36. 'Dr. Fisher and his publishers... have gratiously consented to our use of material', Wallace and Snedecor, *Correlation and Machine Calculation*, p. 4.

37. See endnote No. 2. Also, Berry, Dominic: 'The Resisted Rise of Randomization in Experimental Design: British Agricultural Science, c. 1910–1930', *History and Philosophy of the Life Sciences,* No. 37 (2015), pp. 242–260.
38. Cox and Homeyer, 'George W. Snedecor'.
39. Porter, 'Statistics and Statistical Methods'.
40. Cochran and Cox, *Experimental Designs.*
41. Anonymous: *The Statistical Laboratory of the Iowa State College,* Ames: Iowa State College, 1940, pp. 9–10.
42. Roberts, L., Schaffer, S. and Dear, P. (eds.): *The Mindful Hand,* Amsterdam: KNAW, 2007, p. xix.
43. Gilbert, *Planning Democracy*; Gilbert, Jess: 'Low Modernism and the Agrarian New Deal: A Different Kind of State', in Adams, Jane (ed.): *Fighting for the Farm: Rural America Transformed*, Philadelphia: University of Pennsylvania Press, 2003; Jewett, 'The Social Sciences'; Olsson, *Agrarian Crossings*; Kirkendall, Richard S.: *Social Scientists and Farm Politics in the Age of Roosevelt,* Columbia: University of Missouri Press, 1966; Philip, Sarah T.: *This Land, This Nation: Conservation, Rural America, and the New Deal*, New York: Cambridge University Press, 2007.
44. Cox, Gertrude M.: 'Index number of Iowa farm products prices', *Bulletin (Iowa Agricultural Experiment Station)*, 29.1 (1935), pp. 299–328.
45. Skocpol, Theda and Finegold, Kenneth: 'State Capacity and Economic Intervention in the Early New Deal', *Political Science Quarterly*, 9.2 (1982), pp. 255–278; Rodgers, Daniel T.: *Atlantic Crossings: Social Politics in a Progressive Age,* Cambridge, Mass: Harvard University Press, 2000; Gilbert, *Planning Democracy*; Gilbert, Jess and Howe, Carolyn: 'Beyond 'State vs. Society': Theories of the State and New Deal Agricultural Policies', *American Sociological Review* 56.2 (1991), pp. 204–220.
46. Wallace, Henry A.: 'The Year in Agriculture. The Secretary's Report to the President', in *Yearbook of Agriculture, 1934*, Washington D.C: Government Printing Office, 1934, p. 25.
47. Anonymous, *The Statistical Laboratory.*
48. Ezekiel, Mordecai and Bean, Louis H.: *Economic Bases for the Agricultural Adjustment Act*, Washington D. C.: Government Printing Office, 1933.
49. Lambert, Roger C.: 'The Illusion of Participatory Democracy: The AAA organizes the corn-hog producers', *The Annals of Iowa* 42.6 (1974), pp. 468–477.

50. Dewey quoted in Menand, Louis: *The Metaphysical Club: A Story of Ideas in America*, New York: Farrar, Straus and Giroux, 2001.
51. Saloutos, Theodore: 'New Deal Agricultural Policy: An Evaluation', *Journal of American History*, 61.2 (1974), pp. 394–416; Conrad, David Eugene: *Forgotten Farmers: the Story of Sharecroppers in the New Deal*, Urbana: University of Illinois Press, 1965; Foley, Neil: *The White Scourge: Mexicans, Blacks, and Poor Whites in Texas Cotton Culture*, Berkeley: University of California Press, 1997.
52. Cowie, Jefferson and Salvatore, Nick: 'The long exception: Rethinking the place of the New Deal in American history', *International Labor and Working-Class History*, 74 (2008), pp. 1–32.
53. Walker, Samuel: 'Henry A. Wallace as Agrarian Isolationist 1921–1930', *Agricultural History*, 49.3 (1975), pp. 532–548.
54. Wallace, Henry A.: 'Farm Economists and Agricultural Planning', *Journal of Farm Economics*, 18.1 (1936), pp. 1–11.
55. See namely, Olsson, *Agrarian Crossings*; Garfield, *Searching for the Amazon*. See also the journal *Agriculture in the Americas*, which started to be published in 1941 as an expression of USDA's commitment to pan-American endeavors.
56. Quoted by Olsson, *Agrarian Crossings*, p. 67.
57. 'Henry Wallace – our mutual friend', *Agriculture in the Americas* (February 1941).
58. Gilbert, *Planning for Democracy*.
59. Wallace, *The Century of the Common Man*.
60. Crunden, Robert M.: *Ministers of Reform. The Progressives' Achievement in American Civilization, 1889–1920*, Urbana-Champaign: University of Illinois Press, 1985.
61. Phillips, Ralph Wesley: *FAO: Its Origins, Formation, and Evolution, 1945–1981*, Rome: FAO, 1981.
62. Ezekiel, 'Henry A. Wallace, Statistician', p. 801.
63. Kleinman, *A World of Hope, a World of Fear*.
64. Olsson, *Agrarian Crossings*; Borgwardt, *A New Deal for the World*.
65. Immerwahr, *Thinking Small*; Scott, James C.: *Seeing like a State How Certain Schemes to Improve the Human Condition Have Failed*, New Haven: Yale University Press, 1998.
66. Cox and Homeyer, 'George W. Snedecor': pp. 3–4.
67. Anderson, 'Gertrude Mary Cox', p. 123.
68. Billiard, 'The Impact of the United Nations'.

69. Ibid.
70. Cox, Gertrude M.: 'Statistical Frontiers', *Journal of the American Statistical Association*, 52 (1957), pp. 1–12, p. 3.
71. Hunter, 'Cox in Egypt', pp. 54–56.
72. Anderson, 'Gertrude Mary Cox'.
73. 'Statistical Development in Certain Countries and Possible Remedial Measures', *Statistical Commission, United Nations Economic Council* (9 March 1945).
74. Mazower, *No Enchanted Palace*, p. 17.
75. Cox, Gertrude M.: 'Opportunities for Teaching and Research', *Journal of the American Statistical Association*, No. 40 (1945), pp. 71–74.
76. Sarhan, Ammar E.: 'Teaching of Statistics in Egypt', *The American Statistician,* 11.1 (1957), pp. 15–17; 'Statistical Development in Certain Countries'; 'An International Programme for Education and Training in Statistics', Statistical Commission, United Nations Economic Council (25 April 1949).
77. Cox, Gertrude M.: 'The Organization and Functions of the Institute of Statistics of the University of North Carolina', *Journal of the Royal Statistical Society. Series B (Methodological*), 12.1 (1950), pp. 1–18. The reference to American scientific hegemony is taken from Krige, John: *American Hegemony and the Postwar Reconstruction of Science in Europe*, Cambridge, Mass: MIT Press, 2008.
78. 'An International Programme for Education and Training in Statistics', p. 10.
79. Ibid., pp. 4–5.
80. Ibid., p. 14.
81. Abbate, Janet: *Women's Changing Participation in Computing*, Cambridge, MA: MIT Press, 2012; Slaton, Amy E.: *Race, Rigor and Selectivity in U.S. Engineering*, Cambridge, MA: Harvard University Press, 2010; Ensmenger, Nathan: *The Computer Boys Take Over*, Cambridge, MA: MIT Press, 2012; Marie Hicks: *Programmed Inequality*, Cambridge, MA: MIT Press, 2016.
82. Ross, Dorothy: *The Origins of American Social Science*, Cambridge: Cambridge University Press, 1996.
83. Rossiter, *Women Scientists*; Hunter, 'Cox in Egypt'.
84. Chandra Mohanty: *Feminism Without Borders*, Durham: Duke University Press, 2003, p. 55.

85. Mohanty, *Feminism Without Borders*, p. 59; Anderson, Warwick: *Colonial Pathologies*, Durham: Duke University Press, 2006.
86. Britton, Dana M.: 'The Epistemology of the Gendered Organization,' *Gender & Society*, 14 (2000), pp. 418–434 and pp. 428–429.

10

The Bona Fide Contracts: An Engineering Company in Wartime Shanghai, 1937–1945

Carles Brasó Broggi

Introduction

The industrialization of Shanghai during the first half of the twentieth century was led by Chinese private companies that imported foreign technologies. Foreign companies and foreign goods had privileges in China that were granted under the so-called treaty port system, in force from the opium wars until the clash of the Pacific War: low import taxes, extraterritoriality and a wide range of special concessions that were granted to foreign interests in territories such as Shanghai, by far the most important treaty port.[1] Foreign engineers and engineering companies played a key role in transferring industrial machinery to Chinese merchants. Chinese industrialists had often amassed experience working in foreign companies before building their own industrial firms (they were called *compradores*).[2] For instance, a community of Ningbo *compradores* living in Shanghai created innovative companies in steamships, textiles, matches, concrete, flour, tobacco, rubber, finance and insurance.[3] Even

C. Brasó Broggi (✉)
Universitat Oberta de Catalunya, Barcelona, Spain
e-mail: cbraso@uoc.edu

© The Author(s) 2018

D. Pretel, L. Camprubí (eds.), *Technology and Globalisation*, Palgrave Studies in Economic History, https://doi.org/10.1007/978-3-319-75450-5_10

though some of these companies appeared before 1911, it was in the three decades that followed when the city of Shanghai and its surroundings experienced a process of industrialization that had no parallel in the history of China.[4] The business networks between Chinese industrialists and foreign engineers became a safety net when Japan started its full-fledged occupation of China in 1937.

The full-scale invasion of China had Shanghai as one of the most important battlefronts. It was the world's sixth most populated city and the most important industrial center of China (and one of the main industrial places in Asia).[5] Politically, the city was divided into a complex patchwork of administrations: the Chinese old city, the International Concession, the French Concession and the outskirt districts, administered by the Chinese local governments. The battle of Shanghai was one of the longest and deadliest of the entire Japanese occupation, with more than 250,000 direct deaths and the destruction of the entire industrial zones of Zhabei, Baoshan and other districts.[6] After the battle of Shanghai, the Japanese attacked Nanjing, the capital of China, causing one of the worst massacres of the Second World War and forcing the Nationalist government, led by General Chiang Kai-shek, to retreat to the interior of China, to the city of Chongqing, the new capital until 1945.[7] Meanwhile, between 1937 and 1941, the French Concession and the International Concession of Shanghai (except the territories north of the Suzhou river) were kept as neutral zones and became known as "isolated islands" (*gudao*). Millions of refugees fled to the concessions and around 150 factories were transported from the outskirts of the city to the neutral zones. At the end of 1937, it was estimated that around 400 industrial plants had been destroyed in Shanghai while 570 remained: among them around 200 reached some kind of deal with the Japanese occupiers (they were occupied, undersold or merged with other Japanese concerns).[8] However, other companies sought a transnational strategy to survive inside the protection of the neutral zones.

This chapter aims to analyze the details of this transnational strategy through the collaboration agreements reached between a transnational engineering company called China Engineers Ltd and several Chinese textile companies during the wartime period. The founder of China Engineers was a Eurasian (with a British mother and a Chinese father) of

British nationality named William Charles Gomersall (1895–1960). Eurasians were often hidden in the racist society of China's treaty ports, so Gomersall was educated with his mother's family in Britain, where he studied electrical engineering at the Polytechnic of Regent Street and Manchester College of Technology while working in an apprenticeship at Lancashire Dynamo and Motor Co. at Trafford Park. However, he soon returned to China, where he worked in the engineering department of the British firm Jardine Matheson & Co. He then created his own company called China Engineers in 1928. He also joined the Institution of Electrical Engineers in 1915 as a student; he was accepted as an associate member in 1922 and as a full member in 1940.[9]

The shareholder structure of China Engineers was diverse, comprising British citizens and engineers, *compradores* from Ningbo and merchants from different countries (Parsees from South Asia and Sephardic Jews from the Middle East).[10] China Engineers was thus a transnational company that does not fit into the national categories that often define the playground of China's economic history or the treaty port system, but instead into the transnational and cosmopolitan environment of Shanghai.[11] The company was created by three founding partners, William Charles Gomersall, the British engineer Eric Shaftesbury Elliston and the Chinese tycoon Philip Z.T. Lee. Lee came from a wealthy family from Ningbo and was educated in France and England. He returned to China in 1912, just after the fall of the Qing Dynasty and the rise of Sun Yatsen's Republican nationalist government. He came with two airplanes he had bought from Austrian manufacturer Etrich Monoplanes. He was then asked to create the first aviation department in the history of China.[12] Sun Yatsen enforced the idea to develop industry to save the country (*shiye qiuguo*). According to the Chinese leader, there were four elements that contributed to the development of a country: labor, capital, managerial talent and markets. China had plenty of labor and markets but lacked talent and capital.[13] He then identified capital with machinery, when he argued that "capital is machinery and machinery, capital" and gave facilities for all industrial undertakings.[14] Even though his rule in China was short, his ideas were influential and China experienced a golden age of industrial development during the 1920s and 1930s.

China Engineers was established with the three founding members making up the board of directors (with Gomersall as director for life) and with their powers entrusted to a managing director. However, as the capital of the company increased, the board of directors was expanded (with a limitation of five members) but Gomersall kept control of the company until he died in 1960, except between 1942 and 1945 when he was imprisoned by the Japanese.[15] The company was thus clearly led by one person, but it integrated a varied range of shareholders from different nationalities. From its foundation in 1928, the company was registered in Hong Kong but its main business developed locally in Shanghai. China Engineers provided all kinds of textile machinery, components and engineering services to Chinese firms.[16] However, the Japanese occupation of China and the beginning of the Second World War brought the entire machinery trade to a standstill, forcing the company to diversify its business.

In a previous publication I have analyzed the transnational networks of China Engineers from the perspective of Chinese textile industrialists.[17] For them, adopting the transnational strategy was taking a stance against the political directives implemented by Sun Yatsen's successor, Chiang Kai-shek, which called for Chinese capitalists to move their factories to the interior, where the Nationalist government was in power between 1938 and 1945. The disenchantment between the Shanghai capitalist class and the nationalist mainstream after Sun Yatsen (who died in 1925) has been discussed by many scholars, who have focused on major capitalists such the Liu, Rong and Guo families.[18] Previous research has also noticed the sudden change of nationality of some Chinese firms in 1937, with flags of different countries being hung on the mills in Shanghai and other cities of the Yangzi Delta.[19] However, if sometimes this phenomenon has been treated as anecdotal, this chapter argues that there was a powerful logic of economic globalization behind this reaction.

This chapter aims to focus on the business strategy of China Engineers Ltd facing the collapse of machinery imports and the destruction of factories. The chapter analyzes how China Engineers diversified its operations from its original core business of machinery import and engineering services to equity investment, support in the internationalization of other firms, trading services and auditing. The chapter uses the archives of

China Engineers and a publication of the same firm, the *Quarterly Review of the China Engineers Ltd.*, that is scattered in several archives and libraries of China. It also relies on primary sources from its customers, the Chinese firms, and some press articles. The objective of the chapter is to understand how the company transitioned from a transnational business character in the pre-war period into a true multinational after 1945.

Engineers as Investors

One of the main customers of China Engineers was a textile industrial group called Dafeng. This was a group of three companies controlled by the same body of shareholders, mainly merchants from Ningbo: the group owned two spinning and weaving companies (Zhentai and Baoxing) and one weaving and dyeing company (Dafeng), all in the surroundings of Shanghai (Caojiadu and Baoshan districts).[20] On the eve of the full-scale Japanese occupation of China, these companies represented 1.5% of Chinese-owned cotton spindles and 5% of Chinese-owned cotton looms. It was one of the few Chinese industrial groups that could spin, weave and dye at the same time, having acquired from China Engineers a complete set of modern dyeing technology that had been imported from the United Kingdom in the late 1920s.[21] Some middle-sized groups such as Dafeng had managed to find a niche in the competitive market of textiles in Shanghai by upgrading their production with new technological devices such as printing and dyeing equipment.

In the summer of 1937, the war between the Japanese and the Nationalist troops intensified in the northern part of the Suzhou River, where Dafeng, Zhentai and Baoxing stood. The three factories were situated outside the safety zones and stopped all operations in August.[22] Total destruction could be expected from the Chinese assets that stood out of the neutrality of the concessions, so China Engineers and the board of directors of these Chinese companies decided on a merger to save the factories. In early September, a new British company was registered in Hong Kong that would buy one third of Dafeng. Meanwhile, another company also registered in Hong Kong as a British concern took over the entire assets of Baoxing and Zhentai.[23] The decision was made by the board of directors, but without convening a shareholder meeting as the situation was totally exceptional.[24] China Engineers participated in the merger and became shareholder of the new firms, while William Charles Gomersall was appointed to the board of directors. Before that, the companies had only been managed by Chinese personnel and the shareholders were all Chinese.

In October 1937, British flags hung on the mills, while some of the most expensive machinery sets of Zhentai and Dafeng crossed the Suzhou Creek to seek safety in the International Concession. At that time, the International Concession was crowded by the influx of refugees and other textile companies that were seeking shelter.[25] The transport of the machineries from Baoxing, in the Baoshan district, was hazardous as it was located further away in the north, on the battlefield. The factory was destroyed by a fire but some machines were saved and taken to the International Concession.[26] Luckily most of the stocks of yarns and cloths were hidden in warehouses inside the neutral zone by a loyal group of Chinese workers and managers. Meanwhile, British engineers stayed at the mills of Zhentai and Dafeng, close to the battlefield, to defend the remaining assets.[27]

In November, the Japanese firm Toyoda, backed by the Japanese army, was ready to enter the two mills. The British delegates from China Engineers that stayed in the factory argued that it was a British firm, but the Japanese replied that the merger was only a cover and that no arrangements or sale deeds were acceptable after the start of the conflict. The British engineers answered that the mills and especially the machinery had become British by law, according to the agreements that were signed by the two parts in a peaceful and bona fide manner.[28] According to Gomersall, "If the British authorities were satisfied as to the bona fides of the case, that was enough and their decision could not be questioned."[29] The Japanese authorities took months to investigate the case and in particular the bona fide character of the contracts, which were, according to the press, the most important point to be discussed.[30] Finally, the British engineers withdrew, the British flag was taken down and the two mills were occupied by Toyoda and the Japanese army.[31] The victims gave publicity to the fact that the company was being illegally occupied by the Japanese and Gomersall sent letters to the British and Japanese consuls emphasizing the infringement of the transnational character of the firm, defining it in general terms as "Chinese and foreign interests," a character that was essential to the neutrality of the International Concession and to their business strategy as well.[32]

Meanwhile, prices of textile products rose dramatically. A fourfold price increase in dyed fabrics and other stock sold by Dafeng inside the International Concession picked up profits and piqued the interest of hundreds of entrepreneurs, who established dyeing factories and workshops inside the neutral zones. At the end of 1938, more than 400 companies (mostly workshops but also small factories) produced dyed fabrics in Shanghai by using secondhand machinery and all the capital that had been saved from destruction. It was perhaps the first economic sector that forged ahead to pre-war production levels.[33] China Engineers and the Chinese shareholders of the group decided to build a new factory inside the International Concession with the machines that had been saved (20,000 spindles and 200 looms), and started producing in February 1939.[34]

In May 1939, after long negotiations between the British Consulate and the Japanese authorities, the two original spinning and dyeing factories that had been occupied by the Japanese were transferred back to their owners.[35] An important part of the machinery, motors and boilers, was missing but 20,000 spindles and 640 looms of Zhentai were intact. To avoid the risk of future occupations, another merger was planned and a new company called China Cotton Mills Ltd was created under the Hong Kong ordinances.[36] In practice, the two former spinning and weaving companies (Zhentai and Baoxing) were liquidated and administered directly by the board of the new transnational group called China Cotton Mills. Meanwhile, Dafeng, the original company founded by the Ningbo merchants in 1912, kept control of the dyeing business with no interference from China Engineers.

The board of China Cotton Mills met in December 1939 and declared a year's profit of 100,000 yuan. The next year was even better and the profits multiplied by four, which enabled it to pay to the shareholders dividends of 25% of the paid up capital.[37] The production was surprisingly high: between 1937 and 1940 the dyeing production of Dafeng Co. averaged 500,000 pieces per year, a half of the pre-war normal quantity; meanwhile, China Cotton Mills employed around 3000 workers.[38] It was a period of optimism despite the fact that the neutral concessions of Shanghai were besieged by a situation of total war. It also proved that the transnational strategy had been successful.

In September 1941, China Cotton Mills took a step further into the internationalization of its capital. The partners envisaged a way to increase liquidity by issuing shares for public purchase.[39] Chinese and foreign merchants, Indian traders, refugees from Europe and US businessmen were trapped in the concessions with their savings but with few possibilities to invest. The insecurity of the banking system and hyperinflation impelled merchants to invest in shares, especially if companies were running profits. Therefore, the company's issuance of shares was a total success and a record was established by the heavy over-subscription of China Cotton Mills shares.[40] Even though the majority of new shares were still in Chinese hands, the rest of the capital was distributed among investors from more than ten different countries, and from the two sides of the war.[41] In addition, China Cotton Mills received a loan from the HSBC bank to finish the new factory and the purchase of new spindles when the machinery trade reopened briefly before the beginning of the Pacific War.[42]

However, on December 7, 1941, after the Pearl Harbor incident, Japanese troops crossed the border of the Suzhou Creek and took control of the city's administration. The neutrality of the International Concession ended abruptly. On January 9, Japanese supervisors arrived at the offices of China Cotton Mills and made an inventory of the machinery, stocks and accounts. Over the following days, the three factories were sealed and all production had to stop.[43] The company claimed that its stocks and capital could not be transferred or alienated, as they were mortgaged as guarantee for a loan they had received from the HSBC bank.[44] But now the situation was very different from 1937: there were no neutral zones left and all the companies registered in Hong Kong were considered as enemy property by the Japanese authorities.[45]

Engineers as Transnational Traders

Between 1936 and 1937 Chinese textile companies placed huge amounts of machinery purchase orders for cotton spinning and weaving with England and the United States. It seems that Chinese industrialists were optimistic, despite the continuous threats of Japan and the late impact of the world economic crisis that affected China in 1934–1935.[46] When the war between China and Japan started in July 1937, most of these orders were still pending, on transit or waiting customs. China Engineers complained that the problems of their customers were "magnified" by the quantity of cargo that was affected and by the non-collaboration of the international shipping, insurance and wharf companies that dumped cargoes in the nearest port available, normally Singapore or Hong Kong, and declined responsibility, claiming "war causes," covered by the bill of lading conditions.[47] From the other side, the Chinese customers were not willing to pay surcharges on transport fees that could reach 25% of the original value, and China Engineers had to face a big loss to complete the business. To avoid further complications, the company decided to enter progressively into trade, shipping and insurance.

Lixin was a company of merchants from Jiangsu and another important customer of China Engineers. During the 1920s and 1930s, the company built vertically integrated cotton and woolen mills in the city of Wuxi. Most of the machinery was bought from China Engineers but the last order of 20,000 spindles was not fully paid for by July 1937. The machines arrived in twelve different ships in the second half of 1937, and most of the cargo was unloaded in Shanghai with delays and extra charges. Meanwhile, the plant in Wuxi was occupied by Japanese forces, even though China Engineers claimed British interests, something that was granted by the British Consulate, and British flags flew in the factories of Lixin. In contrast to Dafeng, Lixin had no stocks in a safe place to sell, and therefore was not able to pay back the machinery orders as stated in the contracts. The Chinese company was thus facing bankruptcy, which would surely affect the balance of China Engineers. Gomersall and China Engineers claimed to be "in complete sympathy" with the Chinese dealers who were facing the occupation of their factories; meanwhile, they accused the shipping, insurance and wharf companies of attempting to "make a profit of our losses."[48] However, this situation did not stop China Engineers from taking action against Lixin in the Chinese court and claiming the machinery and the installations of the plant in Wuxi, when they were already occupied by Japan.

William Charles Gomersall visited the factories of Wuxi in December 1937 as a neutral citizen who was examining his proprieties. According to him, all other factories in Wuxi, one of the biggest industrial centers of China, were systematically burnt except for Lixin, where the British flag was still raised. However, the machines inside had been purposely destroyed by the enemy, impeding any attempt to resume production, as the shareholders of Lixin, by allying with China Engineers, had refused to collaborate with Japan. From that point, no further visit was allowed, and China Engineers claimed some kind of compensation from the Japanese authorities if the mills were not going to be given back.[49] The Japanese authorities answered that all further damage would be avoided and that they would act to stop any other attempts to rob or damage the properties of China Engineers, accusing the Japanese company that was managing the installations. By this announcement the Japanese authorities lent credibility to China Engineers' statement regarding the ownership of the textile machinery. The Japanese military garrison stayed in Lixin for four more years and the machinery was not repaired until the end of the war.[50]

The relationship between China Engineers and Lixin did not end with this loss, as the Chinese part kept their side of the bargain by paying back the debt. As in Dafeng's case, but without merging with China Engineers, the board of Lixin embarked upon the formation of another textile company in the International Concession of Shanghai with the remaining machinery (that which had arrived in several shipments in 1937) and a space that was leased by China Engineers. The new company, Changxing Spinning and Weaving Company, was a subsidiary of Lixin and was created in 1940. This new company put to work three small factories in the International and French concessions of Shanghai.[51] China Engineers expressed their admiration for the board of directors of Lixin, especially for its founder, Tang Xiangting, "who might well be inclined to sit down in Shanghai, and adopt a life of ease on the interest of whatever capital was still remaining."[52] Thanks to this new company, China Engineers and Lixin had their debts settled.

However, to receive revenue, Changxing had to produce and sell their production abroad, as the market in raw materials was controlled by Japan and by foreign firms that had access to foreign markets, while communications between Shanghai and the interior of China were affected by war and transport restrictions. In October 1939, China Engineers complained that all Chinese ports were inaccessible owing to the Japanese blockades so Chinese companies could only turn to foreign markets to survive.[53] Even though there was no mention of it, smugglers were also an important part of the business. In other words, if China Engineers wanted the machinery to be paid for, it had to ensure that Changxing had access to raw materials and that they could also find a market for their finished products. Therefore, William Charles Gomersall started to investigate the markets for raw materials in different Asian markets (Manila, Batavia, Singapore, Penang, Ipoh, Kuala-Lumpur, Bangkok and Saigon) and reached agreements with international raw cotton suppliers from the USA and Europe. In this way, China Engineers diversified its business by entering the raw cotton trade.

According to China Engineers, this strategy "fits admirably with our main activities, as the buyers are the same as those who purchase machinery from us."[54] A new company called Raw Cotton Traders soon appeared under Gomersall's leadership to solve the scarcity of raw cotton that was affecting the mills of Shanghai. It was a partnership between Gomersall, the Indian cotton trader and some textile companies and investors, such as China Cotton Mills and Shanghai Worsted Mill. The latter was owned by the Ningbo merchant Li Shuxiong, aka James H. Lee, a partner of China Engineers and younger brother of the founding partner Philip Z.T. Lee.[55] In November 1941, the Indian trader Umrigar left Shanghai to explore the Southeast Asian markets, especially India and Burma, to secure raw cotton supplies for the Chinese mills that were stuck on the "lonely island" of Shanghai.[56] The members of China Engineers were also involved in the wool top trade, a high quality wool that was exported to China in the late 1930s, to cover a demand of the woolen and worsted companies that emerged in Shanghai.[57] In January 1941, it was announced that a former shareholder and member of the board of directors, F.R. Smith, had retired to Australia, and he was replaced by George Ernest Marden (1892–1966). Marden, a British citizen of German descent, was a well-known trader and shipowner of Hong Kong who was director of the important shipping company Wheelock & Co. Ltd.[58] China Engineers became more and more involved in the trading and shipping business.

Furthermore, between 1940 and 1941 China Engineers acted on behalf of Lixin to purchase more than 4000 bales of raw cotton from international trading companies—mainly from the USA—that bought raw cotton in Brazil, Argentina and Paraguay to be transported to Shanghai on Japanese ships.[59] China Engineers paid in advance with foreign currency reserves that it kept in Hong Kong, and the Chinese textile company made a profit from the high price that textile products were sold at in Shanghai. China Engineers earned a trade commission of 3% and a margin (that oscillated between 20 and 30%) on the price of raw cotton sold to Chinese companies.[60] Those that had stocks of textile goods could earn a profit of 200% of its capital during the years between the start of the European war in September 1939 and the beginning of the Pacific War in December 1941.[61] China Engineers considered that "no better time ever presented itself to develop industry in this country."[62]

Sometimes these cargoes were lost during transport and the company had to bear the loss, as happened to a shipment of wool tops that were being transported from Australia to Italy just when the war started in Europe in September 1939. China Engineers bought the cargo when it was at the Suez Canal and shipped it to Singapore; but it was lost between Singapore and Hong Kong. A cargo of American raw cotton was also lost in Manila after the beginning of the Pacific War.[63] Another loss for China Engineers was 2000 bales of cotton yarn that were destined for Hong Kong to execute a contract for British equipment for the War Supplies Board; these were either lost or never paid for.[64] According to the company review, tracing missing and diverted cargoes became one of the core activities of China Engineers. It was a risky business but it was better than accepting the surcharges of the transport companies, as happened in 1937.[65]

The dangers of transport and the high price of war insurances drove China Engineers to get involved in the insurance business as well, although "at first sight it may appear that insurance business and engineering business are far apart." However, according to William Charles Gomersall it was also a "logical step in the development of engineering business."[66] China Engineers was not only financing the machinery and raw cotton purchases of the Chinese textile industrialists, but was also taking care of the transport and insurance, reaching an agreement with the Union Insurance Society in 1940. Some shareholders of China Engineers, such as Ningbo merchant Li Shuxiong (James H. Lee), had already been involved in the insurance business—in his case as chair of China's Insurance Association between 1928 and 1935.[67] So it was another logical step in the transition of China Engineers from offering services in Shanghai to the internationalization of its operations.

China Engineers diversified its business from the original core business of machinery import to trade and insurance. In January 1941, after years of paralysis, the machinery trade experienced a short comeback, and some machinery contracts were signed by China Engineers in Shanghai and Hong Kong to provide machinery to Shanghai and to the zone that was dominated by the Nationalist government in the province of Sichuan. The company, after selling 6000 spindles that would be transported from the Vietnamese border to Xi'an by road—a "herculean work" indeed—stated that the Chongqing government "pursues its fixed policy of resistance, which extraordinary resolution and determination."[68] However, in December 1941, all the assets of China Engineers were seized by the Japanese and William Charles Gomersall was interned in an imprisonment camp for more than two years.[69] The clash of the Pacific War also meant the end of the Treaty Port System and the world of the concessions, which were abolished by the Japanese occupation and collaborationist regimes and also by the government of Chiang Kai-shek in Chongqing.

Engineers as Auditors

The role of China Engineers as auditors began with the mergers of Dafeng, Zhentai and Baoxing in September 1937, when all the assets were calculated in detail.[70] These companies had been appraising their profits and losses using the traditional "four columns system" (*sizhufa*), which was the standard method for Chinese companies. This system accounted for income, fixed costs and costs of production but did not exactly take into account the capital depreciation, as the system was conceived for workshops without major investments in machinery.[71] However, when China Cotton Mills merged Zhentai and Baoxing in 1939, the double-entry system was introduced.[72] China Engineers knew the purchasing price of every textile machine that had been saved from destruction and the merger was calculated according to the market prices and a year on year depreciation of 5%.[73] Having expertise on machinery valuations, China Engineers offered this auditing service to other companies, even to the bigger industrial groups of China, and finally to the Nationalist government as well.

China Engineers had also signed contracts with China's biggest textile groups, Yong'an (also known as Wing On) and Shenxin. In 1934, China Engineers signed a contract to provide finishing machines to Yong'an, but this did not end well. Yong'an complained about the chemical and acid pumps used in the mercerization production chain and an engineer was sent from the original manufacturer in England, Sir James Farmer & Sons. After three months, Yong'an wanted the engineer to stay longer, and there was a discussion about who should bear the expense of this. Yong'an wanted to pay the manufacturer direct, skipping the mediation of China Engineers, which broke the confidence and trust that had been established between these companies. As a result, China Engineers stopped representing Yong'an in China and the business with them came to an end in 1935.[74]

In contrast, the relationship with Shenxin worked better. According to China Engineers, the founder of Shenxin, Rong Zongjing (1873–1938), one of the wealthiest and most famous tycoons in the history of Republican China, stood "without rival as the most important figure of the Chinese industry."[75] Regarding contracts, he did not need to look at a Chinese translation to sign an agreement, as he was confident that China Engineers and William Charles Gomersall would not cheat him, even though the contract was for £60,000 of machinery, more than 20,000 spindles: this was placed in 1936.[76] Shenxin controlled around 20% of all the Chinese spindles that were working in the mid-1930s, competing directly with the most important Japanese industrial groups.[77] Years later, at a public conference, Gomersall showed much respect for Rong Zongjing, and used this anecdote to explain that there was a kind of trust that could only be found at that time in China when making machinery deals with Chinese merchants.[78]

The purchase of machinery from Shenxin by China Engineers was extended in 1937, with the purchase of almost 50,000 spindles. These had to be paid for in installments according to the following terms: after the signature of the contract (10%), on the arrival of half of the machinery (another 10%) and then in twelve equal installments during the year.[79] China Engineers had better payment terms than their competitors. By taking the financial burden of the machinery purchase, they gained trust among Chinese merchants and strong links were consolidated between them. However, the practice of supplying machinery on credit was extended when the war broke out, causing a high degree of stress and worries in the board of the engineering company. The war also caused major losses for Shenxin, which was highly indebted, while Rong Zongjing died in Hong Kong in February 1938, raising tensions between the family and the different mills. The business empire was dismantled and divided between his relatives and other beneficiaries, while the factories were scattered to Shanghai, Hong Kong, Wuxi and Chongqing.

Shenxin was divided into different groups that sought to reach an agreement regarding the previous assets (nine cotton mills, though some had been destroyed or occupied by Japan) and liabilities (the debt of Shenxin and the flour group Fuxin reached 80 million yuan in 1936).[80] Therefore, when in 1939 a group of shareholders of Shenxin founded a new company in Shanghai called Hefeng, they needed a neutral audit to register its capital and break it off from the original group. In October 1940, China Engineers sent the results of the valuation of the mill to Rong Zongjing's son, Rong Hongyuan (who was still the nominal general director of Shenxin) as if it was to be sold as a going concern. The company raised its capital to 750,000 yuan according to the machinery valuation and considering the high rates of inflation. According to China Engineers the machinery was particularly difficult to evaluate, as replacement costs were "abnormally high."[81] This was just one of the deals that China Engineers had with Shenxin.

After December 1941, the Chinese cotton sector almost disappeared and only some textile mills survived in Chongqing, where the Nationalist government resisted the occupation. However, the conditions in Chongqing differed from Shanghai, as the Nationalist government aimed to control the cotton sector with state-owned enterprises and the establishment of controls over pricing and distribution.[82] Because hyperinflation was already a serious problem, the government established several price controls after 1939, and especially after 1942, with the Agricultural

Credit Administration and the attempt to establish price ceilings for raw cotton and yarn, which provoked a reaction from the industrial owners who had moved to Chongqing, such as the other part of the Rong family, who were reluctant to give up their margins.[83]

After the war, the Nationalist government took over the assets that were left abandoned by the Japanese. Some of the factories had been owned by Japan from the beginning but others had been occupied or handed over by their original Chinese proprietors. In 1947, the Kuomintang government asked for a valuation of this enormous industrial capital, which reached 1.7 million spindles, meaning around 40% of all existing textile industrial capital in China.[84] China Engineers, having acquired experience as neutral auditors during the war for companies such as Shenxin, made valuations in some factories that had been taken over by the Nationalist government. These valuations took place in 1947, and China Engineers showed how the industrial equipment of the former Japanese factories was far more advanced than that of the Chinese textile mills.[85] They established a benchmark cost of a 10,000 spindle factory with Japanese, American and British machines using prices of 1937, 1941 and 1947 (replacement prices of 1947 being three times those of 1937).[86] Using this valuation, the nationalist government built the China Textile Reconstruction Company, a state-owned enterprise that would control the business of cotton textiles in post-war China.

The Japanese mills that were transformed into state-owned enterprises had not moved their machinery during the war, and their condition was better than the Chinese mills, with their machinery that had been manufactured before 1937. The machinery trade resumed in 1945, but the civil war between Nationalists and Communists in 1946, hyperinflation and trade barriers drove the machinery trade to focus on the Hong Kong market, where China Engineers played a leading role.[87] China Engineers recovered most of the exclusive agency contracts that had been signed with Western (mainly British) manufacturers, but most of the deliveries of the postwar period would take place in Hong Kong instead of Shanghai. The company increased its agency contracts covering other Asian countries as the China market became more difficult during the late 1940s. In 1941, China Engineers acted as sole agent for fourteen British manufacturers, while in 1953 the company was representing twelve Western power and electrical energy firms, twenty-six textile manufacturers (such as, again, Sir James Farmer & Sons) and twenty-five general manufacturers (ranging from machinery tools, motors and stainless steel to railway materials and bicycles.).[88] That year the main project was in Malaya.[89] By then, China Engineers had become a multinational.

Conclusions

When China Engineers protested against the occupation of the factories from the concessions in 1937, the Japanese embassy in Shanghai answered that the bona fide character of the deals would be "the principal element to be studied."[90] The idea behind this discussion was that these agreements were not a collaborationist action nor a Nationalist stance, but a neutral agreement between Chinese and foreign interests. In Chinese eyes, this strategy that emphasized the transnational character of industrial companies and mills that had been registered as Chinese could only be considered neutral in the context of China's war with Japan and the following collapse of the treaty port system. These bona fide agreements thus symbolize a way in which engineers and Chinese companies embraced a strategy of globalization, a logic where engineers acted as agents and perpetrators with the collaboration of Chinese companies.

China Engineers was criticized by both British and Chinese media for taking advantage of a difficult situation to reap profits. Its transnational character did not fit into the traditional national narratives. In the British community, William Charles Gomersall was seen as a kind of speculator and someone who was more attached to the world of Chinese merchants than to British interests.[91] Other British citizens had made agreements with Chinese companies to create a British façade and to avoid Japanese occupation, but these were perceived as short-term deals. However, China Engineers kept a long-term business relationship with its Chinese partners, helping them to move to Hong Kong in the late 1940s. In his obituary, it was said that Gomersall "had the rare facility of making personal and business friendship with Chinese industrialists."[92] China Engineers would be one of the last British companies to leave Communist China, as late as 1956. The founder of China Engineers strongly supported the importance of maintaining diplomatic ties between China and the United Kingdom for business reasons. In this context, China Engineers was involved in another merger with British firms that were willing to abandon China.[93] In 1951, a book published by the Marxist economist Wei Zichu (Wei Tsu-chu) about the British investment in China depicted China Engineers as an example of a company that specialized in "opportunistic purchasing" and aimed at capital concentration, following the historical rules of Western monopolistic and imperialistic capitalism.[94]

The strategy of China Engineers from 1937 onwards could well be judged as opportunistic and speculative, in the sense that the business prioritized profits over politics or any national engagement. However, there was a strong political connotation in this strategy that had a major impact in the future distribution of the Chinese industrial base from the Yangzi Delta in the pre-war period to Hong Kong, which was soon to catch up during the Cold War. Besides, China Engineers' modus operandi can also be associated with a certain tradition of Chinese business history, the one that defines "merchants without empire" (*shangren wuzuguo*) as the capacity of certain traders to succeed in their economic activities despite a situation that is not favorable at first sight, such as the situation that faced China Engineers and its partners during the war against Japan or the Hokkien maritime traders in the coast of Fujian during the prohibition of private maritime trade in the Ming dynasty.[95]

The experience of war also demonstrated that industrial companies and mills could be divided into different parts: the mill, the machinery, the stocks of raw materials and finished goods, capital shares and more. It also meant that these parts could be alienated, sold or occupied by another actor—sometimes of another nationality—in a very short period of time. Dafeng, Zhentai and Baoxing were absorbed by a company from Hong Kong, but shortly afterwards the mills were occupied by the Japanese. Between September 1937 and April 1938, China Engineers and the Chinese shareholders became used to running a company without a mill and managing the changing of nationality several times. This state of affairs, extraordinary in times of peace, became common after 1937 and did not suppose the bankruptcy or the end of the firm. On the contrary, China Engineers and the Chinese companies who partnered with them earned high profits between 1937 and 1941. It is therefore understandable that they tried to continue with this strategy when the war was over.

Notes

1. See the classic study of the formation of the treaty port system by Fairbank, John King: *Trade and diplomacy on the China coast, 1842–1854,* Cambridge: Harvard University Press, 1953; see also an alternative recent vision in Brasó Broggi, Carles and Martínez-Robles, David, 'Beyond colonial dichotomies: the deficits of Spain and the peripheral powers in Treaty Port China', *Modern Asian Studies* (forthcoming).
2. On *compradores* see Ma, Xueqiang and Zhang, Xiuli: *Churuyu zhongxi zhijian: Jindai shanghai maiban shehui shenghuo*, Shanghai: Shanghai cishu chubanshe, 2009; and Hao, Yen-p'ing: *The Commercial Revolution in Nineteenth-Century China: The Rise of Sino-Western Mercantile Capitalism*, Berkeley: University of California Press, 1986.
3. See a compilation of biographies of Ningbo merchants in Shanghai and their economic contributions in Lu, Pingyi, ed., *Chuangye shanghaitan*, Shanghai: Shanghai kexue jishu wenxian chubanshe, 2003. For a general overview on Ningbo merchants in Shanghai, see Li, Jian: *Shanghai de Ningboren*, Shanghai: Shanghai renmin chubanshe, 2000. See a relevant case study in Cochran, Sherman and Hsieh, Andrew: *The Lius of Shanghai*, Cambridge: Harvard University Press, 2013.
4. See Bergère, Marie-Claire: *L'âge d'or de la bourgeoisie chinoise, 1911–1937*, Paris: Flammarion, 1986.
5. See Ma, Debin,: 'Economic Growth in the Lower Yangzi Region of China in 1911–1937: A Quantitative and Historical Analysis', *The Journal of Economic History*, 68 (2), (2008), pp. 359–60.
6. Bergère, Marie-Claire: *Historire de Shanghai*, Paris: Fayard, 2002, pp. 302–05.
7. A recent history of China's war with Japan can be found in Mitter, Rana: *China's war with Japan. The struggle for survival*, London: Allen Lane, 2013.
8. Huang, Hanmin and Xu, Xinw: *Shanghai jindai gongyehua*, Shanghai: Shanghai shehui kexueyuan, 1998, p. 224; see also Henriot, Christian: 'Shanghai industries under Japanese occupation. Bombs, boom, and bust (1937–1945)', in Henriot, Christian (ed.): *In the Shadow of the Rising Sun: Shanghai under Japanese Occupation*, Cambridge: Cambridge University Press, 2004, pp. 20–24.
9. 'William Gomersal Obituary', *Journal of the Institution of Electrical Engineers,* 6 (70), (1960), p. 605.

10. Hong Kong Public Records Office (hereafter HKPRO), 'Registrar of China Engineers', November 1928, HKPRO, 111-4-34; See also Shanghai Academy of Social Sciences, Chinese Business History Archives (hereafter CBHA) 'Articles of Association of China Engineers Ltd.', 1928–1948, 7-13-515, 1928.
11. They were transnational in the original sense, that "any movement which attempts to thwart this weaving, or to dye the fabric in any one color or disentangle the threads, of the strands is false to this cosmopolitan vision," quotation from the classic article by Bourne, Randolph: 'The Jew and Trans-National America', *Atlantic Monthly*, No. 118 (July 1916), pp. 86–97. On transnationalism in Shanghai, see Goodman, Bryna: 'Improvisations on a Semicolonial Theme, or, How to Read a Celebration of Transnational Urban Community', *The Journal of Asian Studies*, Vol. 43:4 (November 2000), pp. 889–926.
12. Lee, James H.: *A Half Century of Memories*, Hong Kong: South China Photo-Process Printing, 196?, p. 22.
13. Sun, Yatsen: *The international development of China*, New York: Putnam & Sons, 1922, p. 5.
14. According to the essay written in 1919 by Sun Yatsen 'How China's Industry Should Be Developed', compiled and translated by Wei, Julie Lee (ed.): *Sun Yatsen, 1866–1925*, Stanford: Stanford University Press, 1994, p. 237.
15. CBHA, 'Memorandum and Articles of Association of China Engineers Limited. Incorporated October 16, 1928', CBHA 07-13-515, pp. 51–52; see also 'Obituary', *China Engineers Review*, No. 80 (December 1960), p. 2.
16. CBHA, 'Articles of Association of China Engineers Ltd.', CBHA, 7-13-515, pp. 38–40.
17. Brasó Broggi, Carles: *Trade and Technology Networks in the Chinese Textile Industry. Opening Up Before the Reform*, Basingstoke, Palgrave Macmillan, 2016, Chapter 4: "War and isolation", pp. 67–81.
18. For the Liu family business, see Cochran and Hsieh: *The Lius of Shanghai;* and Chan, Kay-Yiu, *Business Expansion and Structural Change in Pre-war China: Liu Hongsheng and his Enterprises, 1920–1937*, Hong Kong: Hong Kong University Press, 2006; about the Rongs, see Coble, Parks: *Chinese Capitalists in Japan's New World*, Los Angeles: University of California Press, 2001.
19. Köll, Elisabeth: *From Cotton Mill to Business Enterprise: The Emergence of Regional Enterprises in Modern China*, Cambridge: Harvard University Press, 2003, p. 270.

20. In early 1937 Shanghai had thirty-one Chinese-owned cotton spinning mills: Seven of them had also weaving and dyeing departments; ten of the total were founded between the two years of 1921 and 1922, when Dafeng and Zhentai were registered. See a chronological evolution of textile mills in Yan, Zhongping: *Zhongguo mianfangzhiye shigao*, Beijing: Kexue Chubanshe, 1955, pp. 357–364.
21. Brasó Broggi, *Trade and Technology Networks*, pp. 55–61; see also *Jindai Shanghai gongshangye gailan,* Shanghai: Shanghai Dang'anguanbian, 1992, pp. 66–96; and Abe, Takeshi: 'The Chinese Market for Japanese Cotton Textile Goods, 1914–1930', in Sugihara, Kaoru (ed.): *Japan, China, and the Growth of the Asian International Economy, 1850–1949*, New York: Oxford University Press, 2005, p. 88.
22. About the sudden stop of the three factories, see Shanghai Municipal Archives (hereafter SMA), 'Dafeng Group (China Cotton Mills), A report on war events, 1943', SMA Q199-3-48, pp. 1–5; and SMA, 'Dafeng archive (China Cotton Mills), A report on the construction of the factory at Yanping Road, 1938', SMA Q199-20-94. See all merger documents in Dafeng Group, agreements between China Cotton Mills, Dafeng, Zhentai and Baoxing, September 1937', SMA Q199-3-173, pp. 1–35.
23. Ibid.
24. According to the explanations given in 1941, when the shareholders of Dafeng met again and they were told that it was the only way to save the factories from destruction, SMA, 'Dafeng archives, Shareholder meeting, Dafeng, October 1941', SMA Q199-3-3, p. 5.
25. Henriot, Christian: 'Shanghai Industries under Japanese Occupation.: Bombs, Boom, and Bust (1937–1945)', in Henriot, Christian: *In the Shadow*, pp. 23–30.
26. SMA, 'Dafeng Group, Board of directors Dafeng, Zhentai and Baoxing, special meeting October 1937', SMA Q199-3-82, pp. 7–14; see also SMA, 'Baoxing Co., Board of directors, December 1938', SMA, Q199-3-32, p. 10.
27. SMA, 'Dafeng archive, Shareholder meeting, November 1937', SMA Q199-3-8, pp. 123–127. See also SMA, 'Dafeng archive, A report on the construction of the factory at Yanping Road, 1938', SMA Q199–20-94, pp. 1–2.
28. *North China Herald* (20 July 1938), p. 103.
29. *The China Engineers Quarterly Review*, No. 28 (January 1939).
30. *North China Herald* (20 July 1938), p. 103.

31. SMA, 'Dafeng archive, A report on war events, 1943', SMA Q199-3-48, pp. 1–5.
32. *North China Herald* (20 July 1938), p. 103.
33. Henriot: 'Shanghai industries', pp. 22–28; see also Huang and Xu: *Shanghai jindai gongyehua*, p. 235. Production increased above pre-war production levels only inside the safety net of the concessions.
34. SMA, 'China Cotton Mills, Board of directors, February 1939', SMA Q199-3-164, p. 23. Minutes were written in English for the first time. SMA, 'Dafeng Co., A diary of the board of directors, October 1937', SMA Q199-20-90, pp. 1–5; see also a letter from William Charles Gomersall in *North China Herald* (31 May 1939), p. 381.
35. According to the British Consulate, Dafeng and Zhentai (Chun Tah Cotton Mill and the former China Dyeing Works) were China Engineers' two subsidiary companies. See 'Consul General Phillips to Sir. A. Clark Kerr, 13-8-1939' F. 10,416/62/10, in *Shanghai. Political and Economic Reports, 1842–1943. Vol. 18: 1936–1943,* London: British Government Records from the International City, Archives Edition, 2008, p. 230. See also SMA, 'Dafeng archives (China Cotton Mills), Wartime documents, 1941–1945', SMA Q199-3-189, p. 3.
36. Name in Chinese *Yingshang Zhongfang shachang gufen youxian gongsi*, English name: China Cotton Mills Ltd. The total cost for constructing the dyeing plant in the International Concession of Shanghai (Yanping Road) was calculated as 500,000 yuan. SMA, 'Dafeng Group (China Cotton Mills), Mergers and acquisitions between China Cotton Mills, Dafeng, Zhentai and Baoxing, February–March 1939', SMA Q199-3-182.
37. SMA, 'China Cotton Mills, Board of directors, June 1941', SMA, Q199-3-164, p. 14.
38. SMA, 'Dafeng Co., Shareholder meeting, October 1941", SMA, Q199-3-3, p. 4; SMA, 'China Cotton Mills, Board of directors, November 1940', SMA Q199-3-164, pp. 27–28.
39. SMA, "China Cotton Mills, Board of directors, February 1941", SMA, Q199-3-164, pp. 19–20.
40. Quoted from an article from *North China Daily News* (September 5, 1941), SMA, 'China Cotton Mills, Newspaper clippings', SMA Q199-3-189, p. 16.
41. See a full shareholder list in SMA, 'Dafeng and China Cotton Mills archives, shareholder lists, commercial brands and other documents, 1930–1946', SMA Q199-3-129, pp. 86–151.

42. SMA, 'China Cotton Mills, Board of directors, November 11, 1941', SMA Q199-3-164, pp. 27–28. The agreement of the loan (at annual interest rate of 7.5 per cent) between China Cotton Mills and the HSBC, in SMA, 'China Cotton Mills, Business documents', SMA Q199-3-129, pp. 5–6; See also *Shanghai* Times (September 5th, 1941), in SMA, 'Newsletter clippings', SMA Q199-3-189, p. 11.
43. SMA, 'China Cotton Mills, General report on the war, 1941', in Dafeng Group (China Cotton Mills), Business documents 1937–1945, SMA Q199-3-177, p. 21; see also SMA, 'China Cotton Mills, List of requisitioned goods', SMA Q199-3-177, p. 2. China Cotton Mills had made an inventory statement as per December 8, 1941 and valuated the plant and machinery (three factories) at 42,186,288 *fabi*, SMA 'China Cotton Mills archives, financial documents, 1937–1945', SMA Q199-3-177, p. 8.
44. SMA 'Dafeng Group (China Cotton Mills), Business documents 1930–1946', SMA Q199-3-129, pp. 5–6.
45. SMA 'Dafeng Group (China Cotton Mills), Board of directors, January 1942', SMA Q199-3-164, pp. 19–20.
46. Shiroyama, Tomoko: *China during the Great Depression.: Market, State and the World Economy, 1929–1937*, Cambridge: Harvard University Press, 2008, pp. 37–59.
47. 'Some aspects of the war situation', *The China Engineers Quarterly Review*, No. 25 (April 1938), p. 1–2.
48. 'Some aspects of the war situation', *The China Engineers Quarterly Review*, No. 25 (April 1938), p. 1–2.
49. *The China Engineers Quarterly Review*, No. 28, January 1939.
50. SMA, 'Lixin archives, Letter from Mr. T. H. Bower, Chief of Industrial Rehabilitation Division, U. N. R. R. A, 2-5-1946. Report of the damages of Lixin', SMA Q195-1-809, p. 1–4.
51. SMA, 'Lixin archives, Documents of Changxing with the Shanghai Municipal Council, 1938–39', SMA Q195-1-712, pp. 1–20.
52. *The China Engineers Quarterly Review*, No. 31 (October 1939), p. 2.
53. *The China Engineers Quarterly Review*, No. 31, (October 1939), p. 2.
54. *The China Engineers Quarterly Review*, No. 33 (April 1940)
55. See also 'New China Textile Company', *North China Herald* (March 20, 1940), p. 460.
56. SMA, 'Anda archives, A history of Raw Cotton Traders Ltd., 1947', SMA Q196-1-491.

57. As investors, China Engineers also participated in a worsted company called Shanghai Worsted Mill, 'Register of the company and the equity share of Shanghai Worsted Mills', Hong Kong Public Records Office, HKPRO, 111-4-54.
58. HKPRO, The China Engineers Limited, 'Allotment of shares 27-12-1940', HKRS 111-4-34; see also *China Engineers Quarterly Review*, No. 36 (January 1941), p. 4; About Marden and Wheelock Co. Ltd. See Lo, York: 'Unsung Kingmakers – the low-key Song Brothers who conquered the Shanghai Bund and Victoria Harbor and backed the development of several key industries in post-War Hong Kong', available at The Industrial History of Hong Kong Group (Last accessed August 30, 2017). URL: http://industrialhistoryhk.org/unsung-kingmakers-the-low-key-song-brothers-who-conquered-the-shanghai-bund-and-the-victoria-harbor-and-backed-the-development-of-several-key-industries-in-post-war-hong-kong-by-york-lo/.
59. SMA, 'Lixin archives, contracts and purchase orders with American company Anderson Clayton, 1940–1941 No. 1', SMA Q195-1-754; and SMA, 'Lixin archives, contracts and purchase orders with American company Anderson Clayton, 1940–1941 No. 2', SMA Q195-1-755.
60. SMA, 'Raw cotton purchase contracts', 1940–1941, in SMA Q195-1-754, p. 4; and SMA Q195-1-759.
61. *The China Engineers Quarterly Review*, No. 33 (April 1940).
62. *The China Engineers Quarterly Review*, No. 33 (April 1940); *The China Engineers Quarterly Review*, No. 75 (April 1958), p. 2.
63. SMA, 'Lixin archives, cotton contracts for Brazilian cotton purchase through China Engineers, 1941–1946', SMA Q195-1-718.
64. HKPRO, 'The accountant General, HSBC, to the War Supplies Board, 30 November 1946', HKPRO, S41-1-2305.
65. 'The China Engineers Limited in China', *Far Eastern Economic Review* (10 May 1956), p. 591.
66. *The China Engineers Quarterly Review*, No. 33 (April 1940).
67. Lee: *A Half Century of Memories*, p. 85.
68. *The China Engineers Quarterly Review*, No. 33 (April 1940), p. 4.
69. 'The China Engineers Limited in China', *Far Eastern Economic Review* (10 May 1956), p. 591.
70. SMA, 'Dafeng Archives, merger documents, 1937', SMA Q199-3-173.
71. Gardella, Robert: 'Squaring accounts: commercial bookkeeping methods and capitalist rationalism in late Qing and Republican China', *The Journal of Asian* Studies, Vol. 51, No. 2 (1992), p. 324; See a study of the

accounts of Dafeng in 1928 in Brasó Broggi, Carles: *Shanghai y la industrialización algodonera en China. El caso de la empresa Dafeng*, PhD Unpublished Thesis, UPF, 2010, pp. 325–333.

72. SMA, 'Shanghai Cooperative Credit Service: an investigation on Dafeng', SMA Q78-2-12,417; SMA, 'Dafeng archives, bookkeeping materials, 1941', SMA Q199-3-162, pp. 5–12;
73. SMA,' Dafeng archives, land contracts with our three factories, 1939', SMA, Q199-3-181.
74. SMA, Yong'an papers, SMA Q197-1-2375, p. 43.
75. *The China Engineers Quarterly Review*, No. 21 (April 1937), p. 2.
76. It was a considerable amount of money at that time. The anecdote is repeated several times; see Gomersall, William Charles: 'Reminiscences of a China hand', *The China Engineers Quarterly Review*, No. 78 (December 1959), p. 13.
77. See Coble, *Chinese capitalists,* pp. 114–16.
78. Gomersall: 'Reminiscences of a China hand', p. 13.
    See Coble, *Chinese capitalists,* pp. 114–16.
79. SMA, 'Shenxin archives, contracts and machinery purchase orders between Shenxin and China Engineers, 1937', SMA Q193-1-2754.
80. Coble: *Shanghai capitalists,* p. 55.
81. SMA, "Shenxin archives, Insurance, machinery and other contracts of Hefeng, 1941", SMA Q193–1-2758.
82. Bian, Morris L. Bian: *The Making of the State Enterprise System in Modern China: The Dynamics of Institutional Change*, Cambridge: Harvard University Press, 2005.
83. Kubo, Toru, *20 seiki chūgoku keizai shi no tankyū*, Nagano: Shinshu University, 2009, pp. 48–59.
84. Jin, Zhihuan: *Zhongguo fangzhi jianshe gongsi yanjiu (1945–1950)*, Shanghai: Fudan daxue chubanshe, 2006, p. 56.
85. SMA, "Shanghai Cotton Textiles Industry Association (Shanghai mianfangzhi gongye tongye gonghui), director of the Committee Wang Qiyu, 1946–1949", S30-2-217, microfilm archives.
86. SMA, 'Shanghai Cotton Textiles Industry Association (Shanghai mianfangzhi gongye tongye gonghui), director of the Committee Wang Qiyu, 1946–1949', S30-2-217, microfilm archives, p. 41.
87. Brasó Broggi, *Trade and technology,* pp. 119–21.
88. Data from 1940 is taken from *The China Engineers Quarterly Review,* No. 36 (January 1941), p. 4. The 14 companies were the following: David Bentley, Cook, Co, John Dixon & Sons, European wool, P&C Garnett,

George Hattersley & Sons, James Kenyon & Sons, Prince Smith & Stells, Riley Ltd., Sandoz Chemicals, James Taylor & Sons, A & H. Simonett, Tweedales & Smalley, Whitehead & Poole, Wilcock.

89. *China Engineers Quarterly Review*, No. 65 (December 1953), p. 8.
90. *North China Herald*, 20-7-1938, p. 103.
91. Howlett, Jonathan J.: 'Accelerated Transition', *European Journal of East Asian Studies*, 13 (2014), p. 184.
92. "Obituary. William Charles Gomersall", *Electrical Engineers* (October 1960), p. 605.
93. See Howlett, 'Accelerated Transition', pp. 184–87.
94. Quoted in *Far Eastern Economic Review* (June 1952), p. 729.
95. Wang, Gungwu: 'Merchants without empire: the Hokkien sojourning communities', in Tracy, James D. (ed.): *The Rise of Merchant Empires: Long-distance Trade in the Early Modern World, 1350–1750*, Cambridge: Cambridge University Press, 1990, p. 419; I would like to thank Professor Lee Pui-tak for choosing this concept that illuminated this chapter, in a workshop that took place in the Chinese University of Hong Kong in June 2013, that was called "Merchants in migration: Transnational networks, Japanese colonialism and China's Civil War, 1930s–1940s". That draft paper was the basis for this chapter.

11

Dutch Irrigation Engineers and Their (Post-) Colonial Irrigation Networks

Maurits W. Ertsen

Introduction

In 1920, at the age of twenty-four, student Paul de Gruyter graduated as civil engineer from Delft Polytechnic (nowadays Delft University of Technology) in the Netherlands. On 7 September of the same year, he was appointed as engineer in the Netherlands East Indian Department of Public Works. The fresh engineer was added to the staff of the Head of the Irrigation Department Tjimanoek in Cheribon, Western Java. In this chapter, the career steps and decisions of De Gruyter allow us to discuss three issues that explore relations between the local and the global for Dutch irrigation engineers in the twentieth century. Using De Gruyter's career and his moves around the globe as the central narrative, this chapter will discuss how the Dutch irrigation engineering network managed to emerge and continue by defining what its members considered 'good practice'.

M. W. Ertsen (✉)
Water Resources Management, Delft University of Technology,
Delft, Netherlands
e-mail: m.w.ertsen@tudelft.nl

© The Author(s) 2018

D. Pretel, L. Camprubí (eds.), *Technology and Globalisation*, Palgrave Studies in Economic History, https://doi.org/10.1007/978-3-319-75450-5_11

Three features of De Gruyter's career allow the building of a larger narrative about locally based Dutch irrigation expertise in a changing global world:

First, De Gruyter's education in Delft allowed him to enter the working hierarchy of the East Indian Department of Public Works. Unlike the first generations of East Indian engineers, De Gruyter was specifically trained and educated for working in the colony. As with most of his colleagues within the East Indian Public Works Department, De Gruyter became a member of the Association of East Indian Civil Engineers (Vereniging voor Waterstaatsingenieurs, VWI). In 1925 he became the secretary of the Sub-Association for Cheribon in the VWI. In 1926 he became the treasurer of the VWI itself. What exactly happened in the months after De Gruyter returned to Java in 1928 is not clear, but in 1929 he resigned as member of the VWI.

Second, his resignation could be related to a major disappointment in his irrigation-based career. Once in the Netherlands East Indies, De Gruyter became involved in a key debate on water distribution. However, his proposals were not readily accepted and he seems to have left the irrigation field. In 1933, De Gruyter returns in the archival record as an engineer with the Javanese State Railway Company (JSRC). In April 1934, he became the Head of the Experimental Bureau for Rolling Stock of JSRC. Although his particular move will appear as non-typical, it was possible because other engineering fields had emerged as well.

Third, a little over a year later, in August 1935, Paul de Gruyter returned to the world of water, but in The Netherlands: he was appointed as Head of the Technical Service of the Rijnland Water Board (one of the largest and oldest of such boards in the country). He remained the Chief Engineer of Rijnland until 1958, when he moved to a position with the United Nations in Syria. In 1961, he moved to Cyprus for the same organization. After 1945, colonial engineers had become—actually had been made, as I will show—international experts, which allowed De Gruyter to return to the international field.

Good practices entered Dutch engineering education, with new professional practices only partially restructuring that same education. It will become clear that after 1945, when Dutch local irrigation engineering went global, that same irrigation engineering remained strongly based on what had become undisputable practices—which were the approaches that had become the standard in East Indian practice. Even when the Dutch way of doing things was challenged in the new international arena of irrigation development—as the engineers from France and Britain had gone through a similar process of decolonization—Dutch post-1945 irrigation practice and education remained firmly rooted in East Indian practice.

Paul de Gruyter retired in 1964, when Dutch international irrigation expertise was well regarded and much sought. He passed away in The Netherlands in May 1975, at the age of seventy-nine. De Gruyter was both a typical and non-typical colonial engineer of his time. Going from graduating to colonial practice was something about one third of Delft's graduates did—De Gruyter was in good company there. Moving from irrigation to trains, however, was not very common. De Gruyter's subsequent move to a Dutch water board was not typical either. However, becoming active in the international world of engineering when most colonies had become independent nations again was something many Dutch engineers did. As such, De Gruyter not only represents his own choices, but also those of his many colleagues that shaped Dutch colonial irrigation practices. Without claiming that De Gruyter was one of the major players in East India—although he was very active—he was clearly one of the individuals who built Dutch colonial and international engineering.

Education and Institutions

Almost a century before Paul de Gruyter came to East India, colonial irrigation engineering made its first moves. The first colonial irrigation efforts—which were actually by Dutch civil servants, as engineers were not available yet—emerged in about 1830, focusing on regulating water distribution for sugar cane cultivation on Java. Sugar cane was the most important irrigated commercial crop by far; it was cultivated on Javanese fields by European factories. When in the 1840s and 1850s several famines occurred on Java, the colonial state included irrigation for rice farming in its policies as well. As a result, the colonial irrigation discourse has been dominated by the relation between sugar cane versus rice—a topic that will return below.[1]

The establishment of a Bureau of Public Works in 1854 was a political recognition of the potential role of engineers and technical support in colonial irrigation development. The engineers remained subordinate to the Civil Service, however. In 1885, the Bureau of Public Works became independent from the general Civil Service. The new Department of Public Works became the centre of irrigation activities. In 1890, the General Irrigation Plan for Java defined nineteen irrigation projects to be developed; a few other projects were included in 1907.

Although the East Indian context confronted the engineers with severe drainage and flooding problems, the East Indies demanded a struggle for water quite different from the mother country. In this respect the engineers had to start pretty much from scratch. The East Indies were strange, in terms of natural aspects, geography, distances, population and so on.[2] As engineers lacked knowledge of the East Indian situation, they constructed one of their first canals with locks, as if the canal were in The Netherlands. Until well into the twentieth century these locks in Demak, Central Java, waited for ships that never came.

The colonial engineers were raised in the Dutch water tradition: they had been trained to fight against water. In 1879, an Irrigation Commission made a plea for dedicated irrigation education for colonial engineers.[3] Establishing an irrigation course, to be taught in Delft by an experienced engineer from the Indies, was seen as an absolute necessity. The report acknowledged that some irrigation design guidebooks were available, most of them in French, but these were not aimed at Java, although the 'Cours d'agriculture et d'hydraulique agricole' included some discussions on rice.[4]

Dedicated educational programmes for colonial civil servants had been available from the early nineteenth century.[5] In 1842 Baud, Minister for Colonial Affairs, decreed that all higher positions within the colonial service were reserved for candidates who had been educated in the Netherlands. By royal decision of 8 January 1842 the Royal Academy in Delft was formally established.[6] Only a diploma from this institution opened up appointment to the higher ranks in the colonial civil service. Next to colonial civil servants, the Academy educated engineers. Five different programmes of four years were offered: civil engineering, mining engineering for the colonies, building engineering, mechanical engineering and commercial engineering.[7] In the first two years of the study programme, the prospective engineers and civil servants shared several courses.

In its first year, the Academy counted forty-six students: ten students were studying to enter the civil service and thirty-six received their training within the Department of General Studies, which hosted the engineering programmes. In 1843 the Academy hosted 117 students, 142 in 1844 and 170 in 1845.[8] Within the engineering programmes the civil engineering study was the most popular by far: in 1864, 183 from the total 207 graduates in the Department of General Studies had chosen civil engineering.[9] With the application of the new Dutch educational law on 30 June 1864 the Academy was abolished. The Department of General Studies became the Polytechnic School. The monopoly of Delft on educating civil servants disappeared, as programmes were established in Leiden and Batavia as well; furthermore, the civil service programme in Delft was no longer part of the Polytechnic.[10] Until 1905, the Polytechnic School was formally registered as secondary professional education, but in that year the School gained academic status and a programme of five years.[11]

Even though specific attention on East Indian issues was not included in the engineering programmes, the Netherlands East Indian Department of Public Works connected itself clearly with the Delft engineering school. In 1874 it decreed that anyone seeking employment with the Department for civil engineering needed a degree from Delft.[12] All in all, some 25 to 30% of Delft graduates went to the Indies.[13] Although the engineers from Delft did not perform too badly in irrigation development from a technical point of view before the establishment of an irrigation course, the call for dedicated attention in the Delft curriculum for hydraulic engineering issues from the East Indies, particularly 'irrigation and other Works in non-flat and tropical regions', was strong from the beginning.[14]

The pleas for dedicated attention basically emphasized differences in technical (e.g. soils, climate, building materials) and social (e.g. laws and regulations, language, administration) issues between the motherland and the colony. An added argument was that a civil engineer in the East Indies had to work on his own. He was not participating in a larger team or guided by senior colleagues; he was often the single authority in a large region and responsible for many more tasks than his fellow engineer would be in the Netherlands. The Dutch engineers were no particular admirers of the quality of education for colonial engineers in Britain, but British attention to dedicated preparations for service in the colony of their engineers was considered an example to follow.[15]

In 1906, negotiations between the Ministry for Colonial Affairs and Delft Polytechnic resulted in nominating a civil engineer on leave from the colony to give the required dedicated course on hydraulic engineering on Java; the course was not obligatory for all civil engineering students. Engineer Grinwis Plaat, who had worked on several locations in the colony on a variety of issues, was this first lecturer. In 1908 his temporary nomination was changed into a permanent position as extraordinary professor in hydraulic engineering. On 1 January 1910 Grinwis Plaat retired as professor. In 1919, after several short-term professorships, Haringhuizen was the first regular professor who stayed for a longer time, until 1938.

In 1920, the same year De Gruyter graduated, the Polytechnic School for the Netherlands East Indies in Bandoeng, Java, was opened, with civil engineering and irrigation as major subjects. The institution was designed to suit East Indian needs. The engineers trained in Bandoeng were supposed to find employment in the colony. The educational system and programme of Delft were taken as the models to follow, but adaptations were made. In Bandoeng the whole programme was obligatory, which was only a small contrast to Delft with its few optional courses. The subcommittee thought that the drawing courses would offer some choice to students in Bandoeng; they could draw those subjects they were most interested in. 'Such freedom would typically be more beneficial than damaging to the programme.'[16]

Hierarchy and Water Distribution

Unlike his predecessors, De Gruyter was trained specifically for his work as irrigation engineer in the East Indies. At the time of his graduation from Delft, attention for irrigation engineering was firmly settled within both the Delft and Bandoeng engineering schools. During his stay in Delft, De Gruyter would have taken some of his senior courses with the new professor Haringhuizen. Very soon after graduating, on 7 September 1920, engineer De Gruyter was appointed in the Netherlands East Indian Department of Public Works, more specifically to the staff of the Head of the Irrigation Department Tjimanoek in Cheribon, Western Java. His engineering career seemed to go well. In November 1925, he was promoted to engineer second class and moved to a new position within the Central Water Board Office for the Princely States in Soerakarta, Central Java.

As an engineer of the third class, De Gruyter's duties would have focused on making technical drawings, and possibly travelling around to map and report on the area. Even after 1925, when he became engineer of the second class,[17] his work would still have been closely controlled by his superiors, as much as the work of the regional office was to be controlled by the Department of Public Works in Batavia. When designs and associated design documents were to be finalized in the Netherlands East Indies, the engineer designer sent the document to the Department. After approval, the documents were sent to the Director of the Department of Public Works, who asked for a governmental decision. Such a procedure was not always appreciated by engineers in the field. 'The project was approved [by Public Works] but with the announcement that it had to be completely rewritten. [...] One of my engineers worked on it for half a year. [...] Another remark concerning that project was that in future I had to send such projects for approval in pencil first. [...] one was exceptionally surprised there that I reacted somewhat annoyed; one of the subordinated engineers even signified that actually all projects should be send in pencil to Batavia; probably to be niggled on by him.'[18]

There was much to be discussed in the 1920s regarding Javanese colonial irrigation, as the new impulse for irrigation—expressed in expenditure for irrigation in Fig. 11.1—by the colonial state created much new work

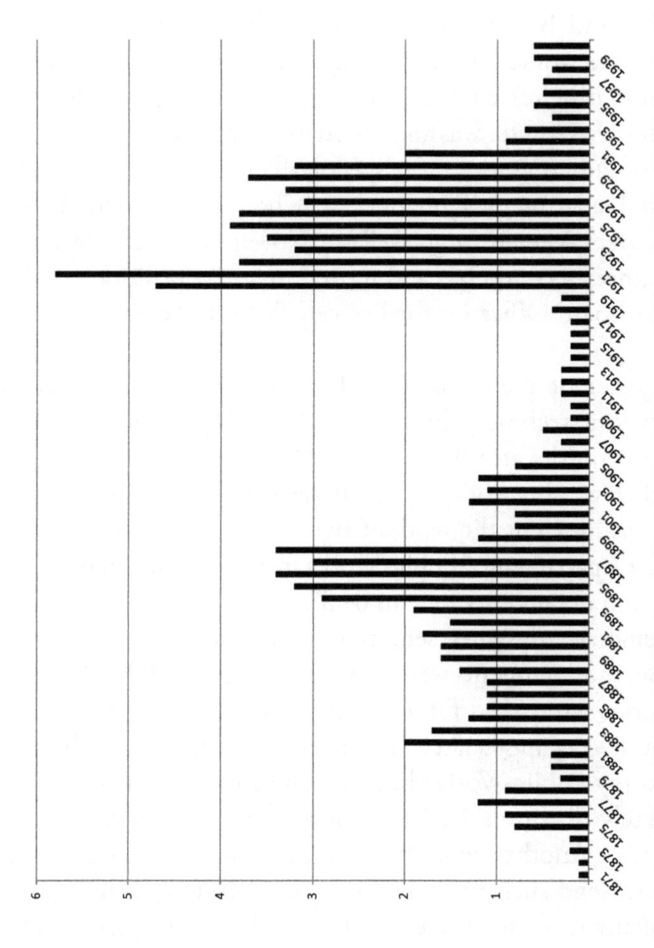

Fig. 11.1 Irrigation expenditure in the Netherlands East Indies between 1871 and 1940. (Millions of guilders per year corrected for inflation (Data from Ravesteijn, W.: *De zegenrijke heeren der wateren. Irrigatie en staat op Java, 1832–1942*, Delft: Delft University Press, 1997, Annex H, p. 360))

for engineers. Those same renewed efforts after the First World War I created the space for engineers to renew a debate they had been engaged in since the start of colonial irrigation efforts: how to distribute irrigation water between the users in irrigation systems. As we will see, De Gruyter did not only enter the Javanese irrigation scene when the debate was renewed. He did much more than that, as he made proposals of his own about how to arrange water distribution and defended them strongly in engineering circles. We will also find out, however, that his proposals were not readily accepted by his peers, and why this was the case.

Well before De Gruyter's efforts, in the first half of the nineteenth century, Dutch colonial engineering irrigation design focused on constructing head works, such as barrages in rivers (weirs) and main canals diverting water from those rivers to existing irrigated areas and fields. Towards the end of the nineteenth century, engineering attention gradually moved to designing complete irrigation systems from intake to field drains. As such, engineers had to deal with a growing number of related issues, from stability of intakes to ways of delivering water to fields. A major duty of Dutch irrigation engineers was to ensure that Java's irrigation systems supplied water to a diversity of crops.

Javanese farmers grew rice on their fields in the wet West Monsoon—between October and April. In the East Monsoon, with much lower water availability from rain or rivers, Javanese peasants grew dry crops (*polowidjo*) such as root crops, beans and ground nuts on part of their land. The other main crop in need of irrigation was sugar cane, the major cash crop grown on Java.[19] The importance of cash crops had been created between 1830 and (about) 1870, when Javanese farmers had to cultivate certain cash crops within the Cultivation System.[20] Two of the crops that were enforced by the colonial government needed irrigation: sugar cane, to produce sugar, and indigo. The income generated by the enforced cultivation of cash crops was considerable: in the period between 1851 and 1860, almost a third of the income of the Dutch state came from the Netherlands East Indies.[21] Under the Cultivation System, sugar companies were under direct control of the colonial government. After the gradual abolishment of the Cultivation System in the second half of the nineteenth century, it was private sugar companies that produced the sugar from the cane. These companies exploited factories on Java, but did not actually own the land on which the sugar cane itself grew. The companies rented the land for a period of three years from Javanese landowners.[22]

Sugar cane was the most important irrigated commercial crop by far, with its sugar serving the world market. Sugar cane was particularly grown in East Java. The famine in 1850/1851 in Demak, central Java, stimulated criticism of the Cultivation System. This criticism coincided with a general call for liberalization of economic activities in favour of private initiatives. The famines were seen as resulting from sugar cane being a major constraint for peasant production to reach a sufficient level. Sugar cane claimed valuable irrigation water, both for the crop and for driving the watermills to crush the cane. Furthermore, sugar cane was removed rather late from the fields, basically when the West Monsoon had already started and growing rice was again hampered. As long as the Cultivation System allowed the profits to go to the colonial government, it was hardly interested in such problems. With the sugar industry still largely under governmental control, rice interests came second.[23]

After the gradual abolishment of the Cultivation System and the increase of private economic activities, the government no longer benefited directly from cash crops. This change in interest coincided with—or allowed—a growing concern about the welfare of the Javanese peasant. One of the first measures the government took was aimed at decreasing water use by watermills belonging to sugar factories (1870). Another measure concerned the cropping dates of sugar cane (1871). In the West Monsoon, all irrigation water was to be made available for rice after 15 October.[24] Despite these measures, conflicts between rice and sugar cane—which were grown within the same irrigation systems—remained. Rice did not compete with sugar cane for irrigation water during the West Monsoon, as sugar cane did not require irrigation at that time. However, the cane required much of the scarce irrigation water in the dry East Monsoon and did compete directly with *polowidjo* crops for water. Furthermore, the issue of the effect of late removal of sugar cane on rice harvests on those fields remained.

How to deal with these competing demands became a major issue of Dutch colonial irrigation water management. Growing sugar cane was seen as very valuable and a very good example of maximization of the value of the water. *Polowidjo* would allow two families to make a living, whereas the same area with sugar cane provided income for up to six families.[25] At the same time, even its high profits should not mean the unlimited growth of sugar cane: Javanese agriculture had to be able to exist and grow too. The Javanese had to eat, but exporting rice could perhaps become an interesting future economic option as well. These considerations led to a water distribution policy that strived to maximize the economic value of irrigation water—a value that in the Netherlands East Indies was expressed in the value of the total crop harvested on a certain surface area.

The total value of a crop, already mentioned in the report of 1879 mentioned above, could be expressed in monetary terms and was created by applying the correct amount of water.[26] Irrigation water could be valued differently in terms of location and the time when water was available. Irrigating a crop that had just received water or irrigating fallow land did not create additional value, or could even be harmful. Irrigating crops in need of water created a very high value. From all these different values of crops and fields, a theoretical total value for all water at a moment in time could be calculated. This total value and how to reach it varied from moment to moment, depending on the actual situation of crops and water availability.

This importance of the actual crop harvest was not only a direct inheritance from the Cultivation System, when the cash crops that were sold on the world market produced huge amounts of money for the Dutch government, but was also related to the colonial tax system on Java. The tax to be paid by farmers and companies in the Netherlands East Indies was based on the actual crop that was harvested on a standard area of land—cash or food crops alike. The higher the harvest, the higher the tax would be. As I have explained in more detail elsewhere, this Dutch colonial taxation system contrasted rather strongly with the system in British India.[27] In India, irrigation was also important to the colonial state, but much more to ensure food security. Whether land could be irrigated to prevent crop failure was more important than irrigating for maximum harvest. Not coincidentally, land taxes in British India were higher for fields that could be irrigated, irrespective of how water was used.

Puddings and Trains

On Java, with cash crops under private management after the Cultivation System had been abolished, land taxes based on crop production had become the main source of income for the colonial government. Irrigation had to ensure maximum harvests and thus maximum monetary values per surface area by applying the correct amount of water to crops at the correct moment.[28] Adaptation to actual circumstances and demands was a key element to ensure maximum values for water use at all times and at all places within irrigated areas. This created the necessity to adapt, measure and control flows of water to different parts of the irrigation system, units, fields or even crops, to allow for changing demands for water at different places in the system at different times. In 1893, the colonial government invited the Director of Public Works to 'design, in consultation with his colleague from the Civil Service, regulations for distribution of irrigation water in the areas of Pekalen and Pategoean, respectively in the residencies of Probolinggo and Pasoeroean, which could be applied as a test, which, when they prove to perform well in practice, can be definitively established later'.[29]

An experiment in these two small irrigation systems on East Java had to suggest the most appropriate water management regulation. Lamminga was involved in the design of the Pekalen works. Even though there was a preference for the Pategoean method expressed by the colonial government, the water distribution regulation developed in the Pemali irrigation area around Tegal in Central Java became the standard approach. The Pemali works and its regulation were designed by Lamminga and his team; the Pemali system is credited for setting the standard for later systems in terms of water distribution.[30] The influence of this system is also explained when taking into account the debate in the Netherlands East Indies at the turn of the twentieth century on the (negative) impacts of Dutch colonial rule on the Javanese. Dutch colonial rule would put too much pressure on the Javanese population and keep the farmers in a continuing situation of poverty. Irrigation was seen as a major solution for farmers' poverty. Around the same time, the Pemali system was seen as one of the better functioning irrigation systems, which was a major reason for it to become the standard design and implementation for irrigation on Java.[31]

Over the years, more defined design standards developed, particularly how to compute canal dimensions and which division structures to use. The designs of the huge Krawang works—irrigable area of 100,000 acres on West Java, opened in November 1925—provided much of the standards for De Gruyter's work.[32] Originally, the Krawang standards prescribed the so-called Cipoletti and Thomson weirs as measuring structures. These devices were accurate and could be easily moved around—an asset as early canals did not have permanent measuring devices and the water distributed to sugar cane was measured just before the water entered the field(s).[33] In later designs, permanent Cipoletti weirs were placed downstream of intakes of all canals leading to groups of fields. This allowed permanent water measurement, but was also seen as somewhat cumbersome, because inflow control through the intake gate and flow measurement at the Cipoletti weir were separate actions at separate locations. A device that could both control and measure irrigation water would save much time and effort.

Dutch engineers discussed their preferred solution in their own colonial engineering journal *De Waterstaatsingenieur*, typically referring to experiences elsewhere. It was Dutch irrigation engineer S.H.A. Begemann who introduced such an external structure when he applied a Venturi structure in the Penewon area, just south of Modjokerto in East Java in 1923.[34] His Venturimeter became the new standard for discharge measurement structures in the East Indies. Begemann and his colleagues had known Venturis for many years, as they had been applied in the Netherlands East Indies as measuring devices in the closed conduits of drinking water systems. After reading about experiences in British India and the USA with Venturis in irrigation networks, Begemann decided to build them himself.[35] The results from his designs allowed Begemann to prove that his Venturis worked well. They could measure water flows and were easy to manage. Furthermore, they did not need huge water level differences between upstream and downstream sides of the structure—a huge advantage in the flat areas on the Javanese north coast where the Dutch built most of their new systems after 1920.

As a result, the second half of the 1920s saw Venturi structures appear as popular discharge measurement structures in new and existing irrigation systems on Java. In the Krawang system, the Cipoletti weirs that were originally planned were replaced with Venturi structures; in Demak, Venturi structures gradually replaced existing structures. This popularity of the Venturi did not mean that the discussion about the perfect water control structure stopped in colonial irrigation circles. As late as the 1930s, reports continued to document experiments with different measuring devices.[36] Already in 1925, in an attempt to promote another alternative, De Gruyter proposed a structure designed by himself in an extensive overview of possible discharge measurement structures. His structure was 'a hotch-potch of the broad crested weir, the regular sluice and the Venturi-flume. It possesses such particularly suited qualities, that in my view it is the most ideal type by far.'[37]

Similar to Begemann, who used experience from elsewhere in selecting the Venturi, De Gruyter had adapted an existing irrigation structure from British India 'from Mr. Crump of the Punjab Irrigation'.[38] De Gruyter knew that the irrigation systems on the Indian subcontinent had long canals and many outlets. As already explained, British colonial irrigation was based on the principle that it was much more important that many fields could be irrigated. Exact water measurement was not very important. Furthermore, with British indirect colonial rule, operation of the large irrigation systems had to be simple, by as few people as possible, and cheap. What the British were looking for was an irrigation system, which would function without management, by itself as it were. In 1922, Crump introduced his own invention in irrigation systems in the Punjab: the Adjustable Proportional Module (APM). The structure consisted of a narrow throat with a sloping sill and rounded roof-block on top of the throat to create an orifice. Its hydraulic properties ensured that the water flow through the structure was relatively constant. As such, an APM provided a predictable amount of water without much need for actual management from engineers or other irrigation staff.

What De Gruyter proposed was replacing Crump's original fixed roof-block with an adjustable sliding gate.[39] This would change the APM from a device that gave stable water flows into an adjustable water control structure that could be used for discharge measurement and regulation.

De Gruyter had clearly studied British Indian irrigation. His personal copy of an important British irrigation handbook contains notes and sketches on many pages.[40] In his 1925 paper introducing his idea, De Gruyter had good hopes 'to convince the reader about the eminent properties of this construction'.[41] What he could not yet show was results of his own proposals regarding existing irrigation on Java. Therefore he proposed to experiment with his structure. Perhaps a secondary canal in the Tjipoenegara works close to Pamanoekan could be used? Unfortunately, the Cipoletti weirs in this system were replaced with the new and popular Venturimeters.

De Gruyter had to look for other options to test his own idea. After being given the opportunity to experiment with his structure in the Irrigation Department Tjimanoek, the area around Indramajoe, he provided an extensive description of the structure, especially its hydraulic behaviour; the experiments showed him that the structure was even better than he had thought! Nevertheless, the apple of his eye was never really considered by his peers from civil engineering as a breakthrough in colonial irrigation on Java. It is very likely that the way the structure was introduced into the debate influenced its success. De Gruyter could not show results from Javanese irrigation practice, so he proposed a structure based on theoretical considerations. This experimental status was hard to overcome in the practice-oriented network of civil irrigation engineers, as De Gruyter himself acknowledges. 'The writer fully acknowledges the general need, before importing new constructions, not to take chances, not to ban the old before the new has sufficiently proven in practice to possess bigger advantages than the old.'[42] But he continued, somewhat disappointed, '[n]evertheless one should still give the new a chance and this can only be achieved by constructing it'.[43]

This last argument was right on the money. In East Indian irrigation, the proof of the pudding was in the eating. Whether candidate structures came from areas close to Java or from further away, and irrespective of whether they had to be changed much, successes of proposals had to be shown in practice. De Gruyter could not show results from Javanese irrigation practice for his modified Crump flume. His position at that time within the Central Office of the Vorstenlanden would not have helped. The Vorstenlanden, the area associated with the Sultan of Jogjakarta, were semi-independent Javanese kingdoms within the colonial state. The Central Office served as a kind of Ministry of Public Works; De Gruyter would have done mainly deskwork. This position made it difficult for him to enter the practice-oriented network of irrigation engineers. Begemann could propose the Venturi structure based on his own trials and experiences in 'his own' Javanese irrigation system. De Gruyter had no real-time irrigation systems to show.

For De Gruyter 1926 was a year of debate. We do not encounter anything from him in 1927, but we know that he took a well-deserved nine months of leave in The Netherlands. At the end of 1928, De Gruyter returned to Java. Again, we do not encounter much of him in the archives, but his resignation in 1929 as member of the VWI suggests that his disappointment in his fellow irrigation engineers' reception of his ideas had not yet changed. Only in 1933 do we find De Gruyter back in the archival record, which shows that he had become an engineer with the Javanese State Railway Company (JSRC). In April 1934, he even became the Head of the Experimental Bureau for Rolling Stock of JSRC. Railways were a relatively new field of engineering in the East Indies and became closely related to the field of hydropower. From the late nineteenth century onwards, hydropower exploitation started as a private initiative—starting in the city of Bandoeng in 1906. In 1912, the Governmental Railways became interested in electrifying part of their network and established a hydropower unit. In 1917, the railway hydropower unit merged with the governmental Electrical System Unit to form the Service for Hydropower and Electricity, with its head office in Bandoeng.[44]

The Hydropower Service employed civil, mechanical and electrotechnical engineers, with quite a few coming from other countries than the Netherlands; many came from Switzerland. After all, Dutch engineers were not really known for their experience with hydropower generation. Hydropower had 'something of the unknown, of the mystique, which attracts the normal Dutchman with so much power to the mountains'.[45] Colonial reliance on foreign engineers is also reflected in the numbers of engineers hired (Fig. 11.2). That same figure also shows that the number of foreign engineers declined in the late 1920s. This reflects the Dutch desire to become less dependent on foreign experts and to replace them with experienced Dutch engineers. Finding (experienced)

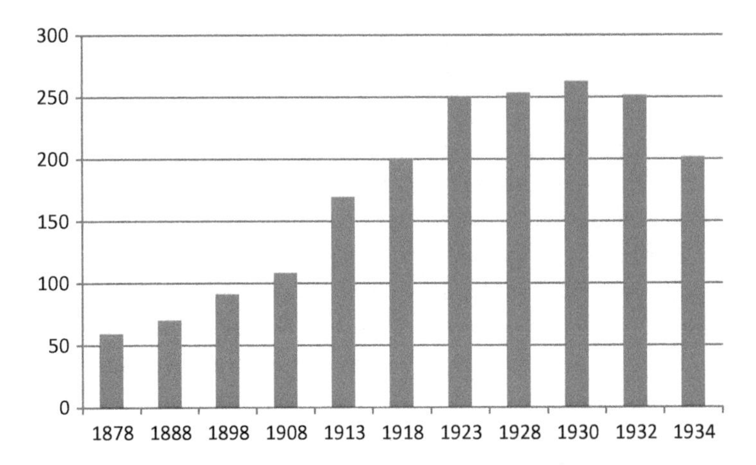

Fig. 11.2 Numbers of engineers and their training origin in the Netherlands East Indies. (Numbers from Ravesteijn: *Zegenrijke heeren*, Annex F, p. 358)

irrigation engineers to work in hydropower was not that easy, as experienced engineers were not always found—or were perhaps not attracted to the field. For De Gruyter, however, with his recent disappointments, a move from irrigation to another field may have been very attractive. We can only speculate, as the archives provide no evidence, but it is likely that De Gruyter replaced one of those foreign engineers in colonial government service.

Local Made Global

Perhaps the colonial railways were not that attractive or perhaps De Gruyter had had enough of the colonial world, but whatever the case, a little over one year later after his appointment as head, in August 1935, Paul de Gruyter returned to the world of water. He did so in The Netherlands, as he was appointed as Head of the Technical Service of the Rijnland Water Board (one of the largest and oldest such boards in the country). Again, we do not find too much of his work and career there. What we do know is that he remained the Chief Engineer of Rijnland until 1958, when he took up a position for the United Nations in Syria.

This move reflects the changing position of Dutch engineering within the emerging international realities after the Second World War and its decolonization trends.

Indonesia's independence in the 1940s effectively put a halt to the massive Dutch engineering efforts in the colony. The Second World War forms a line of demarcation for the Netherlands East Indian irrigation context. Indonesia disappeared as a secure field of practice. This disruption of colonial realities after 1945 was not unique for Indonesia and the Netherlands; within twenty years most colonies gained their independence and were redefined as 'developing countries'. In this new reality Dutch irrigation engineers started to work in countries worldwide; the first generation did so with their experience in the Indies embodied in their persons, while the second generation did so based on their education and training in Delft. Dutch engineers started working in other tropical regions and engineers from different countries started to work in independent Indonesia.

Before the Second World War, about 280 (mainly Dutch) engineers had been employed by the Netherlands East Indian Department of Public Works. After 1945, some 100 engineers remained active in the areas under Dutch control. Irrigation received relatively little attention; restoring and building transportation infrastructure was of much higher importance. In 1949, formal negotiations between Dutch and Indonesian representatives confirmed what had been coming for many years: Indonesia became an independent republic. Despite the expectation of the Dutch—including many engineers who perceived their work as being of great value for Indonesia—that a Dutch presence would continue, relations with the new Republic of Indonesia cooled down considerably in the 1950s. At the end of the 1960s, however, relations improved again.[46] In 1967, the Indonesian government applied for international assistance, focusing on the rehabilitation of existing irrigation infrastructure. The Dutch offer for assistance was accepted by Indonesia, as we will discuss below.

In the meantime, the post-Second World War development aid programmes, which replaced most colonial development programmes, had become a niche for activities of European and American engineers in Africa, Asia and Latin America, including most former colonial areas.[47]

International development organizations under the wings of the United Nations, such as the Food and Agricultural Organization, the World Bank and the Asian Development Bank, required engineering expertise. Irrigation development was one of the key fields for increasing global food production. In 1951, in response to these developments, Dutch engineering firms united themselves in the umbrella organization Netherlands Engineering Consultants (NEDCO). Geographically, the employment opportunities for Dutch irrigation engineers increased; their horizon expanded to countries such as Bangladesh, India, Pakistan, Nigeria and Colombia. As such, the colonial engineers were turned into international experts, who now lived in and worked from the Netherlands but did not work in the Netherlands.

How to translate this new international status of Dutch irrigation engineers was not immediately evident. Once Indonesia's independence effectively put a halt to Dutch engineering intervention in this territory and the East Indies as an employment option had disappeared, the direct connection between education and employment had been cut and the usefulness of specific colonial educational programmes was discussed immediately. The Agricultural School in Wageningen—the other key education institute for the colony—kept offering specializations directed at non-Dutch regions, but renamed them 'tropical', not 'colonial'.[48] Within Delft, similar discussions could not be avoided; the focus of the discussions was on its own graduates, Bandoeng was ignored.[49] After 1945, Delft students could continue to graduate in irrigation. Where they could choose between two main graduation specializations in 1930 ('Dutch' civil engineering and 'East-Indian' civil engineering, with just a few courses being different), the 1955 programme offered a prospective engineer seven specializations: (1) general hydraulic engineering, (2) polders, (3) irrigation and hydropower, (4) bridges and roads, (5) utility buildings, (6) sanitary engineering and (7) theoretical subjects.[50] From these options, the specialization for irrigation and hydropower, the 'non-native branches of civil engineering practice', was the successor of the East-Indian programme.[51]

Armed with the continued attention for irrigation after 1945, the professors responsible for irrigation education at Dutch institutes were among the first to emphasize the need to broaden the perspective of irrigation engineering from the East Indies to all tropical regions. As a Netherlands East-Indian specialist and major player in the international water orientation, Eijsvoogel—who had become irrigation professor in 1946 at the Agricultural School in Wageningen—stressed the attention of the Dutch engineers for layout, organization and management of irrigation systems. Leaving aside communist countries and some notable exceptions to the rule, Eijsvoogel distinguished between 'distribution to the individual farmer where properties have a reasonable size (thus e.g. in the USA, South Africa, Australia) and distribution to a group of farmers where small landholding prevails (that is in the tropics)'.[52] In just a few pages, 'tropics' had been made equal to 'Javanese', when Eijsvoogel stated that Lamminga had thoroughly studied and defined the principles which had to be applied when designing and constructing irrigation systems in 'tropical regions'. In this way, the East Indian irrigation concept, based on the commercial crop sugar cane challenging water distribution to food crops, was redefined (perhaps narrowed down) to distributing water to smallholders.

Some years later, Berkhout, irrigation professor in Delft since 1954, emphasized the strategic aspect related to attention for irrigation in the Netherlands. Allowing the irrigation domain to disappear from the Delft curriculum would mean that the Netherlands could not maintain its leading position in the international aid programmes that emerged worldwide. Global recognition of the importance of irrigation for food production was increasing.[53] It was observed that the expertise available in the Netherlands because of its East Indian past should not be wasted. Apart from its colonies at that time, Surinam and New Guinea, and the UN-guided development programmes, Berkhout saw possibilities for cooperation within the context of Benelux.[54] Dutch engineers could assist Belgium to develop irrigation in its African colony Congo. Involving Dutch irrigation engineers in the British and French territories would be an option too, as was a return to Indonesia sometime in the future. All in all, Berkhout was very pleased that irrigation had not only maintained a professorship, but also that a condensed course in irrigation was obligatory for all civil engineering students.[55]

In 1958, De Gruyter returned to the international community with his two-year appointment as water management expert in Syria. What his tasks were exactly remains unclear, but his appointment confirms that Netherlands East Indian irrigation expertise could be successfully redefined as international water expertise. Interestingly, Indonesian practice suggests a different move. Where the Dutch managed to rephrase their colonial expertise as international expertise, in Indonesia that same colonial expertise was redefined as Indonesian expertise—at least partially by the Dutch themselves.

The Dutch engineers returned to Indonesia in 1967, when an Indonesian large-scale programme of rehabilitation and/or upgrading of some 780,000 hectares—with 90% on Java—was launched. The Dutch government offered assistance as well, and Dutch engineers started to work on such projects as the Tarum canals on West Java and the Djratunseluna area on Central Java.[56] Next to rehabilitation, some new irrigation facilities were included as well, although new designs were very often based on, or directly related to, colonial plans and reports. Colonization of Javanese on the other islands remained an important policy for independent Indonesia. Relabelled 'transmigration', resettlement programmes were resumed, and between 1950 and 1974 in total about 500,000 people were resettled in the Outer Islands. In 1980, about a million people had been resettled, where the third five-year development plan (Repelita III, planning period 1979–1984) achieved the resettlement of almost 1.5 million people.[57] As in colonial times, developing irrigation was nearly always included in transmigration projects, as in Lampung—an important target area for transmigration.[58]

Up to the early 2000s, when I visited the area, new irrigation systems had been constructed in the Lampung area. In 2003, the responsible Japanese consulting designer had just selected and constructed Crump–De Gruyter gates for off-takes to canals—De Gruyter would have been very pleased with that choice. Relevant for this chapter is that the Japanese design documents explicitly refer to the Irrigation Design Standards (IDS) of Indonesia. These were published in 1986 by the Indonesian government in a successful attempt to standardize irrigation design in Indonesia and effectively limit the number of different types of structures applied in Indonesia. The IDS series includes thirteen volumes and was produced with the assistance of a Dutch consulting firm. It presents an overview of the irrigation design process, with available options for different elements of an irrigation system in a 'how-to-do-it' approach. The first volume stresses that it includes only the '[i]rrigation systems and methods as commonly used in Indonesia'.[59]

Each of the design criteria volumes, apart from Volume VII (Drawing Standards) contains a list of references. Original material from the Netherlands East Indies is quoted in three volumes, dealing with the general approach to irrigation design, canal sedimentation and control structures. Measuring and regulating flows remains a top priority, but a designer can now select from a group of distribution structures, including a Crump–De Gruyter gate. The influence of the standards is considerable, not only because irrigation designers (need to) apply them; the standards are also used as lecture material in educating Indonesian civil engineers, providing the design rules for future Indonesian irrigation engineers. The Dutch consultants who coauthored the Irrigation IDS will probably have recognized (perhaps even used) their own lecture notes from Delft in the writing process.

De Gruyter and Dutch Irrigation

This chapter has presented Paul de Gruyter, a Dutch engineer with an international career. His career allowed the discussion of several aspects of Dutch colonial engineers in relation to their knowledges, networks and professional environments. De Gruyter enjoyed his education in Delft. In contrast with the civil engineers who had been trained in Delft before 1910, he had been prepared for working in the East Indies. Where the first Dutch engineers left for Batavia with a thoroughly Dutch-centred education, graduates after 1910 did so with at least a few courses and exercises on technical topics relevant for East India—mainly irrigation. De Gruyter joined the East Indian Department of Public Works, to be exposed to its hierarchy in different ways. First, he was the youngest engineer in the ranks, and promotion was arranged in a strict schedule. Second, all his work had to be checked by his superiors, both in his direct working circle as by the Department itself in Batavia. Third, when De Gruyter shared his own proposals for better water distribution on Java in a series of papers in the periodical of the engineering profession on Java, he discovered that these did not find a warm welcome. He had to compete with proposals from other engineers, who did have more practical experience, could test their own proposals because of their position, and as such could show success in irrigation practice—not just theory. Whether the move to the railways and later The Netherlands was a way to find more recognition is not clear, but it is tempting to argue along those lines. It must have been disappointing for De Gruyter that the Crump–De Gruyter gate did not find the warm welcome he himself thought it deserved.

De Gruyter's return to The Netherlands and his move to Syria in 1958 allowed an exploration of the relations between The Netherlands, its main colony the Indies and the global world of development after 1945. The strong position the Dutch and their engineers held in the growing field of international development projects allowed Paul de Gruyter to move east once more in the late 1950s. There he would have encountered colleagues from other European countries, who, as much as the Dutch did with their experience from the former Netherlands East Indies, used colonial expertise, as the British did with former British India and the French with North Africa.[60]

When former colonies became independent nations, the majority of Dutch, French or British colonial engineers became international experts. As the Dutch did with their Javanese irrigation approach, colonial expertise was translated into an international approach, based on the claim that the (Javanese, Indian, African) context was typical for the international development context—obviously a convenient argument when the aim is to maintain employability for former colonial experts. Maintaining the employability of 'non-Dutch experts' was certainly the wish of Delft Polytechnic, which redefined its colonial programmes into international programmes—even though the training material remained firmly based on the colonial approach. Still, after 1945 the Dutch and international water worlds remained rather separate for many decades. This is shown in Delft education, where prospective water engineers could either take a Dutch-oriented water programme or an internationally oriented programme at relevant universities. Scientific staff at those same universities either taught within the Dutch or the international programme.[61]

Direct relations between the Dutch and colonial engineering contexts were not very strong in terms of knowledge and professional exchange. I have not found many engineers who like De Gruyter started working in Dutch engineering after an East Indian career. Circumstances may have changed, but irrigation design and management practices in the post-colonial period remained largely based on colonial approaches. Although he could not witness it himself, De Gruyter probably would have liked the idea that the water distribution structure he had proposed in the Netherlands East Indies in the 1920s—which had been rejected by his peers at the time—had become one of the possible applications that international experts could select when designing Indonesian irrigation systems as late as the early 2000s.

Notes

1. Ertsen, M.W.: *Locales of happiness. Colonial irrigation in the Netherlands East Indies and its remains, 1830–1980*, Delft: VSSD Press, 2010. Ertsen, M.W. and Ravesteijn, W.: 'Living water. The development of irrigation technology and waterpower', in Ravesteijn, W. and Kop J. (eds.): *For profit and prosperity. The contribution made by Dutch engineers to Public Works in Indonesia 1800–2000*, Zaltbommel: Aprilis; Leiden: KITLV, 2008, pp. 239–271.
2. Van Doorn, J.A.A.: *De laatste eeuw van Indië; ontwikkeling en ondergang van een koloniaal project*, Amsterdam: Bakker, 1994. Compare with Armytage, W.H.G.: *A social history of engineering*, London: Faber and Faber, 1976.
3. *Rapport omtrent het irrigatiewezen op Java en Madoera*, Batavia: Landsdrukkerij, 1879.
4. Nadault de Buffon, B.: *Cours d'agriculture et d'hydraulique agricole comprenant les principes generaux de l'economie rurale et les divers travaux etc.*, Paris: Victor Dalmont, 1858.
5. Ertsen, M.W.: 'Indigenous or international. The evolution and significance of East Indian civil engineering', in Ravesteijn and Kop: *For profit and prosperity*, pp. 381–401; This was late compared with England. In 1806, the big rival and example colonial power England had already established the East Indian College to organize the education of its

future civil servants in the British Indies, Van Leur, J.W.L. and Ammerlaan, R.P.M.: *De Indische instelling te Delft; meer dan een opleiding tot bestuursambtenaar*, Delft: Volkenkundig Museum Nusantera, 1990; In 1794, the Engineering College at Madras had already started to train surveyors; in 1859 this college was transformed into a civil engineering school, Ambirajan, S.: 'Science and Technology Education in South India', in MacLeod, R. and Kumar, D. (eds): *Technology and the Raj. Western technology and Technical Transfers to India 1700–1947*, New Delhi: Sage, 1995, pp. 113–133). In 1847 an engineering college had been started at Roorkee (India); later colleges included Calcutta (1856), Bombay (1888), Sibpur (1880) and Poona (1854), Derbyshire, I.: 'The building of India's railways', in MacLeod and Kumar: *Technology and the Raj*, pp. 177–215.

6. Van Leur and Ammerlaan, *Indische instelling*. This date is relatively late in comparison with similar institutes established in Germany, including Karlsruhe (1825), München (1827), Dresden (1828), Stuttgart (1829), Hannover (1831) and Darmstadt (1836), Groen, M.: *Het wetenschappelijk onderwijs in Nederland van 1815 tot 1980. Een onderwijskundig overzicht. II. Wis- en Natuurkunde, Letteren, Technische Wetenschappen, Landbouwwetenschappen*, Eindhoven, 1998).
7. Groen, *Onderwijskundig overzicht*. With the establishment of the Academy, a separation between military and civil technical education was realized in the Netherlands (Van Leur and Ammerlaan, *Indische instelling*. Van Doorn, *Laatste eeuw*).
8. Van Leur and Ammerlaan, *Indische instelling*.
9. Schippers, H.: *Van tusschenlieden tot ingenieurs. De geschiedenis van het Hoger Technisch Onderwijs in Nederland*, Hilversum: Verloren, 1989.
10. De Jong, J.: *De Waaier van het fortuin. De Nederlanders in Azië en de Indonesische Archipel 1595–1950*, Den Haag: SDU, 1998; Van Leur and Ammerlaan, *Indische instelling*.
11. Higher technical education was at an equal level with university education; engineering had become an academic profession in 1905 (Van Doorn, *Laatste eeuw*).
12. Van Doorn, *Laatste eeuw*.
13. Van Doorn, *Laatste eeuw*. Some of them returned to Delft to become professor.
14. Van Sandick, R.A.: 'Ter herinnering aan P.Th.L. Grinwis Plaat c.i.', *De Ingenieur*, Vol. 26, No. 4 (2011), p. 202.

15. For examples of non-admiration, see Snethlage, R.A.I.: 'De opleiding onzer Indische Ingenieurs', *De Ingenieur*, Vol. 5, No. 46 (1890), p. 433; 'Openbare werken in Britsch-Indië', De Ingenieur, Vol. 7, No. 3 (1892), p. 25. Compare with Fasseur, C.: *De Indologen; ambtenaren voor de Oost, 1825–1950*, Amsterdam: Bakker, 1993.
16. Hoogewerff, S., Weys, C.W. and Van Sandick: *Het leerplan der op te richten Nederlandsch-Indische Technische Hoogeschool*, Den Haag: Belinfante, 1918, p. 21.
17. *De Waterstaatsingenieur*, 1925.
18. Van Oort, M.A.: 'Reorganisatie van den Waterstaatsdienst', *De Waterstaatsingenieur*, Vol. 7, No. 1 (1919), pp. 12–13.
19. See Bosma, U., Giuisti-Cordero, J. and Knight, R. (eds): *Sugarlandia revisited. Sugar and colonialism in Asia and the Americas, 1800 to 1940*, New York: Berghahn Books, 2007.
20. Although the termination of the arrangements of the Cultivation System differed per crop type and lasted in some cases up to 1915.
21. Fasseur, C.: *Kultuurstelsel en koloniale baten. De Nederlandse exploitative van Java 1840–1860*, Leiden: Universitaire Pers Leiden, 1985.
22. See Elson, R. E.: *Javanese peasants and the colonial sugar industry: impact and change in an East Java residency, 1830–1940*, Singapore: Oxford University Press, 1984. The cropping cycle of sugar cane was three years: first year planting, second and third year maturing and several times harvesting.
23. *Onderzoek naar de mindere welvaart der Inlandschen bevolking op Java en Madoera, deel 7: Irrigatie*, 's Gravenhage: Staatsdrukkerij, 1910, p. 87.
24. *Onderzoek irrigatie.*
25. *Onderzoek naar de mindere welvaart der Inlandschen bevolking op Java en Madoera. Besluiten en voorstellen*, 's Gravenhage: Staatsdrukkerij, 1914, 32.
26. Weijs, C.W.: 'Grondslagen eener regeling van het gebruik van bevloeiingswater'. *Handelingen van het tweede Congres van het Algemeen Syndicaat van Suikerfabrikanten op Java*, Soerabaja, 1898, pp. 164–212.
27. Ertsen, *Locales of happiness.*
28. Weijs, *Grondslagen*; See Ertsen, *Locales of happiness*, for an extended discussion on water management principles in British India and the Netherlands East Indies.
29. *Onderzoek irrigatie*, p. 93. In discussions on these experiments, other irrigation systems are regularly discussed too (such as the Kening and

Madioen systems). Available detailed information, however, on these other systems is at best fragmented.

30. Weijs, C.W.: 'Ir. A.G. Lamminga', *De Ingenieur*, Vol. 36, No. 20 (1921), pp. 372–379; Ravesteijn, *Zegenrijke heeren*. Lamminga's works have been regarded as examples for irrigation design. Lamminga himself gave the credits to the 'push for erecting a separate service for irrigation in 1885 with the reorganization of the Service for Public Works'. Lamminga, A.G.: *Beschouwingen over den tegenwoordigen stand van het irrigatiewezen in Nederlandsch-Indië*,'s-Gravenhage: Gebrs. J. & H. van Langenhuysen, 1910, p. 5.
31. *Onderzoek irrigatie*. See on this so-called ethical debate and its later versions Moon, S.: *Technology and ethical idealism. A history of development in the Netherlands East Indies*, Leiden: CNWS Publications, 2007; Mrázek, R.: *Engineers of happy land. Technology and nationalism in a colony*, Princeton: Princeton University Press, 2002.
32. The history of irrigation in the Krawang area started in the late nineteenth century with plans to develop smaller-scale irrigation works (Bakhoven, H.G.A: 'De bevloeiing van de vlakte van Noord-Krawang uit de Tjitaroem', *De Ingenieur in Nederlands Indië* 3 (7 (1936), pp. VI.107–135.
33. All losses in the canals before the measuring point were not taken into account. The losses, however, could be considerable and constrained water availability for peasant crops.
34. Begemann, S.H.A.: 'Toepassing van Venturimeters voor bevloeiïngsleidingen met gebruik van differentiaal peilschalen', *De Waterstaatsingenieur*, Vol. 12, No. 11 (1924), pp. 325–330. Van Maanen, Th.D.: *Irrigatie in Nederlandsch-Indië. Een handleiding bij het ontwerpen van irrigatiewerken ten dienste van studeerenden en practici*, Batavia: Boekhandel Visser en Co, 1931 (first print 1924).
35. Begemann, *Toepassing*.
36. Ertsen, *Locales of happiness*.
37. De Gruyter, P.: 'Beschouwingen over aftapsluizen en meetinrichtingen voor bevloeiingswerken', *De Waterstaatsingenieur*, Vol. 13, No. 3 (1925), p. 70.
38. De Gruyter, *Beschouwingen*, p. 70. The original note is Crump, E.S.: *A note. Dated 15th of June 1922, by Mr. E.S. Crump, executive engineer, on the moduling of irrigation channels*, Typescript, 1922.
39. De Gruyter: *Beschouwingen*; De Gruyter, P.: 'Een nieuwe aftap- tevens meetsluis en de resultaten van een proef met een dergelijk kunstwerk',

De Waterstaatsingenieur, Vol. 14, No. 12 (1926), pp. 391–408; Vol. 15, No. 1 (1927), p. 1–15; Vol. 15, No. 2 (1927), pp. 25–34. Bos, M.G. (ed.): *Discharge measurement structures*, Wageningen: International Institute for Land Reclamation and Improvement, 1990. The APM block was 'adjustable' in the sense that it could be changed, but only on a seasonal basis, as adjustable meant 'use force to remove and then construct again'.

40. Buckley, R.B.: *Irrigation pocket book or, facts, figures and formulae, for irrigation engineers: being a series of notes on miscellaneous subjects connected with irrigation*, London, Spon, 1911; See: Ali, I: *The Punjab under imperialism 1885–1947*, Princeton: Princeton University Press, 1988; Bolding, A., Mollinga, P.P. and Van Straaten, K.: 'Modules for modernisation: colonial irrigation in India and the technological dimension of agrarian change', *Journal of Development Studies*, Vol. 31, No. 6 (1995), pp. 805–844; Stone, I.: *Canal Irrigation in British India: Perspectives on technological change in a peasant society* (Cambridge: Cambridge University Press, 1984.
41. De Gruyter, *Beschouwingen*, p. 70.
42. De Gruyter, *Een nieuwe sluis*, p. 392.
43. De Gruyter, *Een nieuwe sluis*, p. 392.
44. Groothoff, A.: 'Eenige mededeelingen over de waterkrachtindustrie in Scandinavië en over het waterkrachtvraagstuk in Nederlandsch-Indië', *De Ingenieur*, Vol. 33, No. 2 (1918), pp. 18–33.
45. Groothoff, *Eenige mededeelingen*, p. 32.
46. Which obviously relates to the (rather violent) change in government in Indonesia in the 1960s.
47. See Ertsen, M.W.: *Improvising planned development on the Gezira plain, Sudan, 1900–1980*, New York: Palgrave Macmillan, 2016; Hodge, J.M: *Triumph of the expert. Agrarian doctrines of development and the legacies of British colonialism*, Athens: Ohio University Press, 2007; Hodge, J.M., Hödl, G. and Kopf, M. (eds): *Developing Africa. Concepts and practices in twentieth-century colonialism*, Manchester: Manchester University Press, 2014; Kohlrausch, M. and Trischler, H.: *Building Europe on expertise. Innovators, organizers, networks*, Basingstoke: Palgrave Macmillan, 2014; Mehos, D. and Moon, S.: 'The uses of portability: circulating experts in the technopolitics of Cold War and decolonization, in Hecht, G. (ed.): *Entangled Geographies: Empire and Technopolitics in the Global Cold War*, Cambridge: MIT Press, 2011, pp. 43–74.

48. Tropical rural engineering ('tropische cultuurtechniek', the MSc programme I did myself between 1986 and 1993) was a mixture of a new programme on rural engineering and old irrigation-based courses. Other elements of the old programme survived as tropical crop science ('tropische plantenteelt').
49. Baudet, H.: *De lange weg naar de Technische Universiteit Delft. 1. De Delftse ingenieursschool en haar voorgeschiedenis*, Den Haag: SDU, 1992.
50. Groen, *Onderwijskundig overzicht*, 263.
51. Brouwer, A.R.H.: *Waterkracht perspectieven*, Technische Hogeschool Delft: Inaugurele rede, 1955, p. 18.
52. Eysvoogel, W.F.: Eenige aspecten van de moderne irrigatie-techniek in Indonesië, *Voordrachten van het Koninklijk Instituut van Ingenieurs*, No. 1 (1950), p. 341; See also Eysvoogel, W.F.: *De verbetering van den oostmoessonbevloeiingstoestand op Java*, Landbouwhogeschool Wageningen: Inaugurele rede, 1946.
53. As for example discussed in Clawson, M. (ed.): *Natural resources and international development*, Baltimore: John Hopkins Press, 1964.
54. The partnership between Belgium, the Netherlands and Luxemburg.
55. Berkhout, F.M.C.: *De waarde van kennis van irrigatie voor de Nederlandse civiel-ingenieur*, Technische Hogeschool Delft: Inaugurele rede, 1954, p. 16.
56. Vivekananthan, M.N.: 'Rehabilitation of irrigation systems in east Java, Indonesia', Rabat: International Commission on Irrigation and Drainage, 1987, pp. 147–173; *Reconnaissance survey Djratunseluna area*, 1968.
57. Gany, A.H.A.: *The irrigation based transmigration program in Indonesia. An interdisciplinary study of population settlement and related strategies.* PhD thesis University of Manitoba, Canada, 1993.
58. 'Bevloeiing in de Lampongsche Districten', *De Waterstaatsingenieur*, Vol. 6, No. 7 (1918), p. 296. Pelzer, K.J.: *Pioneer settlement in the Asiatic tropics. Studies in land utilization and agricultural colonization in Southeastern Asia*, New York: American Geographical Society, 1945.
59. *Irrigation design standards; Irrigation design manual; Supporting volume for irrigation design standards*, Jakarta: Ministry of Public Works, KP-01, 1986, p. 3.
60. Ertsen, M.W.: 'The development of irrigation design schools or how history structures human action', *Irrigation and Drainage*, Vol. 56, No. 1 (2007), pp. 1–19; See for US engineers and their international careers Teisch, J.B.: *Engineering Nature. Water, development, and the global spread*

of American environmental expertise, Chapel Hill: University of North Carolina Press, 2011; See for theoretical underpinning Van de Poel, I.: 'The transformation of technological regimes', *Research Policy*, Vol. 32, No.1 (2003), pp. 49–68.

61. Ertsen, *Locales of happiness*.

12

Engineers and Scientists as Commercial Agents of the Spanish Nuclear Programme

Joseba De la Torre, M. d. Mar Rubio-Varas, and Gloria Sanz Lafuente

Within the framework of the Cold War, the military and civil uses of the atom became the largest technological, socio-political and economic impact on both sides of the Iron Curtain. The unique nature of the technological, financial and managerial decisions about the civil uses of nuclear power granted a leading role to engineers and scientists employed

The original version of this chapter title was revised to *Engineers and Scientists as Commercial Agents of the Spanish Nuclear Programme.*
The correction to this chapter is available at
https://doi.org/10.1007/978-3-319-75450-5_15

Acknowledgements: the Spanish Ministry of Economy and Competitiveness (project ref. HAR2014-53825-R) and the European Commission and Euratom research and training programme 2014–2018 (*History of Nuclear Energy and Society (HoNESt)*, grant agreement N°662268) financed parts of the fieldwork required to compile sources for this chapter.

J. De la Torre • G. Sanz Lafuente
Economics Department, Universidad Pública de Navarra (UPNA), Pamplona, Spain

M. d. M. Rubio-Varas (✉)
INARBE, Universidad Pública de Navarra (UPNA), Pamplona, Spain
e-mail: mar.rubio@unavarra.es

© The Author(s) 2018

D. Pretel, L. Camprubí (eds.), *Technology and Globalisation*, Palgrave Studies in Economic History, https://doi.org/10.1007/978-3-319-75450-5_12

by the state and private companies.[1] That was the case in the USA as it rose to dominate the Western market for nuclear reactors by the mid-1960s. Engineers and scientists also played a crucial role in other advanced economies such as Germany when they began to scout out the sector. But what about underdeveloped economies? We aim at analysing the role of engineers and scientists as agents of economic modernization in Spain, which at the time was an underdeveloped economy ruled by an authoritarian regime.

The Spanish nuclear law of 1964 responded both to the pressure of the private electricity companies to procure the nuclear business for themselves and to the industrialization strategy propelled by the dictatorship.[2] A handful of modernizing engineers and scientists were decisive in the process. Presas,[3] as well as Camprubi,[4] unveiled the key role of Otero Navascués who was directing the Nuclear Energy Board projects. We know far less of the other engineers and scientists who joining international networks, transferred technological/scientific knowledge and trained qualified teams within Spain. They also acted as commercial agents. Acquiring foreign atomic technology required specialized knowledge in order to decide which of the different alternatives available in the market to choose. Engineers and scientists stayed at the core of all commercial negotiations and transactions involving nuclear technology. One of their basic tools was the cultivation of good relationships in order to influence institutions, industry, experts and especially policymakers, who had the last word. Finally, the history of the nuclear programme was determined by a very small group of people for whom mutual knowledge and trust were essential. Making use of new archival sources,[5] this chapter identifies some of these actors, in particular the Spanish industrial engineers Jaime MacVeigh and Manuel Gutiérrez-Cortines and the German scientist Karl Wirtz. These three actors played a strategic role in promoting the Spanish nuclear programme. But in order to understand the role played by these engineers, we first introduce the global market for nuclear reactors from the 1950s to the 1980s and explain that Spain became the principal market for the major international manufacturers of nuclear reactors during the decade between the mid-1960s and the mid-1970s.

The Global Market for Nuclear Reactors 1950s–1980s: Spain as a Major Client

On 20 December 1951, electricity was first generated from nuclear power at the EBR-I (experimental breeder reactor I) in Idaho, USA.[6] Yet the beginning of civil nuclear power is commonly set at President Eisenhower's address to the General Assembly of the United Nations on 8 December 1953, later called the 'Atoms for Peace' speech.[7] Most civil nuclear programmes around the world began and grew from the 1950s to the 1970s. The first nuclear reactors connected to the electricity grid—Obninsk in the Soviet Union by 1954, Calder Hall in the UK by 1956 and Shippingport in the USA by 1957—proved the concept, but were far from being commercially viable. They were small (5 MW, 50 MW and 90 MW respectively), each applying dissimilar technologies and with plenty of unknowns to be solved. Eventually, three types of reactors were commercialized internationally. The light water nuclear power reactor, using low enriched uranium as its fuel and ordinary water as its coolant and moderator, was built originally to a US design in Western countries and to a similar Soviet design in the USSR and Eastern European countries; The gas graphite reactor using natural uranium as its fuel, moderated by graphite and cooled by carbon dioxide, was a technological design favoured by the UK and France. Finally, Canada marketed a quite different nuclear power reactor using natural uranium as its fuel and heavy water as its coolant and moderator. In the end, it was the light-water reactor promoted by the USA that was triumphant over the more expensive gas-cooled reactors built by the British and the French in the 1950s.[8]

However, until the introduction of 'turnkey projects' in the nuclear business in 1962, with a bid for the construction of a plant at Oyster Creek, New Jersey, the commercial market for nuclear reactors remained stagnant. So did the international market: a total of ten reactors were ordered internationally up to 1964, with a few other reactors domestically ordered and under construction in the USA, the UK and the USSR.[9] The turnkey plants were offered by the nuclear reactor building companies at a guaranteed fixed price, set in advance, competitive with coal- and oil-fired alternatives. Even if in the long term it proved to be a bad business, in which the manufacturers lost money,[10] turnkey projects propelled domestic and international sales of nuclear reactors. Up to 1975, some 245 commercial reactors had begun to be built around the world in the thirty or so countries that decided to pursue nuclear energy. Of those, almost one hundred were ordered and were beginning to be built in the USA alone. All the US reactors were domestically built by five American manufacturers (General Electric, Westinghouse, Babcock &Wilcox, Combustion Engineering and Atomic General).

Just over one hundred reactors were ordered internationally (including Soviet sales to Eastern countries). General Electric and Westinghouse captured, with the help of US economic diplomacy and the financial assistance of the Export–Import Bank, almost 80% of the international sales of nuclear reactors to the Western countries up to the mid-1970s. During the second half of the 1970s, other Western manufacturers, which had been gaining experience by building nuclear plants in their countries, came to compete in the international market, mostly the German Kraftwerk Union, the French Framatome and the Canadian AECL. The British mostly failed to market their Magnox reactor to the world, while the Russians imposed their technology on the Eastern Bloc. By 1975, the curve of orders had already passed its peak in the USA.[11] Worldwide, 1976 marked the historical maximum of nuclear power plants beginning to be constructed, with building works being initiated for forty-three new nuclear plants.[12] Coincidentally, 1976 also saw the last governmental authorization for a new nuclear power plant in Spain.

As an early adopter of nuclear energy, by 1973 Spain already ranked third among countries, with the largest share of nuclear electricity over total electricity produced and being the seventh largest producer in the West.[13] In fact, over the first half of the 1970s, the Spanish electricity utilities became the largest nuclear clients of world nuclear manufacturers, just ahead of Japan and South Korea.[14] All the major international nuclear companies attempted to break into the Spanish market, which was the fastest growing nuclear power developer in Europe at the time.[15] The remainder of this chapter aims to expose the role that engineers and scientist played in helping the destitute, internationally ostracized and dictatorial country that was Spain become a relevant player in the global nuclear market.

Nuclear Optimism and Energy Planning by Entrepreneurs and Engineers: MacVeigh and Cortines as Atomic Leaders

In Spain, experience in managing large engineering works, handling credit and establishing contact with the USA and European firms enabled the electrical lobby to take on the nuclear programme. Since the 1940s, the electricity companies had built up groups that comprised highly skilled engineers who were used to collaborating with foreign experts on large engineering projects. Some industrial engineers became the true architects of the Spanish nuclear strategy, taking on the functions of business leadership, project management, public dissemination of energy policy and the orchestration of a lobby. This is our main hypothesis. These engineers had first-hand knowledge of what the USA had been doing since 1945, but also knew of the limitations of Spanish industrial capacities. They frequented the meetings of the Atomic Industrial Forum in New York and reacted to the business expectations that were put forward at the 1955 Geneva Conference. Upon return from Switzerland, a group of these industry captains had a meeting with General Franco to convey their objectives the nuclear programme should be the responsibility of private enterprises.[16] It was unclear at this stage whether the nuclear civil programme was to be a private endeavour in Spain or whether it would continue in the hands of the government's Nuclear Energy Board (JEN).[17] A British report summarized the situation in early 1957: at least 80% of the electrical supply industry was controlled by a few private companies in association with the main Spanish banks; the electrical engineering industry was developing on a fairly satisfactory scale stimulated by the demand from state and private power production schemes; the larger utility firms were closely associated with French, German and US firms under patent agreements.[18] 'The range of small products is increasing rapidly, but there is no plant for production of the massive electrical generating components required in modern base load stations.'[19] The utilities therefore knew their weaknesses and recognized the need to establish strategic alliances.

The next step entailed the creation of two business consortia to build nuclear power stations in regions that were historically controlled by private enterprises, thus enabling them to distribute a substantial part of the future of the nuclear market and reinforce their unequivocal commitment. The Northern Nuclear Plants SA (Nuclenor: Iberduero and Electra Viesgo) and the Centrales Nucleares SA (Cenusa: Unión Eléctrica Madrileña, Hidroeléctrica Española and Sevillana de Electricidad) for the southern region were founded by electrical and financial entrepreneurs but managed by engineers.[20]

Jaime MacVeigh Alfós (191?–1985),[21] and Manuel Gutiérrez-Cortines (1901–1980),[22] were two of them. Both were industrial engineers and had developed contacts with foreign firms. Son of a counsellor of the Hispano-American Bank, MacVeigh finished his studies in the School of Industrial Engineers of Madrid in 1943. Between 1945 and 1950, engenieer MacVeigh had a managing role within the Spanish TALGO project (Tren Articulado Ligero Goicoechea Oriol, Goicoechea-Oriol light articulated train). The construction of the famous articulated train was made in conjunction with American Car & Foundry, taking MacVeigh to some of the nuclear industry sites in Pennsylvania, Delaware and New Jersey. Upon his return to Madrid, he abandoned TALGO and was hired by the Banco Urquijo. MacVeigh went on to specialize in nuclear projects, convinced that they offered 'the solution to Spain's electric problem'.[23] Cortines, in the other hand, was managing director of Standard Electric in Spain and Portugal before the civil war. From 1940, he worked as counsellor in Electra Viesgo, one of the future partners who built Garoña nuclear power plant (NPP) in the 1960s.

Between 1956 and 1962, MacVeigh and Cortines, as technical advisors and counsellors for Cenusa and Nuclenor respectively, developed a very long-term expansion strategy. Together with the most significant executives of the electric companies, both engineers participated in the Assessment Commission of Industrial Reactors (CARDI) at the Spanish Nuclear Energy Board (JEN), where they established alliances with scientists, experts and businessmen.[24] MacVeigh's public presence had been constant since 1955, spreading the idea that Spain should bet on the economy of the atom. His political and business contacts abroad allowed him to know precisely the technological advances that were being made in the USA and in Europe. Indirectly these two engineers contributed to the turn of the governement's industrial policy that left the business of nuclear power plants in the hands of private firms. In short, these two men become a turning point in the commitment to nuclear power and in the maturation of an industrial sector that was unprecedented in Spain. In 1957, MacVeigh collaborated in founding the nuclear engineering company Técnicas Atómicas (Tecnatom SA).[25] Through this company, MacVeigh directed and led the project of the first Spanish nuclear plant on the Tajo river, at Zorita, and meanwhile Cortines initiated the drafts for the so-called Ebro–Bilbao nuclear power station. Simultaneously both leaders networked with the national atomic agencies of the USA and the UK, as much as with the chief multinationals in the atomic business. Last but not least, they negotiated contracts with international banks and suppliers and supervised the building process once it began. They shared learning and commercial trips. As Cortines wrote some years later, one of the reasons why the Spanish programme had adopted three different technologies was the negotiations undertaken by each company. Each reactor belonged to a different utility and 'all of them were offered under excellent economic conditions'.[26]

Why this passion for nuclear energy? Firstly, because it was a political decision, 'Atoms for a dictatorship'.[27] Francoism embraced the nuclear programme for military and economic reasons. Nevertheless, and secondly, we believe that it was possible not only because of the times of atomic optimism in which the Geneva Conference opened, but also because it was technologically one of the main challenges for engineers of the mid-twentieth century. And of course, the electrical utilities saw a business opportunity to conquer and, in Spain, they had excellent political contacts through which to do this. In this sense we can understand why the first non-governmental technical/economic report on the need to incorporate nuclear energy into the Spanish energy matrix was produced by the Research Service of Banco Urquijo, an industrial bank with important interests in electricity, in February 1957.[28] Entitled 'Essay on a nuclear programme for Spain' (*Ensayo sobre un programa de energía nuclear en España)* and written by Jaime MacVeigh, it concluded that before 1970 Spain would face a major deficit of traditional electricity generation that could only be offset by introducing nuclear power. Given the urgent need to meet the excess demand, MacVeigh urged in his essay that work should start on the first commercial nuclear power plants no later than 1964 and continue after that at an exponential rate.[29] MacVeigh's hypothesis on the growth between 5% and 7% annual of the Spanish electric demand from 1950 to 2000 proved to be excessively conservative in the short and medium term, although remarkably precise in the long term. His nuclear projections, however, were extremely optimistic. At the height of the Spanish nuclear programme, in the final phase of the dictatorship between 1971 and 1976, the installed capacity would never have exceeded 33 GW if all the projects that the utilities applied for had become operational.[30] This engineer knew the obstacles that a project such as the one he was proposing would have to face. MacVeigh recognized that the technical problems had been solved but the economic ones had not.[31] The report from the Urquijo Research Service admitted that the cost of nuclear kWh produced at commercial nuclear power plants at that time was no less than twice the current fossil fuel costs. MacVeigh further stated in his essay that the order of magnitude of investments in nuclear power plants from 1965 to 2000 would have to be between 150 and 280 billion pesetas of 1955 (from $14 to $25 billion at the official exchange rate). That was excluding the additional facilities that a nuclear programme would entail. Including those, the full development of the programme could account for 25–40% more.[32] In other words it was a great opportunity to do business and to promote the industrialization of the country.

MacVeigh, Cortines and other businessmen and technocrats were invited to visit the British nuclear installations in 1957. Although in the end the British did not get any atomic contracts from Spain, this contact was very important for the utilities' managers. For instance, a report titled 'Training and Construction Programme for a Nuclear Power Project in Spain' summarized what should be the process of training a team of engineers to build an atomic plant. In 1958, MacVeigh conducted a training course for senior executives that would be very useful in the Zorita project. Briefly, and according to the British experts, the most important consideration was to nominate at the earliest possible moment the chief engineer for the whole project, since this training would be of long duration, and he must be allowed to choose his own staff. The requirements for the chief engineer were that he should have considerable experience in construction of conventional power station plants, as well as strong technical and administrative ability to draw up enquiries for tenders and to supervise construction of the plant. During the first year, the chief engineer's main task would be to recruit his team and to start its training at the Hartwell Reactor School and with the Calder construction teams. The complete team would contain a similar number of engineers experienced in conventional power station practice, covering such areas as turbines, alternators, instrumentation and switchgear. The chief engineer would pay visits to the industrial firms from whom tenders were being sought and would also spend some time at the research station operated by JEN in Spain and the UK Atomic Energy Authority (UKAEA). Construction could be started within three months of planning an order and should be in operation about three-and-a-half years afterwards. The chief engineer would supervise construction and ultimately operation of the station, and he would supervise the contractors with the assistance of his design engineers, the operation superintendent and the maintenance engineer.[33] Eventually, MacVeigh as chief advisor of Tecnatom would apply this very same strategy in the first plant that was built in Spain, but using American firms instead.

On a commercial mission to England in 1960, MacVeigh made a relevant situation diagnosis. His company's plans relied on the government's opinion. He believed that the Spanish industry could participate in a conventional power station by manufacturing turbines, instrumentation, alternators and steam raising equipment, but excluding the fuel cycle. For this reason, he considered the JEN trials to be a big mistake. MacVeigh declared himself a supporter of the plan to import the reactor and stated that he was familiar with the US financial facilities. To achieve this, it was necessary to influence the decision-makers. Gutiérrez-Cortines personally told Minister of Industry Joaquín Planell that the companies 'doubted whether the government intended allowing the private utilities to build nuclear stations'. The answer was 'the Government would support [...] at the appropriate time'.[34] The problem, in fact, was that the people in charge of the autarkist policy were still in their posts and did not understand the meaning of the new economic policy. By late 1961, MacVeigh was more forceful in his strategy and drafted a confidential report wherein he was critical of the policy of the JEN and its President.[35] This report was given to López Bravo, manager director of Spanish Foreign Currency Institute (IEME), and soon after to become Industry Minister. In MacVeigh's opinion, JEN's plan to proceed with the experimental projects was illusory and expensive. Moreover, the fuel cycle would not be viable until 1970 and would be hardly able to cover the needs for uranium; this would delay the rollout of the power stations. Therefore, MacVeigh's report advised supporting the private initiative that was prepared to immediately construct 'a small and standard plant with a future'; it would be of the 'boiling water' type and come at a 'reasonable' cost. Those running the Zorita project firmly believed that they had to speed up the nuclear race by using US technology and collaborating with the Spanish private companies. MacVeigh was right to consider that everything depended on two essential factors: 'the capacity of Spanish industry' to respond to this challenge and 'the (economic) liberalisation' that access to the foreign market would provide.[36] No doubt MacVeigh knew whom he was addressing. López Bravo—'Mr. Efficiency' as he was nicknamed by a well-known banker—would have all the authority and the last word in awarding the nuclear programme and the tenders. IEME's manager was responsible for liberalizing the foreign currency market and foreign transactions, and as the Minister of Industry promoted an industrial policy that was favourable to foreign capital in general terms, and to American firms in particular.

Furthermore, Cortines and MacVeigh collaborated in founding the Spanish Atomic Forum (Fórum Atómico Españo, FAE) at the end of 1961. FAE became the Spanish nuclear lobby, since it was an alliance of industries that sought a market niche in nuclear energy: steel, metal-mechanical, shipbuilding, chemical and electronic companies, in addition to electricity producers and distributors and engineering and consulting firms. FAE had the support of public and semi-official entities and became part of the European Atomic Forum (Foratom), a necessary link for its internationalization and for complying with nuclear safety regulations. In addition, as vice-presidents of the lobby, MacVeigh and Cortines were counted among the instigators of the project for a European fast fusion reactor, necessarily encountering scientists and the atomic business ecosystem in Europe and North America. In fact, FAE and the Spanish Nuclear Board organized several meetings with French, British, North American and Western German industries during the 1960s.

It should be noted that Spanish industry did not start from zero. The experience gained in conventional thermal power plants, at least since 1950, had allowed engineers to conduct major projects and develop mechanical and electrical assembly techniques. In addition, the factories making specialized machinery and auxiliary equipment for the generation and industrial use of electricity, sometimes with foreign patents, had benefited from manufacturing growth during the 1950s.[37] The nuclear programme was identified as an opportunity. These companies had to innovate in product, techniques, knowledge and management. The manufacture of capital goods and industrial assemblies requires large-scale engineering and consulting services. Some of these firms were born at the time and others, which already existed, adapted to the new challenge by diversifying their production lines and building up strategic links with foreign companies. This was a way in which to solve the two great shortcomings of the Spanish companies: capital and knowledge. In MacVeigh's words, nuclear participation meant 'improving, in general, its construction and productivity standards'.[38] The challenge was to build the first nuclear power plants. Zorita and Garoña were the litmus test for MacVeigh and Cortines.

While Cortines served as an executive officer,[39] MacVeigh acted as a chief engineer with some additional management functions. Even though Cortines referred to Garoña as 'my nuclear plant', the execution of the contract was the direct responsibility of the American multinational. General Electric coordinated, controlled and subcontracted all phases of the project through its subsidiary GE Technical Services Co. (GETSCO). So Cortines negotiated capital goods contracts and financial credits, but was not directing the works. On the contrary, Westinghouse Electric International and its subsidiary Westinghouse Atomic Power of Spain relied on Tecnatom, the Spanish consultancy directed by MacVeigh, to supervise the entire Zorita's project of engineering and management. Why did WEICO confer so much responsibility on a local consulting firm that GE disregarded? In 1965, Tecnatom was only a little company, with a small but highly qualified staff, and with more theoretical knowledge than experience in atomic technology. It is true that Garoña was triple the size of Zorita, which made the project more complex. GE had to consider, with good reason, that the Spanish engineering and consultants were still at an embryonic stage, considering the degree of specialization, technical assistance and learning that were required. Our hypothesis is that the main intangible was embodied in the figure of the advisor-delegate MacVeigh, the project leader who had worked in USA, communicated well in English and personally knew WEICO and also possessed the knowledge and contacts to operate in a complicated country such as Franco's Spain. With all this, MacVeigh could convince the Americans and the policymakers and impose his criteria. MacVeigh put into practice a very personal management structure in Zorita NPP. He negotiated import licences and dollars with the Spanish authorities and contracted logistics in Europe or the USA for the transport of the vessel and the reactor. He also supervised the exit and return of the Spanish uranium to be enriched in America, and agreed loans with public and private banks abroad in a variety of different currencies.[40]

Once the power plant programme was in place, the Minister of Industry gave priority to the future fast breeder reactor. López Bravo again sought the support of the JEN and the nuclear industry. MacVeigh and Cortines also played a key role. On this occasion, they sought to approach European partners and use their old networks in the UK, France and West Germany. The point of connection was the supply of uranium for Zorita NPP. At least since 1962, the UKAEA and Tecnatom had held meetings and negotiations on this issue. According to a British memorandum of September 1968, the Spanish government was very interested in advancing this new technological development. The Energy Director in the Ministry for Industry, 'who is from private industry', expressed that Spain needed nuclear power and the fast reactor seemed the most promising way of obtaining it. He asked about the British plan and requested a visit to the Dounreay installations. Some days later MacVeigh confirmed his own interest in the fast breeder. 'He knew in outline' of the British proposal for Spain and confessed that the American Betchell were 'very impressed indeed with the British fast reactor position and were convinced' that 'it would be adopted in the USA'. MacVeigh 'was personally prepared to back it and recommend his organisations, Tecnatom and UEM, to put money into it'. The report concluded by stating that 'as you know MacVeigh is an influential man in Spain and is a very good friend of ours'.[41]

However, MacVeigh's time at Tecnatom was running out. This consulting engineering company had been created to build Zorita. Once completed and connected to the network in 1969, Tecnatom entered a phase of uncertainty. The board of directors was given a two year trial period, after which its future would be evaluated, not excluding the possibility of dissolving the company. This situation was perhaps decisive in MacVeigh's decision to leave the company in late 1969.[42] Banco Atlántico hired him, a firm linked to the finances of Opus Dei and expanding its industrial businesses. Meanwhile, Cortines acceded to the presidency of the Spanish Atomic Forum and to the council of the European Forum, and tried to advance the development of the fast reactor with partners in Europe and America.

From Scientific Knowledge to Entrepreneurial Ecosystem: Karl Wirtz and the Spanish Atomic Programme

The technological and financial challenges entailed by atomic development led to the early appearance of international projects during the 1950s. The idea was to share experiences, costs and financial risks in this new sector, which still had uncertain technological potential. Karl Wirtz, one of the scientists behind the atomic bomb of Hitler's Third Reich, had been considered 'an old friend of Spain' by the JEN since at least 1949. Private firms, entrepreneurs, diplomatic services and political decision-makers played an important role in shaping the Spanish atomic network between the 1950s and 1970s. However, no national or international atomic programme can be understood from reductionist approaches that refuse to interrelate politics, technology, commerce and culture.[43] Nor can such a programme be explained by one sole actor. In contrast, only an investigation of the connections between all social spaces and multiple actors enables the development of a historical explanation. In addition, these actors moved in substantially different political and economic contexts. We need only mention the political distance between Franco's Spain and the economic development policy of the Federal Republic of Germany(West Germany), a country that was democratic, industrialized and fully integrated into Europe.

If we adopt the day-to-day activities of a sales department manager as our point of reference, Karl Wirtz was not, strictly speaking, a sales representative. He was not in charge of managing private businesses. He was not authorized to negotiate contracts or credits. He did not define distribution channels or set prices, loans or interest charges for nuclear services and products. Nevertheless, his trips and meetings outside West Germany formed part of a long chain of public and private relationships that shaped the commercial and scientific ties between countries interested in implementing atomic programmes.[44]

From 1972 to 1979, Wirtz served as an organizer and consultant to the West German federal government regarding the transfer of nuclear technology from West Germany to Spain. The reasons cited for his appointment were his scientific qualifications and his familiarity with the Spanish market. Wirtz's contact with the JEN dated to the creation of the Board, as recalled by Otero Navascués, with whom Wirtz had shared a 'close friendship' since 1949. But who was Karl Wirtz (1910–1994)? Receiving his doctorate in physical chemistry in 1934 as a heavy-water specialist, Wirtz had completed his academic training in the tumultuous context of the purge of Jewish scientists and any other scientist believed to oppose the Nazi dictatorship. In 1937, he became an assistant at the Physics Institute of the Kaiser Wilhelm Society. As late as September 1939 he maintained international contacts with British and American colleagues, which the war would later end.[45]

'I have done that.' 'I conducted the negotiations with the firms.' 'I have done the experimental work.' These were statements that Wirtz made sure he emphasized during the Farm Hall interrogations. A cottage located near Cambridge in England, Farm Hall was the internment centre to which Wirtz was taken in 1945 with the other scientists who had worked on the Third Reich's atom bomb project. Werner Heisenberg, Otto Hahn and Carl Friedrich von Weizsäcker all spent time there. Ambition, lack of humility and a combination of scientific skills and the ability to communicate with firms were elements of the image that Wirtz projected of himself. He seemed optimistic as well, stating that the odds of him and his colleagues making a quick return to Germany were 70%, in contrast with Heisenberg's doubts. Although Wirtz came to believe that he and the others would be taken to the USA to work, in early 1946

Wirtz and the rest of the group returned to Göttingen in the British occupation zone amidst post-war difficulties and at a time when the city was teeming with refugees.[46] The scientific reconstruction of Western Germany began before the establishment of West Germany as a state (1949). For example, the Max Planck Society was formed in the British zone in September 1946 and authorized in the other zones in 1948.[47] Albeit with difficulties, German nuclear scientists resumed their international relationships quickly during the formative years of the international nuclear community. Wirtz underwent a rapid denazification process and travelled to Brighton in September 1948 for the annual meeting of the British Association for the Advancement of Science. In 1949, he was invited to visit the University of Melbourne (Australia). He spent the last few months of 1950 in Argentina invited by the Nuclear Energy Commission of that country, reestablishing contact with former colleagues there who would later return to West Germany. Thanks to the efforts of the French chemist Bertrand Goldschmidt, and not without causing tensions, Wirtz attended a conference on reactor technology in Kjeller (Norway) in 1953. There he met Alwin Weinberg of the American Oak Ridge National Laboratory. In fact, he never cut ties with his former colleagues who joined the American scientific system, and he made his first trip to the USA in 1955. He returned a year later, joined by the head of the DEMAG corporation, which merged with Atomics International to form the Interatom corporation.[48]

Wirtz's international presence played a part in the first Adenauer government's interest in the nuclear economy, despite the fact that the Paris Accords, along with other legal measures, prohibited nuclear production and research in Germany for civil and military uses until 1955 and the enrichment of uranium until 1960.[49] The initial relationships and exchanges created shared social spaces for industrialists and scientists, such as the Society for the Study of Physics (Physikalische Studiengesellschaft, PSG) founded in Düsseldorf in 1954. Although he had already started to form a group of young scientists and engineers in Göttingen, Wirtz leveraged the PSG to assemble a reactor research group, which he moved to Karlsruhe.[50] As a professor and director of the Institute for Neutron Physics and Reactor Technology, he later presided over the scientific council of the Karlsruhe Nuclear Research Centre (Kernforschungzentrum Karlsruhe,

KfK) in Baden-Württemberg until 1978, an institution that with the Jülich Research Centre (Kernforschungsanlage Jülich, KfA) in North Rhein-Westphalia was one of two scientific, academic and incubator sites for research and development in this sector.[51] The advances that emerged from the KfK included a new uranium enrichment technology similar to Erwin W. Becker's nozzle process, and utilized by the Steag AG Corporation, and a system for monitoring fuel movements, which was conceived by Wirtz and enhanced by Wolf Häfele. From 1956 to 1971, Wirtz was a member of the German Atomic Commission, and he directed that agency's 'reactor circle' until 1966. He also contributed to designing Germany's first and second atomic programmes. The relevance of this complex institutional, public and private framework lies in the implicit, constant cooperation it represents between the scientific system and the firms that shared the strategic objective of internationalization despite conflicts over development and financing.[52]

In this historical context, Wirtz came to play the role of mediator between scientific institutions and nuclear firms in Spain and West Germany, although Spain was not the only hub of his international activity. The first step along this path consisted of academic collaboration. Relationships with Spain were resumed immediately after the war in Göttingen to enable Spanish physicists to research their doctoral theses.[53] Otero Navascués travelled there in 1949, having taken an interest in the German pioneers of reactor technology. In March 1950, the newspaper *Die Welt* reported that Wirtz would travel to Madrid at the request of the Franco government to assist in the establishment of a cosmic radiation laboratory. The article was referring to the future research centre of the JEN in Moncloa, which was intended to replicate the UK's Harwell centre and France's Saclay centre.[54] Additionally, Wirtz was one of the first foreign lecturers at the Applied Nuclear Physics Course organized by the Spanish National Research Council (Consejo Superior de Investigaciones Cientificas, CSIC) between November 1950 and July 1951. According to Aschmann, the interest of the National Institute of Industry (Instituto Nacional de Industria, INI) in acquiring German scientific and engineering know-how figured in the background of that first trip.[55] In fact, Manuel Espinosa Rodriguez, former naval attaché to the embassy in Berlin, travelled to the three Western occupation zones to rekindle former relationships and recruit specialized technicians. To this end, Lieutenant Ignacio Moyano wired approximately US $30,000 from the Spanish Institute of Foreign Currency (IEME) to the Bank Deutscher Länder in Frankfurt. Ten scientists expressed interest, including Wirtz.[56] In the 1950s, Wirtz made additional trips to Spain and corresponded with Otero Navascués, Maria Aranzazu Vigón, Carlos Sanchez del Río (JEN) and Romero Ortiz (Geological and Mining Institute).[57] At the foundation established with JEN, Spanish scientific experts, politicians, and commercial managers started to closely interact in atomic programmes and national and international forums. During the early stages, the search for uranium and other nuclear fuel was prioritized. Later, between 1955 and 1960, Wirtz advised JEN on test reactor development, arguing in favour of natural uranium to avoid dependence on the American monopoly on enriched uranium, just as France, the UK and West Germany were seeking to do. Wirtz also negotiated with the German Nuclear Affairs Ministry to obtain financing for JEN. In 1958, he attended the inauguration in Moncloa of the experimental reactor acquired from GE.[58]

Wirtz recognized that his interventions in the 1950s were showing results in the scientific realm but that he would only begin to prosper in commercial terms in the long run, specifically in the 1960s. However, Spanish–German relations weakened in the early 1960s primarily because German industry remained incapable of exporting a nuclear power plant. A secondary reason for these weakened relations was likely the fact that the main atomic products at the time, power plants, were controlled by multinationals from other industrial powers. In Spain, while Vandellós was a French project, Zorita and Garoña were the work of American corporations, WESCO and GE, where MacVeigh and Cortines were pioneering the nuclear business in which they would eventually meet Wirtz.

During this period of nuclear optimism, improving the technological capacity of reactors and ensuring a sufficient fuel supply posed new challenges. Thus the fast reactor and the fuel cycle occupied a substantial portion of the scientific, diplomatic and commercial nuclear agenda in the USA and Europe. In fact, Wirtz and Häfele expressed their scientific interest in advanced, or fast, reactor prototypes at the KfK beginning in the late 1950s. Such prototypes were not exclusively a German development but were included in the scientific plans of the leading nations. In 1960, financing for the project was approved by the German Atomic Commission. The effort's financial and technical complexity required multilateral collaboration and support from the European Atomic Energy Community (Euratom).[59] Another business opportunity that drew together scientific research and the nuclear industry was the fuel cycle. Amidst rapid expansion, progress had to be made in enriched uranium manufacturing for existing power plants and particularly for future technologies for electronuclear production, an essential process subject to the US and Soviet duopoly of that time.

The fast breeder and nuclear fuel manufacturing had been prominent among the concerns of Minister of Industry Lopez Bravo since at least 1966, and these matters were addressed by G. Soltenberg and Otero Navascués during their visit to the uranium mines of Andújar that year.[60] At that time, there were already Spanish scientists in the KfA and the KfK in West Germany. The correspondence between Wirtz and Sanchez del Rio, the head of JEN's Nuclear Physics and Chemistry Division, reveals their close collaboration on the fast breeder.[61] In 1968, Wirtz travelled to Madrid with five collaborators to manage JEN's participation in the Projekt Schneller Brüter (PSB).[62] This scientific and technical cooperation was extended in 1970.[63] However, JEN was not the only organization involved in the project. The PSB had also attracted Unión Eléctrica Madrileña, SA (UEM), the developer of the Zorita nuclear power plant and its expansion. Wirtz kept in touch with the old MacVeigh circuit in Tecnatom and UEM.[64]

The second entry point to the atomic entrepreneurial ecosystem was nuclear fuel manufacturing. A September 1968 entry in Wirtz's travel diary described at length his contacts with the leadership of JEN, making note of the individuals he referred to as 'Mir wichtiger Industriemänner', that is 'men of industry who are important to me': López Bravo, Cortines, Oriol, Urquijo, Kaibel, Sendagorta, Mendoza and Millán, among others. In institutional terms, this list referred to the Ministry of Industry, the Atomic Forum and several of the firms involved in nuclear projects: Ibernuclear, Hidroelectrica Española, Sener Consulting, the Spanish Society for Naval Construction, Unesa and Babcock & Wilcox Spain. Wirtz added: 'López Bravo has influence over Cortines' decision about fuel manufacturing.' The scientist was well informed. The minister's efforts resulted in the creation of Ibernuclear, a mixed partnership (70% private and 30% public) that was managed by Cortines, 'the most important man at the time'.[65] This partnership was placed in charge of site studies for a fuel plant for light water natural uranium reactors.[66] In fact, Cortines had contacted Heinrich Mandel of the West Germany electricity utility Rheinisch-Westfälisches Elektrizitätswerk (RWE) and visited Wolfgang, the headquarters of Schimmelbusch's NUKEM (Nuklear-Chemie-Metallurgie).[67] He also visited the headquarters of the KfK, where he was joined by the head of the UEM, Julio Hernández Rubio.[68]

Under the mandate of López de Letona, the Ministry of Industry, the INI and the electricity utilities promoted the National Uranium Enterprise (Empresa Nacional del Uranio, ENUSA) as a fuel manufacturer and expanded contacts with foreign industries that possessed that technological capacity. Such was the case with the Kraftwerk Union (KWU). Product of a merger of the reactor departments of the Allgemeine Elektricitäts-Gesellschaft (AEG) and Siemens, KWU began to submit bids in 1969 to construct reactors in international calls for tenders and presented offers in Spain for the nuclear power plants in Almaraz, Lemoniz, Ascó and Cofrentes. Managers led these international relationships. However, they did so in close communication with the diplomatic services and institutions such as the KfK through the Society for Atomic Research (Gesellschaft für Kernforschung, GfK).[69] The first agreement between the GfK and JEN was signed in Madrid in 1973. Foreign Minister López Rodó and Wirtz were in attendance, along with Greifeld (GfK), Sánchez del Río, Oltra (JEN) and representatives of Interatom, which still belonged to KWU.[70] However, the challenge of obtaining an important nuclear contract in Spain would remain unmet until 1975. Although Wirtz was not the decisive actor, he represented an important player in a complex process that would eventually enable KWU to construct the Trillo nuclear power plant, the only such plant to be built by a German corporation in a booming market that was practically monopolized by American firms.[71] A 1976 letter to the KWU president in which Wirtz offered contacts in Galicia to help KWU win the call for tenders for the Regodola nuclear power plant reveals that the context had changed.[72] By the mid-1970s, business managers had become the key decision-makers in megaprojects.

Not every decision passed through Wirtz, and not all interventions were rewarded with success. However, his work as an enduring catalyst of public and private relationships in the international nuclear community was crucial in the rapprochement between German and Spanish firms and scientific institutions in the nuclear field. A bridge between nuclear scientific knowledge and the Spanish and German entrepreneurial ecosystems, Wirtz made these connections part of his career and his activity with the KfK. He did so at times without complete information and aware of the limitations through collecting data and contacts, linking evidence and fostering trust. How did he do it? By introducing himself into the lines of communication between key industrial actors and political decision-makers in Spain, by writing reports that he distributed to leaders of industry, politicians and scientists, providing information and his impressions of these meetings, and by constantly promoting scientific cooperation.

Notes

1. Balogh, B.: *Chain Reaction: Expert Debate and Public Participation in American Commercial Nuclear Power, 1945–1975.* Cambridge: Cambridge University Press, 1991.
2. De la Torre, J. and Rubio-Varas, M. d. M.: 'Nuclear Power for a Dictatorship: State and Business involvement in the Spanish Atomic Program, 1950–85', *Journal of Contemporary History*, Vol. 51. No.2 (2016) 385–411.
3. Presas, Albert: Science on the periphery. The Spanish reception of nuclear energy: An attempt at modernity? *Minerva*, No. 43, Iss. 2 (2005), p. 197–218.
4. Camprubí, L.: *Engineers and the making of the Francoist regime.* Cambridge, MA: MIT Press, 2014.
5. Through this chapter we make use of materials from the following archives: Archivo Histórico de la Sociedad Estatal de Participaciones Industriales [ASEPI] (Madrid, Spain); Archivo Histórico del Banco de Bilbao Vizcaya Argentaria [AHBBVA] (Bilbao, Spain); Export-Import Bank Archives [EXIM] (College Park, Maryland, USA); Generallandesarchiv Karlsruhe [GLA] (Karlsruhe, Germany); Politisches Archiv des Auswärtigen Amts [PA AA] (Berlin, Germany); The National Archives of United Kingdom [TNA] (Kew, United Kingdom).

6. Rubio-Varas, M. d. M. and De la Torre, J.: 'How did Spain become the major U.S. nuclear client?', In Rubio-Varas, M. d. M. and J. De la Torre, (eds.), *The Economic History of Nuclear Energy in Spain: Governance, Business and Finance*. London: Palgrave Macmillan, 2017, pp. 116–54.
7. Eisenhower, Dwight: 'Atoms for Peace Speech', 1953. Available at: https://www.iaea.org/about/history/atoms-for-peace-speech. See also Drogan, Mara: 'The Nuclear Imperative: Atoms for Peace and the Development of U.S. Policy on Exporting Nuclear Power, 1953–1955', *Diplomatic History*, Vol. 40, No. 5 (2016), pp. 948–74.
8. United States, General Accounting Office, Comptroller General Report to the Congress, 'U.S. Nuclear Non-Proliferation Policy: Impact on Exports and Nuclear Industry Could Not Be Determined', Washington D.C: US Government Printing Office, 1980, pp. 8–9.
9. Rubio-Varas, M. d. M. and De la Torre, J.: 'Spain—Eximbank's Billion Dollar Client: The Role of the US Financing the Spanish Nuclear Program', in Beltran A., Laborie L., Lanthier, P. and Le Gallic, S. (Eds.), *Electric Worlds / Mondes électriques: Creations, Circulations, Tensions, Transitions (19th–21st C)*. London: Peter Lang, 2016, pp. 245–70.
10. Burness, H. Stuart, Montgomery, W. David and Quirk, James P.: 'The Turnkey Era in Nuclear Power', *Land Economics*, Vol. 56, No. 2 (1980), pp. 188–202.
11. Cohn, S.: *Too Cheap to Meter: An Economic and Philosophical Analysis of the Nuclear Dream*, New York: State University of New York Press, 1997, p. 127.
12. International Atomic Energy Agency (IAEA): *Nuclear Power Reactors in the World*, Vienna: IAEA, 2016, p. 79.
13. Central Intelligence Agency (CIA): *Nuclear Power and the Demand for Uranium Enrichment Services*, Washington: Government Printing Office, 1974. Table 1., p. 4. [CIA Document Number CIA-RDP85T00875R001900030].
14. Rubio-Varas and De la Torre, 'Spain—Eximbank's Billion Dollar Client', p. 239.
15. Ibidem.
16. Gutiérrez Cortines, Manuel: 'Las centrales atómicas en los programas de construcción de las empresas eléctricas', *Conferencia Pronunciada al Círculo de La Unión Mercantil e Industrial de Madrid*. Madrid: Círculo de la Unión Mercantil e Industrial de Madrid, 1958.

17. The Junta de Energía Nuclear (JEN, Nuclear Energy Board) created by the Spanish government in 1951 (springing from preexisting units for atom investigation) generated similar institutional structures for the development of nuclear civil uses as those of other Western governments at the time.
18. TNA. Foreign and Commonwealth Office [FCO). H.R. Jonson, Technical Policy Branch, Notes 7th January 1957).
19. TNA. Foreign and Commonwealth Office [FCO). H.R. Jonson, Technical Policy Branch, Notes 7th January 1957).
20. Romero de Pablo, A. and Sánchez-Ron, J.M.: *Energía nuclear en España. De la JEN al CIEMAT*. CIEMAT, Madrid: Ediciones Doce Calles, 2001.
21. Valbuena, Pablo: *Historia de la Escuela Técnica Superior de Ingenieros Industriales de Madrid desde 1901 hasta 1972.* Proyecto fin de carrera (1996). De la Torre, J: 'Who was Who in the Making of Spanish Nuclear Programme, c.1950–1985', In Rubio-Varas, M. d. M. and J. De la Torre, (eds.), *The Economic History of Nuclear Energy in Spain: Governance, Business and Finance.* London: Palgrave Macmillan, 2017, pp. 33–65.
22. Alzugaray, J.J.: *Reflexiones de un ingeniero,* Madrid: Eds. Encuentro. 1999, pp. 73–4.
23. This was natural since the financier who promoted the TALGO first and the nuclear bet for Hidroeléctrica Española later was Jose María Oriol y Urquijo. For more on this see Ballestero, Alfonso: *José Mª de Oriol y Urquijo*. Madrid: Lid, 2014.
24. De la Torre, J., and Rubio-Varas, M. d. M.: *La financiación exterior del desarrollo industrial español a través del IEME (1950–1982)*, Banco de España, Madrid, 2015, pp. 99–105.
25. Álvaro, Adoración: 'The Globalization of Knowledge-Based Services: Engineering Consulting in Spain, 1953–1975', *Business History Review* 88 (2014), pp. 681–707.
26. Gutiérrez Cortines, Manuel: 'Nuclear Industry in Spain', *Nuclear Engineering International* 17, no. 188 (1972), pp. 31–32.
27. De la Torre and Rubio-Varas: 'Nuclear Power for a Dictatorship'; Otero Navascués, José Mª: Nuclear Energy in Spain, *Nuclear Engineering International* 17, no. 188 (1972), pp. 25–28.
28. MacVeigh, J.: *Ensayo sobre un Programa de energía Nuclear en España*, Madrid: Banco Urquijo, 1957; Puig, N. and Torres, E.: *Banco Urquijo. Un banco con historia,* Madrid: Turner, 2008.

29. According to the hypotheses of the essay, nuclear power installed in 1970 should be 0.35 GW, peaking by the year 2000 between a minimum of 21 GW and a maximum of 39 GW of nuclear power generation; the equivalent of a minimum of 106 to a maximum of 194 nuclear reactors of 200 MW operating in the year 2000. MacVeigh, *Ensayo sobre un Programa de energía Nuclear en España*, p. 13.
30. De la Torre and Rubio-Varas, 'Nuclear Power for a Dictatorship'.
31. ABC, 11 April 1956, p. 41.
32. MacVeigh, *Ensayo sobre un Programa de energía Nuclear en España.*
33. TNA, AB 61105, Notes visit, January 1957.
34. De la Torre, 'Who was who'.
35. De la Torre and Rubio-Varas, *La financiación exterior del desarrollo industrial español a través del IEME*, pp. 109–110.
36. TNA. FO. AB 61105. Visit of Señor MacVeigh to Atomic Construction Limited (14/8/1959); and AB 6591. British Embassy Report (23/11/1962).
37. Catalan, J.: 'La ruptura de la posguerra y la industrialización, 1939–1975', in Nadal, J. (Dir.) *Atlas de la industrialización de España*, Barcelona: Crítica, 2003, pp. 233–88.
38. MacVeigh, J.: 'Posibilidades de la industria española en la construcción de centrales nucleares', in *Conferencia Mundial de la Energía.* Madrid, 5–9 June 1960.
39. For instance, in 1964 he participated in the annual meeting of the Atomic Industrial Forum in San Francisco, 'in which he acted as spokesman for European opinions on the new North American legislation regarding the private ownership of nuclear fuels and the enrichment of uranium under canon'. AHBBVA, Nuclenor Activity Report, 1965.
40. De la Torre, Joseba and Rubio-Varas, M.del Mar, 'Learning by doing: the first Spanish nuclear plant, *Business History Review*. 92, issue 1 (Spring 2018).
41. TNA, AB. 38,185. (27/09/1968). From 1962 to 1968 several meetings were held between the UKAEA and Tecnatom on business reprocessing. Ibidem. AB 38280.
42. ASEPI, Report about Tecnatom (8/3/1971). Tecnatom (2007).
43. It is well known that neither the competitive market nor technical practicality led to the dominance of American-made light-water reactors in electricity, isotope and plutonium production. Radkau, Joachim: 'Kernenergie-Entwicklung in der Bundesrepublik: ein Lernprozess. Die ungeplante Durchsetzung des Leichwasserreaktors und die Krise der

gesellschaftlichen Kontrolle über die Atomwirtschaft, GG 4 (1978), pp. 195–222. Winnacker, Karl and Wirtz, Karl: *Das unverstandene Wunder,* Düsseldorf: Econ Verlag, 1975, pp. 192–193.

44. Wirtz's international activity in the 1960s and 1970s was intense. For example, in 1963, Wirtz participated in the Japan Atomic Industrial Forum and visited the Tokai nuclear research centre. That same year, he took part in the meeting of the United States Atomic Energy Commission (USAEC), GE and the Karlsruhe Nuclear Research Center (Kernforschungszentrum Karlsruhe; KfK), which were held in Arkansas. In 1966 and 1967, he attended the International Atomic Energy Association (IAEA) and the European Atomic Energy Society conferences in Sweden and Portugal. In 1969–1970, he spent a sabbatical at the University of Washington (Seattle). He regularly participated in the meetings of the American Nuclear Society, and in 1974 he was the coeditor of three specialized American journals. In 1975, he delivered a lecture in Mexico City and visited the Mexican National Institute of Nuclear Energy (Instituto Mexicano de Energía Nuclear). In addition, he maintained contact with experts in Brazil, Chile, Argentina and Venezuela. In 1976, he was preparing for a trip to Iran. Hermann, Armin: *Karl Wirtz- Leben und Werk,* Stuttgart: Schattauer, 2006, pp. 137–138,149,168–169. GLA Abt. 69/Kfk INR Nr. 52. Letter from Carlos Sánchez del Río to Karl Wirtz, 03.05.1967. GLA 69 KfK-INR Nr.168. Letter from K. Wirtz to Hans Frewer 13.07.1976.
45. GLA Abt. KfK, GF-1 Nr. 184. Brief z. Hdn. Karl Wirtz. Staatssekretär (Haunschlid). Bundesministerium für Bildung und Wissenschaft 17.08.1972. Brief (Kopie). z. Hdn. Staatssekretär (Haunschlid). KfK Karl Wirtz. 01.09.1972. Müller, Wolfgang D.: *Geschichte der Kernenergie in der Bundesrepublik Deutschland,* Stuttgart: Schäfer-Poeschel,1996, II. p. 127. GLA 69 KfK- GF1-Nr. 339,1. Letter from the President of the Nuclear Energy Board, Jesús Olivares, to Horst Böhm. 20.02.1979. GLA 69 KfK- GF1-Nr. 339,1. Dr. Nentwich. Internationale Beziehungen. Betr.: Besuch der JEN in der Zeit vom 4. bis 6. April 1979. 12.04.1979.GLA Abt.69 KfK-GF-1 Nr. 184. Letter from J.M. Otero to Hans-Hilger Haunschild. 18.09.1972. GLA Abt 69 KfK-GF 1 Nr. 184. Notiz an Herrn Dr. Greifeld (Prof. Dr. K. Wirtz), 10.04.1972. Hermann, *Karl Wirtz,* pp. 33–42.
46. Hermann, *Karl Wirtz,* pp. 11,23, 65, 69.

47. Stam, Thomas: *Staat und Selbstverwaltung. Die Deutsche Forschung im Wiederaufbau 1945–1965*, Köln: Verlag für Wissenschaft und Politik, 1981, p. 85. Keβler, G.: 'Zum Gedenken an Karl Wirtz', *Physikalische Blätter* 50. No. 9, (1994), p. 867. Hermann, *Karl Wirtz*, p. 77. Radkau, Joachim: *Aufstieg und Krise der deutschen Atomwirtschaft 1945–1975,* Reinbeck bei Hamburg: Rowohlt, 1983, p. 37.
48. Wirtz, Karl: *Im Umkreis der Physik*, Karlsruhe: KfK, 1988, p. 89. Hermann, *Karl Wirtz*, p. 137.
49. Law 25 of 29 April 1946 and Law 22 of 2 March 1950. Müller, Müller, Wolfgang D.:*Geschichte der Kernenergie in der Bundesrepublik Deutschland*, Stuttgart: Schäfer-Poeschel,1990, I. p. 43–54.
50. Radkau, *Aufstieg und Fall,* pp. 46. *Winnacker and Wirtz, Das unverstandene, p. 90. Hermann, Karl Wirtz, p. 92.*
51. Oetzel, Günther: *Forschungspolitik in der Bundesrepubilk Deutschland. Entstehung und Entwicklung einer Institution der Groβforschung am Modell des Kernforschungszentrums Karlsruhe KfK, 1956–1963*, Frankfurt am Main: Peter Lang, 1996, pp. 11.12/24/84.
52. CSIC. International Conference on scientific research and energy problems. Madrid, 14–18 October 1974. Participant CV (Karl Wirtz). The conflicts were with engineers and scientists who served on corporate boards, such as Wolfgang Finkelnburg (Siemens), Rudolf Schulten (BBC/Krupp, later BBK) and Alfred Schuller (AEG). Oetzel, *Forschungspolitik,* p. 139. Müller, *Geschichte*, II. p. 342. Eckert, Michael Osietzki, Maria: *Wissenschaft für Macht und Markt. Kernforschung und Mikroelektronik in der Bundesrepublik Deutschland*, München: C.H. Beck, 1989, p. 84.
53. Sánchez Ron, José María: 'International Relations in Spanish Physics from 1900 to the Cold War', *Historical Studies in the Physical and Biological Sciences*, 33/1, (2002), pp. 3–31. Hermann, *Karl Wirtz*, p. 76–77–82.
54. Romero, Ana: 'Un viaje de José María Otero Navascués. Los inicios de la energía nuclear en España', *Arbor*, 659–660, (2000), pp. 509–526. Presas, Albert, 'La correspondencia entre José M. Otero Navascués y Karl Wirtz, un episodio de las relaciones internacionales de la Junta de Energía Nuclear' Arbor, 659–660, (2000), pp. 527–602. Romero de Pablos and Sanchez-Ron, *De la JEN al CIEMAT,* pp. 34–40,69,114. Hermann, *Karl Wirtz*, p. 81–82.

55. The National Institute of Industry (INI), founded in 1941, was the holding of public companies where the state acted as an entrepreneur.
56. Aschmann, Birgit: *'Treue Freunde…'? Westdeutschand und Spanien 1945–1963*, Stuttgart: Franz Steiner Verlag, 1999, pp. 334–35. On the activities of the Spanish National Institute of Industry in Berlin during the Second World War, San Román, Elena: *Ejército e industria: el nacimiento del INI,* Barcelona: Crítica, 1999, pp. 200–204. On continuity and changes, Viñas, Ángel: Franco, Hitler y el estallido de la guerra civil, Madrid: Alianza, 2001, p. 517.
57. GLA Abt 69 KfK-INR. Nr. 52. High Council for Scientific Research. Alfonso X El Sabio Trust. Daza de Valdez Optics Institute. Applied Nuclear Physics Course. November 1950-July 1951. Madrid. Wirtz, Im Umkreis, pp. 87–88.
58. Wirtz, Im Umkreis, p. 87–88. Hermann, *Karl Wirtz,* p. 124. Aschmann, *'Treue Freunde…'?* pp. 231–233.
59. Overbeck, Helmut: *Der Schnelle Brüter in Kalkar. Planung, Bau, vornukleare Inbetriebnahme, Stilllegung*, Düsseldorf: 1992, p. 48ff. Industrial partners in 1968 were Belgonucléaire, Neratom and Interatom who together created the INB company. Electric power industries (RWE, SEP and Synatom) formed the SBK company. Over the course of the 1960s, the French corporation EDF, the Italian public enterprise ENEL and the British CEGB became part of the project. Oetzel, *Forschungspolitik*, p. 242.Oetzel Günther: *Die geplante Zukunft,* Frankfurt am Main: Peter Lang, p. 99ff.
60. PAA B35 Band 77. Auswärtiges Amt. An den Bundesminister für wissenschaftliche Forschung. Betr: Zusammenarbeit mit Spanien auf dem Gebiet der Kernforschung und Kerntechnik, 12.12.1966.
61. GLA Abt. 69/Kfk INR Nr. 52. Letter from Carlos Sánchez del Río to Karl Wirtz, 03.05.1967. Consideration of Wirtz as an 'old friend of Spain' in communications between López Bravo and Gerhard Stoltenberg in 1966 in Aschmann, Birgit (1999), p. 233, note 222.
62. GLA Abt. 69/KfK INR Nr. 104. Besuch bei der Junta de Energía Nuclear in Madrid vom 10. bis 12. Sept. 1968 zwecks Vorbereitung der deutsch/spanischen Zusammenarbeit auf dem Gebiet der schnellen Brüter.
63. GLA. Abt. 69 KfK. PSB. Nr. 471. Zusammenstellung der durchgeführten Projekte bzw. laufenden Aktivitäten. Argentinien, Chile, Spanien, Pakistan, Rumänien. 15.03.1972.

64. ASEPI. Box 4898. Minutes of the board of directors of Unión Eléctrica Madrileña, S.A (UEM) 29.01.1971/26.02.1971/24.10.1971. ASEPI. Box 4898. Note on the meeting of the board of directors of Unión Eléctrica Madrileña, S.A. (UEM), 20.01.1971/26.01.1971 On the visit of Giacomo Bertolotti Batistoni (Tecnatom) in the SBK in Essen ASEPI. Box 5048. Note on the meeting of the board of directors of Unión Eléctrica Madrileña, S.A. (UEM), 14.11.1972. GLA Abt. KfK, GF-1 Nr. 184. Vertrag zwischen der Projektgesellschaft Schneller Brüter GbR (PSB) und der Unión Eléctrica Madrileña, S.A. (UEM), 08.06.1971 (Madrid/Essen/ unsigned).
65. GLA Abt. 69 KfK INR Nr. 228. Handwritten agenda for Karl Wirtz's trip to Madrid 9.9.1968–12.09.1968. GLA Abt. 69 KfK INR Nr. 455. Spanish Government Nuclear Programme Project. Karl Wirtz. Madrid, 17.03.1969. GLA Abt. 69/KfK INR Nr. 57. Copy. Letter from Karl Wirtz to H. Schimmelbusch 16.09.1968.
66. GLA Abt. 69/KfK INR Nr. 104. Nuclear Programme and activities prepared for European Atomic Energy Society Meeting, Lugano, April 26–30, 1969, Madrid, April 1969, p. 3–4.
67. GLA Abt. 69/KfK INR Nr. 57. Letter from H. Schimmelbusch (NUKEM) to Karl Wirtz (INR), 03.09.1968 (Spanien). GLA Abt. 69/ KfK INR Nr. 104. Letter from Manuel Cortines (Chairman of IBERNUCLEAR S.A) to H.E. Schimmelbusch (NUKEM), 1.06.1968.
68. PA AA 129507 Brief und Anlage. Westdeutsche Landesbank Girozentrale (Düsseldorf) an das Chefs des Protokolls Auswärtiges Amt (Bonn) 4.2.1981. Lebenslauf (Dr. Ing. Julio Hernández Rubio) und Vorschlagsbegründung. GLA Abt. 69 KfK INR Nr. 57. Letter from Carlos Sánchez del Río to Karl Wirtz, 6.5.1969. GLA Notiz (Wirtz). Zum Besuch der spanichen Delegation, 8. juli.1969. GLA Abt. 69 KfK INR Nr. 455 Notiz (Wirtz) Zum Besuch der spanischen Delegation am Dienstag 8. Juli 1969. GLA Internationales Büro. Dr. Schnurr. Dr. Laue. 4.8.1970. Betr: besuch den Herren Dr. Schnurr und Dr. Laue bei der Junta de Energía Nuclear, Madrid vom 15. bis 17.07.1970.
69. GLA Abt 69 KfK-GF 1 Nr. 184. Letter from José María Otero to H.E. Schimmelbusch 16.10.1972. GLA Abt 69 KfK-GF 1 Nr. 184. Bericht über einen Besuch bei der Junta de Energía Nuclear in Madrid, 22-11/24-11-1972. Schimmelbusch (NUKEM), Hildenbrandt (KWU), Warrikow (RGB) and Wirtz (GfK) participated.

70. BNN 2.8.1973 *Das kurze Interview: Es gibt kein Tabu. Geschäftsführer Dr. Rudolf Greifeld über Kooperation in der Kernforschung*. GLA Abt. KfK, GF-1 Nr. 184.Kraftwerk Union Aktiengesellschaft zur GfK mbH z. H. Herrn Prof. Dr. Wirtz. Betreff KWU-Angebot an ENUSA über BE-Zusammenarbeit. Erlangen, 5. März 1973. Hilger, Susanne: *'Amerikanisierung' deutscher Unternehmen. Wettbewerbsstrategien und Unternehmenspolitik bei Henkel, Siemens und Daimler-Benz (1945/49–1975),* Stuttgart: Franz Steiner Verlag, 2004, pp. 61ff.
71. Sanz Lafuente, Gloria: 'The Long Road to the Trillo Nuclear Power Plant' in Rubio-Varas M.d.M. and J. De la Torre (eds.), *The Economic History of Nuclear Energy in Spain: Governance, Business and Finance.* London: Palgrave Macmillan, 2017, pp. 187–215.
72. GLA 69 KfK-INR Nr.168. Letter from K. Wirtz to Hans Frewer 13.07.1976. Letter from Hans Frewer to Karl Wirtz, 18.08.1976.

13

Engineers' Diplomacy: The South American Petroleum Institute, 1941–1950s

María Cecilia Zuleta

In the late 1930s, during a rising wave of nationalism, South American networks of oil experts and regional cooperation among oil-producing and oil-importing countries consolidated. Regional technical exchanges and cooperation in oil refining processes notably increased between 1937 and 1942. Soon after the controversial expropriation of foreign-owned oil companies in Mexico in 1938, the South American Union of Engineering Associations (Union Sudamericana de Asociaciones de Ingenieros, USAI) announced in Lima a public call for ideas to improve the competitiveness of South America's petroleum sector.[1] Three years later, after several meetings and initiatives under the auspices of USAI, the South American Petroleum Institute (Instituto Sudamericano del Petróleo, SAPI) was

The author thanks Carla Xóchitl León Cortés for her research assistance, as well as Jesús Reyes Bautista and María Camila Díaz Casas. Special thanks go to Manuel A. Bautista González, for his precious help in the translation of this chapter, and to the editors of this volume, for their valuable suggestions to improve its previous versions. The usual disclaimers apply.

M. C. Zuleta (✉)
Centre for Historical Studies, El Colegio de México, Mexico City, Mexico
e-mail: mczuleta@colmex.mx

© The Author(s) 2018

D. Pretel, L. Camprubí (eds.), *Technology and Globalisation*, Palgrave Studies in Economic History, https://doi.org/10.1007/978-3-319-75450-5_13

founded in January 1941. The outbreak of the Second World War favoured the development of this organisation and opened new opportunities for South American engineers and technicians to become involved in the exploitation and industrialisation of hydrocarbons in the region.

This chapter studies the rise of SAPI in the 1940s, focusing on its origins, organisation, membership composition and leadership. Two distinct periods can be identified in its existence. The first was the crucial years of the Second World War, in a context of oil rationing and continental cooperation; and the second was during the early years of the cold war in a context of developmentalism and state-led industrialisation, increasingly based on fuel imports substitution. It is worth noting that the extensive historiography of Latin America's oil industry has tended to ignore or underestimate the place of engineers and technicians in the development of this region's petroleum industry. An exception is the work of the historian John Wirth. According to Wirth, the foundation of SAPI allowed South American engineers to have broader opportunities for professional exchange that in turn contributed to the circulation of technical and marketing information at a regional level. In Wirth's words, during the interwar period 'South American engineers had their eyes fixed, exclusively, on the North American Petroleum Industry'. Building on Wirth's pioneering work, this chapter aims to demonstrate the strengthening of Latin American communities of petroleum engineers during the war and post-war.[2] SAPI enabled the professional development and identity of South American engineers as well as the technological transformation of the hydrocarbons industry in the region. This chapter highlights the relationship between the professionalisation of petroleum engineering and the increasing involvement of South American oil engineers and technicians in the transnational politics of energy.

South American Oil Industry in the Interwar Period and Wartime

During the interwar period, the South America petroleum industry developed in a context of rapid energy transition from coal to hydrocarbons. The efforts of South American petroleum engineers to increase regional cooperation also coincided with the take-off of the South American petroleum industry and the progressive import substitution policy in the fuels sector in this region. The whole region experienced a transition to a pattern of energy consumption based on hydrocarbons and electricity; however, each country had different experiences and rhythms.[3] For instance, the most urbanised countries of the Southern cone—Chile, Uruguay and Argentina—had been the prominent consumers of fossil fuels in the region since the late nineteenth century.[4] The increasing exploration of new hydrocarbon deposits, facilitated by foreign direct investment by British and American oil multinationals, marked the birth of petroleum industries in the continent.[5] Meanwhile, after the First World War, the wave of automobile imports in the River Plate countries (Argentina and Uruguay), Chile and Brazil had the effect of an expansion of hydrocarbon's derivatives consumption.[6]

By 1939, on the eve of the Second World War, Latin American fields produced nearly 14.6% of world oil production (Fig. 13.1 summarises the main world oil flows for 1938). The United States, the Soviet Union and Venezuela were the main producers and exporters of petroleum.[7] Venezuela alone produced more than 60% of Latin American crude, and almost 10% of the world's oil. The second most important producer of petroleum in Latin America was Mexico, although this country had reached a peak in its oil exports at the end of the 1920s, well before President Cárdenas's expropriation of 1938. Colombia, Peru, Argentina, Ecuador and Bolivia were other producers. While Venezuela (95%), Colombia (86%), Peru (83%) and Ecuador (82%) exported most of their oil production, Argentina's crude was directed to cover half of that country's domestic fuel needs.[8] In 1940, the South American region had the third largest refining capacity for crude, only after the United States and the Soviet Union.[9] A combination of small distilleries and large modern refineries produced a variety of petroleum products such as kerosene, gasoline, lubricants, paraffin wax and asphalts for both domestic and external markets.

South American energy transition during the interwar period encouraged the regional circulation of capital, technologies, experts, practices and knowledge, just as had occurred with railway expansion during the late nineteenth century.[10] European and US oil corporations transferred

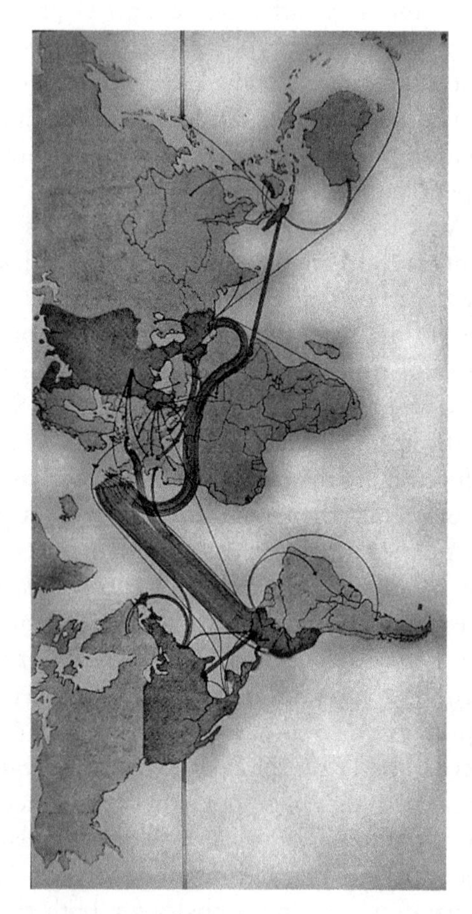

Fig. 13.1 World petroleum transportation, 1938 (Maritime transport of oil: Each line represents 1 million tons). (Source: Jaime Bermejo, "Transporte marítimo de petróleo": *Boletín de informaciones petroleras* (YPF), Buenos Aires, XVIII: 198 (February, 1941), pp. 61–92, Fig. 3)

to South America new technologies—designed at their research laboratories—to prospect, extract and industrialise hydrocarbons. This process of transnational technology transfer was built on global petroleum engineering networks.[11] The expansion of the hydrocarbons industry also required several pressing transformations. The South American petroleum industry demanded professionals, governmental agencies and regulations. Studies of and plans for each nation's energetic consumption were also an urgent need. Therefore, the development of the hydrocarbon sector in South America opened professional opportunities for specialised geologists, chemicals, engineers and technicians. Since the first decade of the century, both in Europe and the United States, the new oil industry had promoted the rise of the profession of petroleum engineering and its legitimation at the international level. Latin America was not, however, an exception. As in other places, the petroleum engineer was considered a professional with knowledge of several scientific and technological fields, such as earth science, industrial chemistry, mining, metallurgy and mineralogy.[12]

During the interwar period, a wave of innovations radically transformed the global oil industry. Petroleum science as a discipline gained academic relevance and petroleum engineers were consolidating their professional identity on a global scale. Advances in subsoil geophysics offered new exploration methods that favoured the discovery of new oil deposits. Offshore extraction operations in the Gulf of Mexico started in the immediate aftermath of the Second World War.[13] The introduction of European and American chemical processes such as cracking and catalysis allowed the so-called 'miracle of molecules',[14] which precluded the birth and development of the petrochemical industry, multiplying the number of petroleum products.[15] Meanwhile, the mass production of new materials made possible the transportation of fuels through transnational networks of oil and gas pipelines.[16]

An institutional innovation, the establishment of state-controlled oil companies, was a vital factor in the development of the oil industry in South America. Latin American state oil companies followed the pioneer model established by Yacimientos Petrolíferos Fiscales (YPF) of Argentina, founded in 1922, the first vertically integrated state oil company, financed with public revenue. After YPF, followed the Administración Nacional de

Combustibles, Alcoholes y Portland (ANCAP) in Uruguay in 1931 and Bolivia's own YPF in 1936.[17]

The foundation of state-owned companies inaugurated a new age in the marketing of fuel and its derivatives in South America, especially among foreign companies. From 1935, market-sharing arrangements, as well as controls on prices and fuel distribution mechanisms, were introduced in Uruguay, Argentina, Chile and Peru. As Bucheli points out, this could be considered evidence of a process of cartelisation in the oil business.[18] By 1940, the regulations became common currency. Tariff controls on crude and derivatives imports were generalised throughout the region, and governments began to set top prices for the different types of hydrocarbons. These new market arrangements provided professional opportunities for petroleum engineers that could resolve complex bureaucratic and technical problems.

Meanwhile, state-owned oil companies were progressively functioning as incubators of innovation, centres of technical training and nodes in the growing networks of petroleum engineers. The state-owned oil companies established scientific laboratories and formalised alliances with higher education institutions in their respective countries. These companies also developed connections with European and American oil companies, which often provided technology and qualified personnel capable of training new generations of local engineers and technicians. As a result, the expertise of foreign engineers was complemented with the knowledge of South American engineers, which was better adapted to local conditions.[19]

The Second World War was a crucial period for South American petroleum engineers and technicians. During those years, these experts gained political and professional legitimacy, to the extent that they started to have a central role in the region's energy diplomacy. The period between 1941 and 1947 was an economic turning point for Latin American economies, with a reorientation of trade towards the pan-American area,[20] and the establishment of cooperation schemes to supply the Allies with primary goods and strategic commodities.[21] As the supply and transportation of fuels was strategic for the war effort, the United States coordinated oil rationing through the authority of the Petroleum Administration for War (PAW), with the advice of the Petroleum Industry War Council.[22]

In South America, the Allies promoted the development of an inter-American supply pool of petroleum. A complex rationing of all kinds of 'essential' supplies was coordinated by Washington through negotiation with Latin American governments. The Committee of Provisions for Latin America was in charge of fixing supply quotas of fuels for each country, at a lower level than previous consumption patterns (for 1942, the quota was 40% of the level of consumption of crude and derivatives in 1941). However, the Argentine YPF refused to join the inter-America petroleum pool.[23]

During the war, PAW and the US State Department, supported by a variety of petroleum experts, inaugurated a new era in the global oil politics and diplomacy. The continental and intercontinental scale of oil rationing during the war demanded nerve-wracking rounds of diplomatic negotiations. The discussion of complex technical questions required the active participation of petroleum experts in these negotiations. In the United States, PAW hired geologists, engineers and scientists working for oil corporations with consulting and executive powers. These experts had responsibilities over the explorations, research, refining, rationing and supply of hydrocarbons at national and global levels. PAW's deputy director and his team also had attachés at the American embassies in South America (such as in Venezuela, Colombia, Trinidad, Ecuador, Peru and Bolivia). During wartime PAW sent so-called technical 'oil missions' to European, Middle and Far Eastern countries, as well as to Latin America (Mexico, Bolivia, Colombia, Ecuador and Peru). Oil experts were the central actors planning and managing the war effort, successfully connecting private and public interests.[24]

The South American Petroleum Institute (SAPI)

Engineers were active in establishing international institutions and professional associations during the interwar period and the 1940s. In 1935, engineers and technicians from the Southern Cone and Brazil created a federation of engineers' associations: USAI, with bases in various South American capital cities. In January 1941, and after two preparatory meetings in July and December 1940, USAI's board decided to create

SAPI.[25] The founders of SAPI followed as models two foreign associations: the British Institute of Petroleum, founded in London in 1914, and the American Petroleum Institute, founded in 1919 with the financial support of large private oil companies and philanthropists.[26]

Originally, SAPI only accepted members from South American countries, although this would eventually change. The new body was closely related to USAI, which had representation in SAPI's executive board.[27] Indeed, the executive board of SAPI was constituted by the Permanent Executive Committee, a delegate from the USAI and the presidents of the national sections.[28] This committee was the main pillar of SAPI, and the one in charge of organising its meetings and congresses. Unlike some US professional associations in the petroleum sector, such as the American Petroleum Institute or the American Association of Petroleum Geologists, SAPI did not develop a comprehensive scientific research agenda.

SAPI was a non-profit corporation and did not undertake business or commercial activities. The Institute raised funding in the form of fees, donations and subsidies. Its administrators secured funds from individual membership fees, regional governments and private multinational oil companies that, like other corporate members, joined the Institute as associates. Public bodies and both public and private oil companies (such as ANCAP, YPF, Shell and Standard Oil) contributed funds to SAPI to regularly publish a bulletin and to organise scientific meetings. The Institute's official bulletin—*Boletín del Instituto Sudamericano del Petróleo*—would be an essential vehicle of knowledge diffusion among petroleum engineers in the region. The bulletin was published practically without interruption until 1951. In it, the Institute published informative articles on South American oilfields, and many scholarly analyses of the petroleum industry in each country of the region. It also published the Spanish translation of several studies conducted by American scientists. SAPI's bulletin was especially relevant for the diffusion and adoption of an Americanised scientific, technical and professional paradigm. SAPI international conferences, in turn, took place every four years, with the objective of furthering the scientific and professional goals of the Institute and increasing its international exposure, although many of these meetings were postponed.

Montevideo was chosen as SAPI's headquarters by the executive board of the Institute. This can be explained by a combination of professional, economic and geopolitical factors. On the one hand, it is a telling sign of the associative strength of the Uruguayan engineers amidst their South American colleagues, as well as the strategic leadership of the state company ANCAP.[29] The localisation of SAPI in Uruguay was a pan-Americanist decision of the engineers associated to USAI in difficult geopolitical circumstances. In a context of pervasive Anglo-American presence in the region,[30] the Uruguayan diplomats led the pro-Allied movement within the inter-American system, strengthening both Uruguayan autonomy vis-à-vis Argentina and Brazil and pan-Americanism in the hemisphere.[31] These tensions between the pan-Americanist and South Americanist views continued to be a source of constant conflict throughout the existence of SAPI.

During the same period, Uruguay had been able to host the Regional Conference of Plata in February 1941, where diplomatic representatives of South American countries agreed on a broad common agenda, including trade cooperation and energy distribution.[32] Even if Buenos Aires concentrated most of the economic and financial power in the Plata region, including the offices of oil companies of neighbouring countries such as the Standard Oil of Bolivia Company,[33] SAPI operated with a certain degree of autonomy vis-à-vis the Argentinean state oil company YPF. The YPF technical prestige and neutralists positions, far away from the Allies' interests, complicated the geopolitics of energetics in the region.[34] In Montevideo the Institute was kept away from the strong diplomatic and scientific pressures of the Axis countries. SAPI's ultimate safeguard was the pro-Allied support of the foreign affairs ministry of Uruguay.

The convention of USAI on January 1941 drafted SAPI's Provisional Statutes, ratified in 1944. They identified three areas of interest for the organisation: scientific, professional and geopolitical. SAPI's main goal was to strengthen the networks of cooperation and interaction between engineers to pursue the professionalisation of petroleum engineering in South America. Moreover, the Statutes established as goals of the Institute support to the circulation of petroleum extraction knowledge, the construction of a network for research collaboration and the training of experts through regional exchanges. In other words, SAPI had the declared objective of encouraging a community of petroleum experts in the region, as well as the advancement of this profession through training and the regulation of professional credentials.[35] The Institute was also explicitly interested in getting involved in the South American geopolitics of hydrocarbons during wartime, through the coordination of intra-regional trade negotiations for fuel supply, with the active support of national governments.[36]

Although Ralph Emery, of the Shell-Mex Uruguay Company, became SAPI's treasurer, the leading figure behind SAPI's foundation was the Uruguayan engineer Carlos Végh Garzón. In Montevideo, in May 1940, in a radio conference that took place as part of the celebrations of the Day of the South American Engineer, Végh Garzón had already urged for the creation of a scientific and professional organisation of experts involved in the petroleum and gas industry of the region.[37] He had in mind a new institute that could foster commercial and knowledge exchanges related to the hydrocarbons industry in South America during wartime. In Végh Garzón's words the 'war demand[ed] a very close coordination of the industrial and commercial activities related to petroleum'.[38]

Having been appointed the president of the Executive Committee of SAPI, Végh Garzón became the leader of the project and played a vital part in its development.[39] Bureaucrat, economist, entrepreneur and politician, Végh Garzón did not study engineering in a military school, but at the School of Engineering of the University of the Republic in Montevideo, from where he graduated in 1926.[40] He combined experience as a corporate director of the state firm ANCAP, with a consulting role in the construction of refineries. Végh Garzon had a broad and extensive professional experience. He had worked in his youth as an engineer in the nationalised railways, overseeing the introduction of equipment and technology from Europe, particularly Belgium and Germany. In 1937, Végh Garzón was in charge for buying British and American equipment for the construction of La Teja refinery in Montevideo. He also collaborated with the Foster Wheeler company in Brazil in 1939 in Bahia and, years later, in the 1950s, in Pernambuco-Recife. Végh Garzón had close ties with the Argentinean-Brazilian company Ipiranga and was a private consultant of Universal Oil Products. After managing ANCAP,

Végh Garzón became Uruguay's Minister of the Treasury (1967) and president of the Bank of the Republic (1968).

Végh Garzón had a comprehensive vision of the hydrocarbon economy and fuel industries in South America and understood the conflicting national interests. Considering the rising number of active petroleum experts in the oilfields of South America, Végh Garzón proposed the foundation of SAPI because he clearly recognised the new professional possibilities opened to petroleum engineers and technicians by the war effort. He had a vision conciliating Uruguay's economic needs with a regional project for articulating the energetic and commercial complementarity of South America. Végh Garzón's education and professional experience, together with his appreciation of the role of engineers in the 'South American hour', made him different from the military and nationalist engineers. For instance, he diverged from strategic military visions behind the performance of the state-owned oil company YPF and the National Oil Council of Brazil (Conselho Nacional do Petróleo). Although Végh Garzón considered that hydrocarbons were key resources for the economic independence of the South American countries, he also supported what the historian John Wirth has called 'the doctrine of large profits from refining', a doctrine first inspired by YPF's experience and policies of capturing the market for energetics.[41]

Being less of a politician than an engineer, Végh Garzón had practical technical ideas rather than a clear political project. Because of this, he advocated for the joint participation of private capital (both domestic and foreign) and governments in the development of the petroleum industry in the region. Therefore, he envisioned the Institute as a professional and scientific organisation that would advance the professional interests of petroleum experts working in both private and public sectors. His ideas on the South American energy complementarity, together with his pan-Americanist vision, were very different from proposals of a more nationalist tone.[42] Even if Végh Garzón's proposals seemed to lose relevance considering the geopolitical transformations of the post-war period, they anticipated future projects and experiences of energetic regional integration.[43] As John Wirth demonstrates, Végh Garzon's ideas on petroleum engineering and the relations between the scientific, technological and entrepreneurial worlds, were truly innovative in the Plata basin, especially his approach to regional and inter-American petroleum geopolitics.

SAPI Membership and National Representation

SAPI organised its members by professional specialisation and nation of origin, although it was not a multilateral organisation. It followed the organisation of the Division of Petroleum Technologists (1922) within the American Institute of Mining Engineers and Metallurgical Engineers. SAPI was a professional organisation that brought together men with different educational, professional, scientific and industrial backgrounds. Its membership was very diverse, ranging from bureaucrats to engineers and technicians working for multinationals, to scientists and academics. It also had corporate members, which represented bureaucratic, political, business, professional, scientific and educational interests. In 1947, according to its bulletin, SAPI had 762 individual members and 116 corporate members (various private companies and associations, eight state-owned companies, seven governmental agencies, six educational and research institutions, and seven public corporations). In 1951, ten years after its foundation, SAPI communicated to the United Nations Economic Commission for Latin America and the Caribbean (ECLAC) that it had 1054 members from twelve countries, including Venezuela, Colombia, Mexico and the United States.[44]

At SAPI's Second Conference of the Permanent International Executive Committee (Montevideo, April 1944) a revision of its statutes was approved in order to allow the admittance of North American members. PEMEX and the magazines *The Equipment Publisher and Company* and *Petróleo Interamericano* eventually joined SAPI as corporate members.[45] Végh Garzon personally invited Mexican state-owned oil company (PEMEX) to become a corporate member of SAPI. As a result, at the end of 1947 eighteen Mexican engineers from Petróleos Mexicanos (PEMEX) became new associates. The same sources reported that ninety-seven US individuals and corporations involved in petroleum extraction, refining and industrialisation joined SAPI.[46] This expansion to North American members strengthened commercial and scientific activities in the hydrocarbon sector across the continent. The presence of US members might be evidence of the linkages between US oil interests and the South American petroleum industry, especially concerning geophysical surveys and refining operations. However, as Table 13.1 shows, Mexico and the USA did not have national sections at the Institute.

Table 13.1 Membership to national sections: South American Petroleum Institute (1941–1951)

		Individual members			Corporate members
Section	Year of constitution	1945	1947	1951	1947
Argentine	1941	283	299	352	26
Bolivia	1942	27	43	57	9
Brazil	1946	–	39	42	9
Chile	1942	104	96	40	9
Colombia	Not formed	–	13	18	–
Ecuador	1942	105	69	104	12
Paraguay	Not formed	–	5	5	–
Peru	1942	48	50	51	13
Uruguay	1941	83	110	112	28
Venezuela	Not formed	–	31	45	1
Mexico	Not formed	–	14	18	–
United States	Not formed	–	75	97	18

Source: Boletín del Instituto Sudamericano del Petróleo, Vol. I, 1943; Vol. V, 1947; Vol. III, 1951; ECLAC, 'Instituto Sudamericano del Petróleo, Nota del secretario ejecutivo', GENERAL E/CN.12/232, 1951–04-09, http://repositorio.cepal.org/

In a little less than a year, between the end of 1941 and the final months of 1942, six national sections were formed and became SAPI's institutional pillar (see Table 13.1). The Institute's international executive committee supported each national section with a subsidy correlated to the number of members.

As Table 13.1 shows, the size and composition of SAPI's national sections were very diverse. The whole set of national sections followed the geography of hydrocarbons in South America: some countries were producers and others importers, revealing the possibilities of energetic integration between suppliers and consumers of petroleum. The first national sections were created in Uruguay, Argentina and Chile, the main consumer economies of hydrocarbons in the region. Venezuela, Latin America's leading exporter, joined SAPI late. The vital importance of Venezuelan oil for the Allied countries (who took over Venezuelan oil industry under the supervision of PAW) might be the most reasonable explanation for its delay in joining SAPI. The delay might also be explained by the politics of petroleum in Venezuela, with the many debates for the approval of legislation that favoured national control of the oil rent and petroleum profits in 1943.[47]

The Colombian section was never created. Nevertheless, SAPI's bulletin published reports on the state of the oil industry in Colombia, studies written by Colombian engineers.[48] The fact that the two main exporters of the region in this period, Venezuela and Colombia, did not join SAPI, reflects disinterest and regional division, since engineers working in those countries preferred alternative organisations more in line with the interests of the multinational oil corporations operating in both countries.

The Argentinean national section had the largest membership. Its composition was very diverse, although its most important base was engineers, scientists and technicians working for the state oil company YPF (50% of the total number of members).[49] Most of these experts had been trained by YPF during the interwar period and had experience in both field production and laboratory research. Experts working for private companies were also members of the Argentinean section. Corporate members went from small private oil firms to multinational corporations and distributors of chemicals, gas and capital goods. Academics from the universities of Buenos Aires, Cuyo (Mendoza) and La Plata, as well as bureaucrats and experts in oil transportation, were represented.

The Second World War strengthened the domestic orientation of the Argentinean petroleum industry, prioritising internal supply over the export of petroleum products to other countries of the region. This situation created confrontation between Argentina and its neighbours, such as Uruguay and Chile, which concentrated on accessing cheap energy and diversified commercial exchanges within the region. Argentina's political standing as a neutral country during the war kept it away from the hemisphere's dominant position, creating tension in the relationship with the United States and the inter-American system. Ultimately, this affected the capacity of the Argentinean engineers to interact with their peers in the international arena.

Despite the weight and prestige of YPF in the region, the Uruguayan national section maintained a position of leadership within SAPI throughout the 1940s.[50] The Uruguayan section headed SAPI's International Relations Commission and published the Institute's bulletin (*Boletín del Instituto Sudamericano del Petróleo*). The membership of the Uruguayan section at SAPI was largely composed by the state fuels company ANCAP, together with the Uruguayan Association of Engineers, and the state companies providing energy transportation and electricity. Commanded by ANCAP and the Uruguayan government, its members shared the objective of promoting a trade policy that could secure energy supplies.[51] The Uruguayan section organised a First National Conference on Supply and Rationing in the Use of Energetics in July 1942. Among the various question discussed in that conference were innovations in chemical processes, the marketing of fuels and alternative energy resources such as wood-gas and hydroelectricity.

On the Pacific coast, the composition of the Peruvian section reflected both the importance of foreign capital in the Peruvian oil industry and the growing power of oil experts and bureaucrats working for the Peruvian Ministry of Development and Public Works (Ministerio de Fomento y Obras Públicas). Particularly visible were the technocrats from the Direction of Mines and Oil, and the Department of Oil of the Mining Engineers Corps, institutions that had created the state oil company Empresa Petrolera Fiscal in 1934. The foundation of the Peruvian section coincided with the renovation of the organisation and curriculum of the National School of Engineering, which had an impact on the professionalisation of engineers in Peru. The core membership of the Peruvian section had been active in the negotiations between the Uruguayan state company ANCAP and the Peruvian oil companies since 1937. Many Peruvian members, such as several geologists, had been connected with Végh Garzón since the First Panamerican Congress of Geology and Mines (January 1942, Santiago de Chile).

Although Bolivia was a founding national section, it was unable to consolidate a significant membership. The section was largely made up of the group of engineers and experts of Yacimientos Petrolíferos Fiscales Bolivianos (YPFB). The Bolivian section also had several technocrats who worked in the Ministry of Mines and Oil. A good example is Juan Pinilla, an engineer active in the foundation of SAPI. Pinilla also participated in the commission that wrote SAPI's societal statutes. He had been director of public works in the Bolivian government and was the technical representative of YPFB in Buenos Aires between 1940 and 1943. He was also involved in the technical and commercial negotiations between YPFB,

YPF and ANCAP during the first half of the 1940s. These negotiations between state-owned oil companies of the Plata region were the immediate origin of the Bolivian section of the Institute. Carlos Végh Garzón himself participated in these negotiations from 1937 onwards.[52] Despite the many expert commissions visiting Bolivia, such as the Merwin Bohan and the YPF missions, the negotiations of oil trade and traffic with Brazil and Argentina proved to be intricate.[53] The many political and diplomatic crises Bolivia faced during these years only made matters worse.[54]

Ecuador was among the smaller hydrocarbon exporters of the region. However, the composition of the Ecuadorian section is quite significant given the large size of its membership, mostly technocrats, university researchers and engineering experts working for two large foreign oil companies, the International Ecuadorian Petroleum Company and the Shell Company of Ecuador. The Ecuadorian section formally participated as a consultant in the debates for the passing of the Ecuadorian oil law of 1937.[55] Although Quito hosted the section's board, engineer Ernesto Escobar Pallares, involved in the business of oil transportation, also established a delegation of the section in the city of Santa Elena.[56] The broad membership of SAPI's Ecuadorean section and its numerous edited publications reveal the geopolitical significance of Ecuador's fuel in times of scarcity. As with the Peruvian and Bolivian cases, Ecuador's export-oriented oil companies had dealt with the Uruguayan company ANCAP through supply contracts since 1937.[57]

The engineering technocracy working for the Chilean state since the 1920s made the most of SAPI's Chilean section.[58] Many came from the Chilean Development Corporation (Corporación Chilena de Fomento, CORFO) and the Chilean Institute of Engineers (Instituto de Ingenieros de Chile). The group of Chilean engineers and entrepreneurs that founded the Oil Company of Chile (Compañía de Petróleos de Chile or COPEC) were also members of the Chilean section. COPEC was a private company with national capital which primarily imported hydrocarbons. The interests of the automobile industry and the multinational oil companies were as well represented. The vice-president of the Chilean section was Hugh Raymond Spilbury Pockok, manager of Royal Dutch Shell in Chile. Between 1942 and 1943, the period with the most severe continental rationing of energy, there were new institutional arrangements to enable the exploration of the oil reserves in the Magallanes area. CORFO and the United Geophysical Company pushed this project, which would become in 1950 the public firm ENAP.[59] Professional and personal interactions of Chilean engineers and geologists with Végh Garzón in the First Panamerican Congress of Mining Engineering and Geology in January 1942 explain the rapid growth of the Chilean section. For instance, Chilean industrialists and engineers were active in fuel rationing between 1941 and 1946, promoting interregional supply policies. Chilean business groups and technocrats linked to the distribution of fuels in Chile saw intraregional trade and cooperation only as a temporary solution to the shortage of fuel in times of war.

It seems clear that political and ideological issues hindered the cohesion of SAPI's community and created problems in basic matters such as funding. Tensions also arose owing to SAPI's division into national sections and the divergent interest of corporate members, limiting the body's scope of action. But this overview is not complete without taking into account the development of other regional engineering professional associations,[60] as well as the informal linkages established between oil producers and oil supplying companies in the River Plate and the Andes.[61] On the other hand, it must be emphasised that the differences in the membership composition of national sections also showed regional asymmetries in petroleum engineering education and professional training.

During the interwar period, South American universities and research centres modernised their courses in geology, petroleum, mining and metallurgic engineering, as well as in industrial chemistry, all of them scientific and technical disciplines essential for the hydrocarbons industry. However, this process was not uniform across countries. There were significant differences within the teaching and research programmes of petroleum engineering across South America, as each country followed different models of technical and geological education (from French, American or German influence) in public institutes, universities, laboratories and state oil companies.

While multinational oil companies and higher education institutions concentrated engineers' education and petroleum-related research in Colombia, Venezuela and Peru, in other countries such as Argentina and Uruguay the state oil companies YPF and ANCAP established partnerships with local universities to provide professional training in chemistry, mineralogy, geology and petroleum engineering.[62] Indeed, on both sides of the River Plate, Uruguay and Argentina developed during the 1930s and 1940s a novel system of education and training through collaboration between academic researchers and professionals working for the oil industry.[63] In Bolivia, Ecuador, Peru and Venezuela petroleum engineering education was established later than in other parts of the region. In these four countries, engineering education heavily relied on practical training at mining corporations or multinational oil companies.[64] In the 1940s, the Bolivian YPFB and the Peruvian government would try to develop a more formal training in petroleum engineering.[65]

Towards an Oil Diplomacy

SAPI members did not have the same ideas concerning the role of the state as a manager of the economy. There was an ongoing tension between the nationalist, technocratic and entrepreneurial factions of SAPI's membership. Engineers educated in military institutions did not agree with their peers trained in universities and working in multinational oil companies on the question of state intervention. Pan-Americanist debates around the conciliation of state companies with private capital were also important. These discussions became particularly relevant after the nationalist expropriation of oil firms in Bolivia (1937) and Mexico (1938) and the consolidation of the state oil companies, specifically YPF.

During the Second World War, SAPI concentrated its efforts in encouraging the coordination of a regional market of fuels, building on the natural energetic complementarity among South American countries. SAPI promoted diplomatic initiatives at a continental scale to discuss and solve the problems of scarcity and rationing of energy during the war. The composition of the national sections of SAPI, which integrated both exporters and importers, demonstrates the conscious planning of a strategy of technical and economic cooperation between providers and importers in the region. It was precisely in these years when intraregional trade in South America grew markedly, with oil playing a strategic role, particularly in Chilean–Peruvian bilateral trade.

The Institute tried to develop linkages with the petroleum sector in the USA. Given the fuels shortage, SAPI made attempts to become a consulting body in the hemispheric planning of supply and rationing of fuels. Its Executive and International Affairs committees established relationships with both PAW and private actors in the US oil sector. With this determination, SAPI promoted its image as a professional organisation of engineers with high technical expertise. It presented itself as a body with practical information on the South American petroleum market and technical expertise in energy production. For this purpose, SAPI employed the engineer Washington P. Bermudez (commercial attaché of the Uruguayan Embassy in Washington during the war) as representative and delegate during the war. In 1947, ANCAP's engineer and president of Uruguayan section, Alfredo Levrero, was SAPI's delegate in the USA. The Institute leveraged its political connections within the region through different means with the objective of becoming an indispensable consulting body in fuel rationing policies. For instance, in 1943, Végh Garzón himself paid a visit to Nelson Rockefeller, the head of the Office of the Coordinator of Inter-American Affairs (CIAA). During the same visit to the United States, Végh Garzón met William R. Boyd Jr, the president of the American Petroleum Institute and the Petroleum Industry War Council. Végh Garzón's 1943 trip to the United States, however, only resulted in the establishment of a formal diplomatic contact between SAPI and Washington.[66]

Another example of SAPI's professional and geopolitical activities was a meeting organised at the end of 1944 by the Ecuadorian section in Santa Elena, a city surrounded by productive oilfields. The Santa Elena meeting was attended by government representatives, professors from the School of Sciences of the Central University of Quito and professionals working for foreign oil companies operating in Ecuador, such as the International Ecuadorian Petroleum Company and the Shell Company of Ecuador. A delegation of technical experts of PAW and the Venezuelan ambassador, engineer Enrique J. Aguerrevere, also attended the Santa Elena meeting.

After a period of stagnation owing to the worsening of the war and the industrial raw materials scarcity of 1944 and 1945, the Executive Committee promoted increasing scientific and technological cooperation, with the objective of supporting hydrocarbons industrialisation through the regional exchange of scientific information and research outputs. From that moment, SAPI also planned to support academic exchange programmes in petroleum engineering, funding scholarships and offering consulting services. Between 1946 and 1950, SAPI tried to establish itself as an institution promoting job allocation for engineering experts in the petroleum industry in South America. This was possible because the Institute had a privileged institutional organisation that encouraged transnational contacts between members working in both multinational and state oil companies. For this endeavour, the Institute's bulletin offered a mechanism to connect experts with the oil industry.

Throughout the 1940s, SAPI was shaped not only by technical transformations and innovations but by the demands of the Allied military-industrial complex. During the war, the Institute faced an uncertain political and economic international scenario that reflected the transition from Good Neighbour diplomacy to the hemispheric cooperation against the Axis powers. This geopolitical transition created tensions in some regional governments, particularly in neutral Argentina.[67] Beyond the war effort and geopolitical tensions, South American refineries increased their production capacity during these years. SAPI operated in a context of accelerated import-substitution policies, which explains this notable expansion of Latin American crude refining capacity.

A radical transformation occurred in the global petroleum industry during the post-war period with the emergence of the exporting oil pole of the Middle East. During the early post-war years, SAPI was impacted by the new ideological divisions and geopolitical clashes that characterised the cold war. This tension and conflicting positions weakened SAPI's influence and initiatives in the region during the period. Despite this, delegates of SAPI organised and participated in various international scientific meetings.[68] In a meeting in Mexico City in June 1951, SAPI applied to become a 'B-type' consulting body for the Economic Commission of Latin American and the Caribbean (ECLAC) of the United Nations. ECLAC accepted the request, as SAPI 'performe[d] activities that are necessary for the economic development of Latin

American countries; henceforth, its collaboration could be useful for the respective bodies and organisations of the United Nations'.[69] However, the post-war developmentalist paradigm was met with controversy in SAPI's community; some members favoured state participation in the petroleum sector while others were against it.

In the bipolar post-war order, engineers became central actors within the global architecture of the new US geopolitics, oriented to resist the expansion of the Soviet sphere of influence, including its scientific and technologic strengths.[70] Nonetheless, the organised efforts of the South American petroleum engineers seemed to fail in the immediate aftermath of the Second World War. The new cold war petroleum order opened a new era of confrontations in South America. Different visions regarding the drivers of economic growth combined with ideological antagonisms divided the professionals who had integrated SAPI during the war, such as engineers, geologists, geophysicists and chemists.[71]

After the war, SAPI entered a period of decline. It was subsumed into the Pan-American Union of Engineering Societies (UPADI), which was founded in 1947, and did not consolidate its role as a consulting body to ECLAC. SAPI also encountered problems in finding an equilibrium between the pro-market tendencies supporting the expansion of private oil companies in South America and the supporters of state oil companies, all within an increasing American influence in the Institute. Finally, SAPI was dismantled in 1957. Its connections with political and business elites had eroded. Domestic and international politics had changed, and new generations of South American engineers confronted professional challenges that were radically different than those of wartime. Post-war developments and the rise of US influence in the region also affected USAI, which became part of UPADI as well.

Final Remarks

During the post-war period, SAPI became primarily an association of professionals that concentrated on the promotion of academic and scientific exchanges across South America. The Institute, however, was not able to maintain its strategic position in the South American geopolitics of oil during the 'age of development'. Newly created institutions—such as the Inter-American Development Bank and ECLAC—came to hold the central roles in regional economic policy and trade integration. In this new context, SAPI concentrated its efforts in promoting knowledge and professional exchanges at a regional level.

In contrast with the post-war situation, during wartime SAPI had been an essential institution not only for the regional exchange of expertise and knowledge but also in the geopolitics of oil. The Institute favoured the professional development of hydrocarbons engineering and encouraged South American technical cooperation, particularly on refining and marketing matters. Indeed, the Institute created new spaces for the circulation of petroleum engineering knowledge during the war. The Institute's bulletin shows the technical maturity attained by the South American petroleum industry and the high level of technical expertise and professionalisation gained by the petroleum engineering community in the subcontinent. SAPI's institutional trajectory and initiatives also reveal its vital role in the formation of regional petroleum expert networks and the development of hydrocarbons engineering in Latin America during the 1940s. However, the question of the education and training of petroleum engineers in the region deserves further comparative research.

The historical evidence shows that SAPI's projects and activities were often negotiated among members with conflicting economic and political interests. Despite these ideological disputes, South American petroleum engineers were able to place themselves at the centre of the regional politics of oil during the Second World War. To make this possible, South American petroleum engineers and technicians interacted with private oil multinationals and state oil-owned companies as well as with various governmental agencies and supranational organisations.

With national bases but regional horizons, SAPI was capable, during the war, of forging and strengthening professional cooperation and expert networks, integrating a diverse community of petroleum scientists, technicians, engineers, business corporations and oil companies. In times of challenges, it not only promoted professional exchanges but also an innovative engineering diplomacy with regional and international scope.

Notes

1. 'Concurso: Política que convendría seguir a los países sudamericanos en materia de Petróleo', *Revista de Ingeniería. Publicación de la Asociación Politécnica del Uruguay*, No. 32 (1938).
2. Wirth, J. D.: *Latin American Oil Companies and the Politics of Energy*, Lincoln: University of Nebraska Press, 1985, pp. 26–28. This is the only and best study of SAPI and has been an important source for the development of this chapter.
3. It is generally assumed that energy transition from biomass to fossil-fuel sources and modern industrialisation are correlated. However, in Latin America countries such as Brazil diverged from this pattern. See Brannstron, Christian: 'Was Brazilian Industrialization Fuelled by Wood? Evaluating the Wood Hypothesis, 1900–1960?', *Environment and History*, Vol. 11, No. 4 (2005), pp. 395–430. An insightful essay on the emergence of global fossils and the hydrocarbons economy is Mitchell, T.: *Carbon Democracy: Political Power in the Age of Oil*, London: Verso, 2011.
4. Rubio, María Del Mar, Yáñez, César, Folchi, Mauricio, and Carreras, Albert: 'Energy as an Indicator of Modernization in Latin America, 1890–1925', *Economic History Review*, 63 (2010), pp. 769–804, tables 3–5, and pp. 783–85. During the last decades, a mature literature on energy transition, employing comparative and econometric methods, has provided insights on the nature of this transition in Latin America.
5. See, for instance, Wilkins, Mira: 'Multinational Oil Companies in South America in the 1920's: Argentina, Bolivia, Brazil, Chile, Colombia, Ecuador and Peru', *Business History Review*, 48, No. 3 (1974), pp. 414–446; Wirth, J.: 'Setting the Brazilian Agenda, 1936–1953', in Wirth, J. (ed.), *Latin American Oil Companies and the Politics of Energy*, Lincoln, NE: University of Nebraska Press, 1985; Bucheli, Marcelo and Aguilera, Ruth: 'Political Survival, Energy Policies, and Multinational Corporations. A Historical Study for Standard Oil of New Jersey in Colombia, Mexico, and Venezuela in the Twentieth Century,' *Management International Review*, 50 (2010), pp. 347–378.
6. Yáñez, César and Badia-Miró, Marc: 'El consumo de automóviles en la América Latina y el Caribe (1902–1930)', *El Trimestre Económico*, Vol. 78 (2), No. 310 (April–June 2011), pp. 317–342, and (November 2005), pp. 395–430.

7. 'The Oil industry in Latin America', *Latin American Memoranda and Notes,* Series 2, No. 8 (March 1943); *The Foreign Trade of Latin America, United States Tariff Commission*, Washington (1942), Part III, 'Selected Latin American Export Commodities', p. 160.
8. 'Conferencia del Sr. Presidente de la Sección Argentina, Ing. Enrique Cánepa. Perspectivas del futuro desarrollo de la Industria del Petróleo en Sud América. Producción, consumo y saldos de importaciones o exportación en Sud América, Cuadro N°2', *Boletín de Informaciones Petroleras*, vol. 22, no. 248 (April 1945), p. 59.
9. 'Division of World Refining Capacity and Nationality of Ownership, Excluding Europe and Far East, 1945', Table 6, in Frey, J. and Ide, H. C. (eds.): A *History of the Petroleum Administration for War, 1941–1945,* Honolulu: University Press of the Pacific, 2005. (originally published in 1946).
10. Drummond, D.: 'Britain Railways Engineers: The First Virtual Global Community', in Herbrechter, S. and Higgins, M. (eds.): *Returning (to) communities. Theory, Culture and Political Practice of the communal, Vol. 28*, Amsterdam: Critical Studies, 2006, pp. 207–221.
11. Mount, H. F.: *Oilfield Revolutionary. The Career of Everette DeGolyer.* College Station, Texas: A&M University Press, 2014. For Bolivia, see Anaya-Giorgis, J. J.: *Estado y Petróleo en Bolivia (1921–2010*), PhD. Dissertation, Buenos Aires: FLACSO, 2015, p. 44, Chart 1, and pp. 394–399.
12. I adopt an expansive concept of petroleum engineering. This field involved 'any technical expertise associated with the production, transportation or refining of crude oil and its products'. Giebelhaus, August W.: 'The Emergence of the Discipline of Petroleum Engineering: An International Comparison', *Icon,* Vol. 2 (1996), p. 110; Constant, Edward: 'Science in Society: Petroleum Engineers and the Oil Fraternity in Texas, 1925–65', *Social Studies of Science*, Vol. 19, No. 3 (1989), pp. 439–472. See also Uren, L. Ch.: *Petroleum Production Engineering, Petroleum Production Economics,* New York: McGraw-Hill Book Company, 1950.
13. Priest, T.: *The Offshore Imperative: Shell Oil's Search for Petroleum in Postwar America*. College Station, TX: Texas A&M University Press, 2007.
14. Enos, J. L.: *Petroleum Progress and Profits. A History of Process of Innovation*, Cambridge, Massachusetts: The M.I.T. Press, 1962; Frey, *History of the Petroleum Administration,* p. 191.

15. Egloff, Gustav: 'Carburantes modernos', *Boletín del Instituto Sudamericano del Petróleo,* vol. 2, No. 5 (February 1945); and Villar, G. E.: 'El petróleo, materia prima de la síntesis orgánica', *Boletín del Instituto Sudamericano del Petróleo,* 2, No. 2 (May 1946).
16. Fitzgerald, Edward P.: 'Business Diplomacy: Walter Teagle, Jersey Standard, and the Anglo-French Pipeline Conflict in the Middle East, 1930–1931', *The Business History Review,* Vol. 67, No. 2 (1993), pp. 207–245.
17. Philip, G.: *Petróleo y política en América Latina: movimientos nacionalistas y compañías estatales,* Mexico City: F.C.E., 1989.
18. Bucheli, Marcelo: 'Multinational Corporations, Business Groups, and Economic Nationalism: Standard Oil (New Jersey), Royal Dutch-Shell, and Energy Politics in Chile 1913–2005', *Enterprise & Society,* no. 11 (2010), pp. 1–50.
19. Matharan, Gabriel A.: 'La investigación industrial en la Argentina: el caso de la industria petrolera de Yacimientos Petrolíferos Fiscales (1925–1942)', *REDES,* Vol. 19, No. 37, (December 2013), pp. 13–42.
20. Badia-Miró, M., Carreras-Marín, A. and Peres, J.: *Intraregional trade in South America, 1913–50. Economic linkages before institutional agreements,* Collecció d'Economia E12/270, Barcelona: Universitat de Barcelona. Facultat d'Economia i Empresa, 2012.
21. *Memoria de la conferencia interamericana sobre sistemas de control económico y financiero,* Washington: Unión Panamericana, 1942.
22. Randall, Stephen J.: 'Harold Ickes and the United States Foreign Petroleum Policy planning, 1939–1945', *Business History Review,* Vol. 52 (1983), pp. 367–387.
23. 'EE.UU de Norteamérica y nuestro petróleo': *Revista Oro Negro. Revista mensual de económica política e industria del petróleo, otros combustibles, óleos vegetales y animales.* Buenos Aires, October–November (1942), pp. 4–9. For an analysis of Argentine diplomacy during World War II, see Rapoport, M.: ¿Aliados o neutrales? *La Argentina frente a la Segunda Guerra Mundial,* Buenos Aires, EUDEBA, 1988.
24. Good examples are Max Thornburg (Standard Oil), Ralph Davies and Everette DeGolyer. On DeGolyer's professional performance see Mount, *Oil Revolutionary,* pp. 174–208.
25. Végh, Carlos R.: 'Sobre la creación del Instituto Sudamericano del Petróleo', *Revista de Ingeniería. Publicación de la Asociación Politécnica del Uruguay,* 34, no. 6 (June 1940), pp. 169–175.

26. Constant, Edward, *Science in Society*, 447–449; Pogue, J. E.: 'The Statistical Work of the American Petroleum Institute', *Journal of American Statistical Association*, (March 1929), pp. 118–120.
27. 'Instituto Sudamericano del Petróleo', in *Boletín de Informaciones Petroleras*, 19, no. 210 (February 1942), p. 19.
28. *Boletín de Informaciones Petroleras*, p. 17.
29. 'Sección técnico-científica. El Uruguay y su problema petrolero. Trabajo presentado al 1er. Congreso Sudamericano de Ingeniería, por el ingeniero Daniel Rey Vercesi, de la ANCAP, octubre de 938', *Revista de Ingeniería*, 33, no. 4 (April 1939), pp. 97–111; Pérez Prins, Ezequiel, 'Situación actual del abastecimiento de combustibles líquidos al Uruguay', *Instituto Sudamericano del Petróleo. Memorias presentadas a la primera conferencia nacional sobre aprovisionamiento y racionalización en el empleo de los combustibles*, Montevideo: Impresora Uruguaya S. A., 1943.
30. Mills, T. C.: *Post War Planning in the Periphery. Anglo-American Economic Diplomacy in South America, 1939–1945*, Edinburgh: Edinburgh University Press, 2012.
31. Figallo, Beatriz: 'Bolivia y la Argentina: los conflictos regionales durante la Segunda Guerra Mundial', *Estudios Interdisciplinarios de América Latina y el Caribe*, vol. 7, no. 1 (1996), pp. 35–70; Figallo, Beatriz: 'Desde la crisis internacional a los conflictos regionales: la Argentina y el Uruguay, 1940–1955', *Anuario del CEH* 1, N° 1 (2001).
32. 'New Course Taken to America's Unit. River Plate Conference Related their Regional Effort to Hemisphere Program', *New York Times* (2 February 1941); *Conferencia regional de los países del Plata (Montevideo, 21 de enero-6 de febrero de 1941). Informe de la Secretaría de la Delegación de Bolivia.* Montevideo, 1941.
33. 'Status de la compañía', *Defraudación, Historia de una empresa petrolera en Bolivia,* New York, 1939, p. 28.
34. See, for instance, Solberg, C. E.: *Petróleo y nacionalismo en la Argentina*, Argentina: Hyspamérica, 1986; Gadano, N.: *Historia del petróleo en la Argentina, 1907–1955: Desde los inicios hasta la caída de Perón,* Buenos Aires: EDHASA, 2006.
35. 'Breve reseña sobre la constitución del Instituto Sudamericano del Petróleo (ISAP)', *Memorias presentadas a la Primera conferencia nacional*, I, pp. 9–10.
36. 'Instituto Sudamericano del Petróleo (ISAP)'. 'Comunicación leída por el Ingeniero Carlos R. Végh Garzón', *Boletín de Informaciones Petroleras* 19, no. 210 (febrero 1942), p. 24.

37. 'Día del Ingeniero Sudamericano', *Revista de Ingeniería. Publicación de la Asociación Politécnica del Uruguay*, 33, no. 7 (July 1939), pp. 254–263; *Boletín Informativo de la USAI, 1935–1946*, Montevideo.
38. *Boletín de Informaciones Petroleras, YPF*, 19, no. 210 (February 1942), p. 24.
39. Wirth, *Setting the Brazilian*, p. 112.
40. 'Hybrid' identities like this can also be found in other national contexts. See, for Spain, Camprubí, L.: *Engineers and the making of the Francoist Regime*, Cambridge, Massachusetts/London: The MIT Press, 2014, p. 157.
41. Wirth, *Setting the Brazilian*, p. 115.
42. For a study of an iconic military engineer, see Solberg, C.: 'Entrepreneurship in Public Enterprise: General Mosconi and the Argentine Petroleum Industry', *Business History Review*, No. 56 (1982), pp. 380–399.
43. See, for instance, Bertoni, R. and Travieso E.: 'Economía política de la energía en clave regional. Una propuesta analítica y un estudio de caso histórico', in Lopes M.A. & Zuleta, M. C.: *Mercados en común. Estudios sobre conexiones transnacionales, negocios y diplomacia en las Américas (siglos XIX y XX)*, México: El Colegio de México, 2016, pp. 543–581.
44. 'Comisión de asuntos internacionales para estudiar a incorporación al Instituto Sudamericano del Petróleo, en calidad de miembros activos, de personas o corporaciones radicadas en EE: UU y México', *Boletín del Instituto Sudamericano del Petróleo*, No. 3 (April 1944), pp. 324–325 and No. 1 (August 1944), pp. 298–299; .
45. 'Instituto Sudamericano del Petróleo (ISAP)', Nota del Secretario Ejecutivo', ECLAC, Digital Repository, E / CN.12 / 253, June 12, 1951, Economic Commission for Latin America, Fourth Period of Sessions, Mexico, D.F., agenda item 10, p. 1, http://hdl.handle.net//11362/14521.
46. 'Nómina de los miembros activos del Instituto Sudamericano del Petróleo', *Boletín del Instituto Sudamericano del Petróleo*, vol. 2: N° 4, (1947), pp. 499–518.
47. Rivas, R.: 'Venezuela, petróleo y la Segunda Guerra Mundial (1939–1945): Un ejemplo histórico para las nuevas generaciones', *Economía*, Vol. 20, No.10 (1995), pp. 163–179.
48. Valderrama, Andrés; Camargo, Juan; Mejía, Idelman; Mejía, Antonio, Lleras, Ernesto and García, Antonio: 'Engineering Education and Identities of Engineers in Colombia, 1887–1972', *Technology and Culture*, Vol. 50 (October 2009), pp. 811–838.

49. In 1945, the headquarters of the Argentine section appointed special representatives at the La Plata Distillery and at Comodoro Rivadavia oilfields, 'Nueva Comisión de la Sección Argentina I.S.A.P.', *Boletín de Informaciones Petroleras*, Vol. XXII: N°256 (December 1945), p. 498.
50. Mataharan, Gabriel A., *La investigación industrial en la Argentina.*
51. Pérez, *Situación actual.*
52. Archivo Cancillería, Bolivia, Yacimientos Petrolíferos Fiscales Bolivianos. Volume 1937–1938, La Paz, March 12, 1938, Muñoz-Reyes to Jorge Eduardo Diez de Medina, Ministro de Relaciones Exteriores y Culto: 'Viaje del Gerente de la Ancap'; Volume YPFB, 1939–1945: 'Juan Pinilla to YPFB's Directory at La Paz 'Sociedad entre YPFB y ANCAP', Buenos Aires, (28 October 1941); See Zuleta, María Cecilia: 'Horizontes, negociaciones y disyuntivas en los tratos de YPFB con YPF, 1937–1945', *Revista de Gestión Pública*, Vol. 2, no. 3, (January–June 2013), pp. 107–143.
53. On the Bohan Mission and its proposals for YPFB, see Zuleta, *Horizontes*, 2013, and Anaya Giorgis, *Estado*, 2015.
54. See Figallo, Beatriz, *Bolivia y la Argentina*; Figallo, *Desde la Crisis Internacional.*
55. 'Memoria anual de la sección ecuatoriana del Instituto Sudamericano del Petróleo, October 1942–December 43', *Boletín del Instituto Sudamericano del Petróleo*, 2 (December 1943), pp. 164–174.
56. *Instituto Sudamericano del Petróleo. Sección Nacional Ecuador. Conferencias y trabajos correspondientes a la Junta General Ordinaria celebrada en Quito, 7–10 de diciembre de 1944.* Montevideo, Instituto Sudamericano del Petróleo, 1945. I thank Oscar Torres Montúfar for his help in locating and consulting this source at the US Geological Survey Library.
57. Administración Nacional de Combustible, Alcoholes y Portland, *Lo que nos mueve es todo un país, 1931–2006. 75 años ANCAP,* Montevideo, ANCAP, 2006.
58. Ibáñez Santamaría, Adolfo: 'Los ingenieros, el Estado y la política en Chile. Del Ministerio de Fomento a la Corporación de Fomento, 1927–1939', *Historia,* Vol. 18, (1983), pp. 45–102.
59. Jofré-Rodríguez, J.: *Forjadores de la actividad petrolera en Chile,* Santiago: Instituto de Ingenieros de Minas de Chile, 1955, and Martinic, M.: *Historia del petróleo en Magallanes,* Punta Arenas: ENAP, 2005.
60. For instance, in the Andes were founded the Instituto de Ingenieros de Chile (1888), the Sociedad Colombiana de Ingenieros (1887), the Sociedad de Ingenieros de Bolivia (1922), the Sociedad de Ingenieros de

Perú (1898) and the Sindicato de Ingenieros Peruanos (1931). In Brazil, the Clube de Engenharia do Rio de Janeiro (1880) and the Federação Brasileira de Associações de Engenheiros (1935). In the River Plate were created the Centro Nacional de Ingenieros (Argentina, 1895), the Asociación Politécnica del Uruguay (1905), and the Asociación de Ingenieros del Uruguay (1922).

61. See Rodríguez, Ana María, et al.: 'El nacionalismo petrolero argentino de la década del 20 y su influencia en el surgimiento de la ANCAP', *Hoy es Historia,* Montevideo, Año 1, No. 3 (1984), pp. 35–50.
62. Soprano, Germán F.: 'Autonomía universitaria e intervención política en la trayectoria de liderazgos y grupos académicos en Ciencias Naturales de la Universidad Nacional de La Plata 1930–1955', *Anuario del Instituto de Historia Argentina*, no. 9 (2009), pp. 97–147.
63. Matharan, *La investigación industrial.*
64. Contreras, Manuel: 'Ingeniería y Estado en Bolivia, durante la primera mitad del siglo XX', Cueto, Marcos (coord.): *Saberes Andinos. Ciencia y tecnología en Bolivia, Ecuador y Perú*, Lima: Instituto de Estudios Peruanos (1996), pp. 127–157.
65. Anaya Giorgis, *Estado,* 2015; and Rodríguez Valencia, K.: *Historia de la Universidad Nacional de Ingeniería. La apertura a espacios nuevos (1930–1955),* Lima: Universidad Nacional de Ingeniería, Proyecto Historia UNI, 1999, III.
66. *Boletín Del Instituto Sudamericano Del Petróleo*, 1, No. 4 (August 1944), pp. 313–314.
67. See Mcpherson, A.: *Intimate Ties, Bitter Struggles: The United States and Latin America Since 1945* (Issues In The History Of American Foreign Relations), Washington: Potomac Books, 2006, pp. 17–44; Tillman, A. And Scarfi, J. P. (Eds.): *Cooperation and Hegemony In US-Latin American Relations. Revisiting the Western Hemisphere Idea*, London: Palgrave, 2016.
68. The more important meeting was the first international SAPI's Congress (*Primer Congreso Sudamericano del Petróleo*, Montevideo, March 1951), which was attended by more than 300 delegates.
69. CEPAL, Repositorio Digital, E/CN.12/253, June 12 1951, Comisión Económica para América Latina, Cuarto Período de Sesiones, México, D.F., tema 10 del orden del día, p. 1, http://archivo.cepal.org/pdfs/1951/S5100314.pdf; https://repositorio.cepal.org/handle/11362/14521; https://repositorio.cepal.org/bitstream/handle/11362/14521/S5100549_fr.pdf?sequence=3&isAllowed=y. Date of publication: 12 June 1956.

70. Aksel, K.: *The Engineering Generation: The Story of the Technicians Who Enabled American Cold War Foreign Policy, 1945–1961*, PhD. Dissertation, Proquest Dissertations Publishing, 2016 (University of Colorado, Department of History); Doel, Ronald E.: 'Constituting the Postwar Sciences. The Military's Influence on the Environmental Sciences in the US After 1945', *Social Studies of Science*, Vol. 33, No. 5 (Oct. 2003), pp. 635–666.
71. Mitchell, *Carbon Democracy*, pp. 141 and 238.

14

Epilogue: Technology's Activists and Global Dynamics

Ian Inkster

Backstage Moves to the Limelight

This volume's editors are firm in their contention that the engineering, legal and managing experts from around the mid-nineteenth century were far more central to the dynamics of global economic and cultural change than earlier analysts allowed. To an extent this perspective is becoming more dominant as a result of the very fine detailed work done in the area of history of science by a wide range of intellectual, social and cultural historians, as well as in the history of technology by new historians who are first recognising and then stressing networks, institutions, intellectual property rights and, perhaps in particular, the intervening power—both limiting and stimulating—of developmental states. It is the latter which ensure that no technique is an island. Where these political economies are basically liberal, capitalist and democratic, then the gatekeepers may be non-authoritarian in any directly institutional or political

I. Inkster (✉)
School of Oriental and African Studies (SOAS), University of London, London, UK
e-mail: ii1@soas.ac.uk

© The Author(s) 2018

D. Pretel, L. Camprubí (eds.), *Technology and Globalisation*, Palgrave Studies in Economic History, https://doi.org/10.1007/978-3-319-75450-5_14

sense, and options may be seemingly open and debatable. Nevertheless, no individual, collectivity or creative site floats freely.

States embroiled in some form of political economy catch-up were in most cases either escaping from formal colonialism (Japan followed by key smaller East Asian nations), emerging from it (India or Indonesia) or recovering from partial escape from it (China, much of Eastern Europe and Latin America). Through the echoes of imperial withdrawal, for most of Africa below the Middle East the detrimental traumas of Western colonialism still inhibit the emergence of formal planning of technological development—corruption within regimes, identity confusion amongst often rich elites, interference by cold war warriors until the 1990s and from then the social and political dimensions of new global technologies of social media and knowledge diffusion. In the Middle East itself, the failure of post-colonial regimes to construct pathways towards new forms of democracy, or to use the wealth from natural resources to build legitimate budgets and then employment-generating technologies from the enormous array of techniques, expertise and institutions that were available, has nurtured the massive divergences of income and wealth within their much-contested borders.

Now, by any standard this is a huge perspective, and historians of technology might well be accused of an overt imperialism of their own. The present volume seeks to advance this approach, nuance it and help fill its data bank. So here we have not merely the developmental state but the colonial state, post-colonialism, the cold war state, the state at war and so on. In chapters by several authors political economies are at the forefront of concern.

From Intellectual Limelight to Fashioning the Backstage

Recent work on a broad front, as well as more specialist work on patents and particular industries, has established the importance of fashioning the backstage of global knowledge and its institutions. Incremental technological progress is essential to the commercialising of so-called 'breakthrough' technologies, and in any significant economic system if the sites that create incremental changes are absent, or if the activity within them is distrusted or just too expensive or tortuous for any commercial evaluation and choice, then such systems are in trouble. Even a very major nation such as the UK is periodically labelled as being scientifically creative and technologically inventive yet somehow also technologically unproductive—the 'British disease'. Although some of this seeming enigma—at work also in much of modern Europe and the USA—might actually be reduced by considering such elements as market failure, commitments amongst producers to existing technological infrastructures or procedures, or slovenly and at times contradictory public sector regulations and funding patterns, the approach in this volume gives greater attention to less established technological systems and to the workings (or relative absence) of networks and institutions often neglected by more economistic approaches.

This 'backstage' of advancement is of central historical importance to this volume. Such a claim becomes more acceptable when we maintain our distinctions between knowledge and information. Chapters in this volume wed themselves to the prime importance in technological advancement of the former rather than the latter. It is true that a text, a manual or a patent (even a lecture) may provide information vital to a technological advance, but this is in most cases more likely in a knowledge context—active agents whether mechanics or entrepreneurs, engineers or patent agents, operated upon internalised knowledge fostered in apprenticeships, colleges and night schools, factories and workshops, and in the networks of the engineering world that range from patent office libraries to associations and innovative city and industrial sites. These places nurtured and fashioned knowledge that becomes vital in the adaptation and usage of the flows of information that moved through them. Information might well flow to and from a site or node on a network; but characteristics of the site of endeavour or micro-environ at key points convert or help convert such information into knowledge.[1] That is, the dynamics and characteristics within any technological site of endeavour (a factory, a lab or a patent library), as well as its degree of connectivity inside and beyond the urban or industrial environment are of great importance in the global process whereby knowledge becomes technique in some places and not others, at some times but not others, and in some directions rather than others. This is probably true even when we replace the notion of 'others' with that of 'not at all'. And here I use the term 'global' both spatially and temporally; even in antiquity and across the large spaces of the eastern Mediterranean and Indian Ocean, knowledge travelled when it was made proximate in trade and more informal commercial connectivity—trading middlemen were also linguistic and cognitive *compradores* of knowledge exchanges.[2] In more recent years the exponential increase in information, especially concerning technologies, has depended on what I have termed for simplicity's sake a 'backstage' of knowledge production and settlement. Of great importance since the 1830s has been the growth of complex systems of property rights that also serve the purposes of information dispersal, knowledge settlement and national competition between great industrial powers.

By the time of the so-called Vienna Patent Congress of 1873, several nations recognised intellectual property systems as integral to purposeful technological change and transfer, for any good patent system 'attracts capital from abroad, which, in the absence of patent protection, will find means of secure investment elsewhere, acts as a global information system, and saves the time and money of inventors'.[3] In this volume David Pretel considers patent agency in the widest possible way in order to blur the simplistic distinctions between foreground or limelight and background or support structures. Of the patent practitioners, Pretel argues that they 'varied geographically, in line with the industrial and institutional disparities from country to country, even region to region'. Whether this arose principally from differences of industrial culture or from patterns of late development remains an open question, and Pretel is more concerned with how the growing body of expertise gave patent agents the authority to shape, interpret and influence patent laws.[4] The author asks us to think of the international patent system 'as a series of interlocking and overlapping networks of national offices, companies, engineers, capitalists and inventors that transcended the boundaries of Western Europe and the United States'.

Science from Limelight to Useful and Reliable Knowledge

One of the techniques used here—very much developed by the historians of science themselves—is to dethrone 'science' as a 'necessary and sufficient' factor in explanations of technological change, whilst offering a more circumstantial account of relations between knowledge and praxis, theorist and practitioners, and individuals and the sites in which they operate. Of course, none of this rejects the notion that Western-based scientific knowledge and methods remained vital in many instances of technological creativity beyond the pale of Western cultures and institutions.[5]

But such contributions as those of Leida Fernandez-Prieto on the Harvard Botanic Station for Tropical Research and Sugar Cane Investigation show just how nuanced and contradictory relations between science knowledge and its technological outcomes could be. Fernandez-Prieto argues that the creation of the Harvard Botanic Station for Tropical Research and Sugar Cane Investigation drew a metaphorical line separating the development of sugar cultivation and the field of tropical economic botany from global, American and local market needs, but with an agenda at least partially set by Edwin Atkin's purely business-generated demand for a high yield and resistant sugarcane in the face of competition from sugar beet. With Atkins combining funding, insight and energy in the emergence of a scientifically innovating eastern Cuban sugar industry (the search for a commercially ideal hybrid sugarcane as its focus), the 'price of bringing Harvard into imperial networks of botanical research and displacing Europe as the global centre of tropical knowledge, included modernising Cuba's sugar industry from an enclave built on slavery'.

The result was a Harvard-dominated experimental site that 'shows the collaboration of networks of imperial gardens and experimental stations throughout the tropics, the interactions of botanists inside transatlantic knowledge circuits and exchanges between productive and scientific groups at different levels (global, regional and local)'.That is, regionally and locally, North American scientists became 'mediating agents' in economic practices. So here the role of science and scientists as knowledge creators and transmitters does embrace the commercially directed research going on, the application of innovations, the construction of a knowledge system and so on. Furthermore, the author sees this productive merging of scientific research and commercial usage within one site and ethos as in fact characteristic of scientific knowledge production in the early twentieth century. Similarly, in their chapter on statistical methods Tiago Saraiva and Amy Slaton—if I am not putting their case too strongly—show how a US methodological technique or frame of the 1940s could itself become an agent of knowledge transfer and legitimacy. They focus on how statistical methods became devices for Americanising scientific practices on a global scale, and thereby open up a whole gamut of questions linking science as method to new practices at the periphery, in this case Egypt and India.

Connectivities Within and Beyond

Here we are thinking first of the institutional, network and site connections fashioned within the world of technology and globalisation in which 'techno-scientific experts' figured centrally, as well as the manner in which underlying socio-economic (especially politico-military) processes impacted on technological communities and forced out new modes or degrees of response. In tune with what has been argued at times in this volume, even the perspective that argues that most of technological globalism and modernity has centred on an historical process of technological transfer from a core to a periphery has now incorporated the notion that this was very often an aggressive, incomplete and inefficient process, or where it succeeded in part did so by creative adaptation and modification as much as by any simple process of adoption.[6] As Lemon and Medina have summarised in a recent account of Latin America, artefacts and ideas 'regularly cross borders, are modified in their travels, and take on multiple meanings in different contexts'.[7] Modifications and the construction of meanings were often in the hands—or passed through the fashioning hands—of engineering and managing experts and agents, and the authors in this present volume are more or less concerned with whether such key institutions and nodes or sites of activity acted principally as gate-keepers or as an essential communal learning or experiential system within a global political economy.

The chapter by Vera on the anti-metric movement in the USA shows how concerted efforts of key engineering experts could halt a major national project that was to all intents and purposes an improved technological system within the leading industrial nation of our time. Certainly, this case highlights how at times gate-keeping and communal learning might be more or less deliberately confused! Owing to the stalwart opposition of American mechanical engineers through their established institutions and networks, led by the American Society of Mechanical Engineers (ASME), the National Association of Manufacturers (NAM) and the National Bureau of Standards, supported by such eminent figures as Andrew Carnegie and the inventor Elihu Thomson, of General Electric, as well as by William Thomson, Lord Kelvin, and using all the media and political networks at their disposal, thereby rendering the 'compulsory introduction of the metric system a highly contested political issue', engineers and manufacturers 'achieved their ultimate goals by undermining the scientific authority of metric proponents and by intimidating politicians to not pass an unpopular law'. This success was in spite of the fact that—clearly enough—the scientists and experts who supported metrication seemed to have both rational science and global commercial practice on their side. What is more, the expert anti-metrics could with ease be branded unfashionably pro-British and insular as against the progressive metricists such as the most celebrated of all British scientists, Lord Kelvin, supporting global rationality and the sweeping away of restrictive measures of past days. In this exemplary case, the expert engineers in support of retaining the old imperial system used explicit arguments from national tradition, from the 'British connection' and from cost rationality—the metricists lived in an ideal world comparing one system as established with another as established, yet the transition costs would be so high as to reduce any commercial benefits flowing from metrication to zero. This argument is a form of engineering argument without doubt, analogous to that used by British industrialists and engineers in rejecting the Le Blanc-Solvay transition for manufacture of sodium carbonate in the alkali industry. British interests who had invested in Leblanc technology from the 1820s argued that in ideal conditions of a level playing field Solvay was a measurable and thus investable improvement, but in the real engineering world the wiping out of established fixed assets would counter any rise in productivity.[8] The great value of Vera's paper is to show how American expert engineers could effectively transcribe such logic from the world of fixed assets to established institutions, standards and forms of transaction as well as to expenses of retooling, and that they did so very publicly using subtle social and cultural imagery.[9]

It is clear that this theme of agency allows for a very wide range of work. For instance, the chapter on the Spanish nuclear programme by Joseba de la Torre, Mar Rubio-Varas and Gloria Sanz Lafuente considers 'the role of engineers and scientist as agents of economic modernisation in Spain, at the time an underdeveloped economy ruled by an authoritarian dictatorship', and illustrates how under such a political situation and with such high-level technological imperatives the engineers dominated the core of all commercial negotiations and transactions involving nuclear technology. It was Spanish private industry and not Franco's state that groomed the Spanish engineer Jaime MacVeigh as a principal technical change agent of great effectiveness. Interestingly, it was the Research Service of Banco Urquijo in 1957 that published MacVeigh's ambitious long-term projection, which would have begun in 1964 and produced the equivalent of a minimum of 106 to a maximum of 194 nuclear reactors of 200 MW operating by the year 2000. It is clear enough that electrical engineers and private interests were in systematic cooperation in leading the nuclear programme, following a relatively speedy phase of learning and network building. In the following years MacVeigh's activity was prodigious internationally, but especially with principal US corporations in terms of both equipment (the first Spanish nuclear plant on the Tajo river, at Zorita, with technology from WESCO) and networking for influence and expertise (General Electric and Westinghouse), as well as with British and American national agencies. In the absence of realistic state aid, he and others such as the German ex-Nazi physicist Karl Wirtz made exhaustive effort to forge contacts and relationships with the emerging international forums (the 'shared spaces') of the 1950s and 1960s. Here the quickly forged engineering 'backstage' could roll over the limelight of Franco's political dominance.

Perhaps an even more forceful case is that provided in Braso's chapter on Shanghai during 1937–1941. It is shown here how technology transfers and their commercial deployment can continue under the most hopeless circumstances of war, regulation and the disappearance of certified property rights when expert agents are determined. In the case of Shanghai textile manufacturing offered by Braso, transnational agency embracing the trapped energies of 'Chinese and foreign merchants, Indian traders, refugees from Europe, and US businessmen' operating within or into the concessions 'with their savings but with few possibilities to invest'. Technological agency was thus confused both perchance and with deliberation, but active *compradores* and technical activists continued to make profit from transferred technologies, to increase their market strength during inflation, under legal and institutional devices that were as innovative and opportunistic as any technical innovation. The history of China Engineers shows just how in extreme exigencies an engineering enterprise could transform itself rapidly into trading, transport and insurance businesses in order to maintain viable networks when normal commercial imperatives were stalled by warfare.

Connections *beyond* those which are nominally technological are in some ways perhaps the major theme for any institutional and critical treatments of technology in history—that is the global connections between technological development and the social, cultural and economic processes both working on technological change and responding to it. Thus, the much-vaunted Second Industrial Revolution straddles the earlier years of our period and is intimately associated with the huge development of an Atlantic-based trading economy carrying men, manuals and machines from advanced technological sites within Great Powers to a wider array of places through aggressive colonialism, profitable trading and investment, and lucrative exchanges of knowledge. Whilst much of this was predicated upon very fundamental technological innovations in military and naval power, in communications and transport, and in the huge civil engineering projects that for the first time allowed the white settlement of millions of Europeans and many Americans into local environs that would have stultified such ambitions in the earlier years of colonialism, it also arose from or within serious long-term changes in the even wider global context. European industrialisation transformed from a set of mechanical engineering advances in factories, mills and workshops into enormous projects populated by expatriate experts, rooted in vast capital and protected and advanced by a dominant gunboat diplomacy, all alchemically transmuted within civil engineering innovations in port and riverine coastal cities that quickly became the most advanced sites of a new modernity.

There is a sense in which the aggressive technological colonialism of the late nineteenth century became a crude transmutation of the much earlier Chinese cultural tendency to see the control of their own vast land empire as being an object of their wide conception of engineering. As Schäfer reminds us, beyond technical efficacy, the Chinese literati had envisaged engineering know-how as civilising 'in its benefits for both the state and its common people, being dependent on the moral application of knowledge, defined in terms of commensurability'. If we read civilising as controlling and measure the distance between Beijing and the southern ports as greater than many of the spans of European empire during industrialisation, then the allusion becomes stronger.

Schäfer's discussion of early China in this book is important as she shows the coexistence there of two features of technological change that are too often seen in Western and modernistic terms—the Chinese authorities tended to trust incremental technological change over breakthroughs for reasons primarily but not entirely social, and they saw that within a notion of technological change, organisational innovation could well substitute for technical solutions and advancements (a la Schumpeter who repeatedly emphasised that any proper definition of innovation must include those of new markets and altered institutions), they were generally cheaper and far less socially disruptive. It is well to be reminded by Schäfer that well before the industrial modernisation of the west in the eighteenth century the elite engineering practitioners and administrators defined their skills and knowledge in terms of how their efforts to regulate natural forces were important for ordering (zhi 治, i.e. governing) the world, thus making it a habitable and 'cultured' (wen 文) place. As such, engineering projects to control water, land and society were mainly pursued by the cultured inhabitants of China's various dynasties, whereas the barbarians outside China were believed to be unaware of such means.

As two of our authors, Diogo and Navarro, emphasise with an entirely frank statement from the (Portuguese) *Journal of Mines and Public Works* in 1899, 'When colonial nations want to take real possession of their territories they send their engineers overseas.'[10] This same reflection and attitude can be seen throughout the pages of such influential British and US engineering forums as *The Engineer*, *Journal of the Society of Chemical Industry* or *Engineering*. Mostly such civil engineering sites began as enclaves of colonialism or of very strong trading compulsions, and many remained so well into the twentieth century, indeed into that 1970s weakening of the older global nexus by further technological advances in brand new industries—portmanteau terms such as biotechnology, information technology, nanotechnology, telecommunications, new materials and blue-sky industries may disguise as much as they enlighten, but this later period certainly witnessed a great loosening of the grip of the older 'machinofacture' or machine and chemical manufacturing technologies of the years approximately 1830 to 1970.[11]We might add here that private-sector machinofacture and the vast expert infrastructure that it evolved in our period was the principal technological feature of the industrialisation process, and that earlier public-sector technological development before say about 1700 had exerted most impact through civil engineering in transport and religious and military control systems (from cathedrals to docks, bridges, shipbuilding, canals and improved roadways), and that from the 1870s the reemergence of huge civil engineering activity in public goods and infrastructures occurred most notably in the colonies and dependencies, being designed principally as muniments of control or means of commerce.

Maria Paula Diogo and Bruno Navarro 'discuss the strategies deployed by Portugal, a peripheral country in Europe, to use its African Empire as a token for asserting its position in the European arena'. They do so by documenting the activity of the engineers behind the great infrastructural—especially railway—projects for their two colonies of Angola and Mozambique. Engineers made concrete and steel of the imperial agenda designed to enhance Portuguese geopolitical status and presence and bring African territory into a global market determined by patterns of colonialism which were themselves governed by new technologies of transport and communications. Covering the period of heightened imperial rivalry of the Great Powers following the Berlin Conference of 1884–1885, this meant for the relatively minor power of Portugal the emergence of technological 'civilising outposts' in her colonies. As with so many of the colonial projects of this sort, and documented severally in this volume, the physical and skill loci were infrastructure and civil engineering.

This chapter also reminds us of the crucial importance of colonial natural resources to further technological advance in the West. Coal, copper, iron, cobalt, gold, diamonds, timber, rubber, coffee and cocoa were valuable commodities that demanded strict management rules across the Portuguese territory, thus granting infrastructures a vital role. Such a list of natural resources for the relatively minor case of Portugal shows that civil engineering projects were not just a package of fixed capitals nor only a site of imperial advancement and power demonstration; they were also intended as crucial to the development of physical sites that linked the acceleration of mechanical, chemical and military engineering (linked of course to further colonialisms) to colonial civil engineering through the fast-rising global demand for key natural resources for manufacturing within the Great Powers themselves. This seems obvious but is far too often neglected as an explanation of state involvement in civil engineering during these years. In such projects, not only were global standards developed and set but also engineering 'expertise shaped the (Portuguese) technopolitical agenda and legitimised the governmental elites' political action'. This is an admirable formulation that fits so many cases of civil engineering enclaves in colonised territories in the years even immediately following the Second World War.

A major theme in this chapter is the building of railways in colonies to 'secure them'. This was common throughout this phase of colonial expansion. Even in cases where canals, roads or coastal traffic might have been better developed, the colonial authorities opted for railways as the best commercial link to ports for export development, as an expert engineering choice that demonstrated technological superiority—not only to subject peoples but competing power, and as means of supplying armed forces to frontier wars or civil uprisings. India under the Raj was of course the case in point, where canals were often the better choice yet far more money and thought went into railways.[12] It was an engineer, F.W. Simms, sent out by the East India Company in the 1850s, who reported that railways 'are not only a great desideratum but with proper attention can be constructed and maintained as perfectly as in any part of Europe', quoting low prices for land, labour, building facilities and parliamentary legislation for colonies![13] For Portugal, as Diogo and Navarro point out clearly, a railroad network with its colonies would ensure that Portugal could no longer be called 'a small country'. In many colonial situations, well-trained civil engineers eagerly abetted the most spectacular of projects. Railways also illustrate well the inappropriate interventions of colonial states in the misuse of manpower and resources as well as the initial suboptimum technological choices and decisions, an outstanding instance being the guarantee system unnecessarily grafted on to railway development in colonial India, which as with most forms of official protection inflated all costs and removed most decisions from the arena of optimum technological choice to that of high-minded bureaucracy.[14]

Even more broadly, of course, the more contemporary applications of technology are clearly forged in complex institutional imperatives flowing across simplistic divisions of the political, the economic or the technological. Accordingly, several of our twentieth-century studies go well beyond such distinctions and show clearly the interplay or even recursions of technological forces and socio-economic events, especially the interventions of the state and of warfare or war conditions. In this context, several of the chapters (Zuleta, Martykanova, Ertsen) focus on central positions held, and roles played, by engineering both as a community and as outstanding individual change agents. The nuances were many, as illustrated by Zuleta in her chapter on the South American Petroleum Institute after 1941. Here it becomes clear that regional expert associations formed from a rising nationalism during the 1930s were jolted into something more potent by the exigencies of the Second World-War, when oil shortages made necessary the regional networking of experts in such a vital field. Zuleta argues that closer networking and cooperation of engineering experts drawn from the diverse cultures of the military and private enterprise 'enabled the professional development and identity of South American engineers as well as the technological transformation of the hydrocarbons industry in the region'.

The contribution by Ian Inkster straddles the two ends of the globalisation of technology, dealing with processes both within and beyond the world of the activist engineers and agencies of technological change. On the one hand it is clear that the camphor-based chemical technology that produced the celluloid, plastics, artificial fibres, new materials and explosives of the Second Industrial Revolution evolved from a system of intellectual property rights operating from the 1860s that depended on hyperactive individual agents, creative technological sites of endeavour and an international knowledge system. On the other hand, it is just as clear that the socio-political outcomes of the production and commercial exploitation of camphor included violence and civil war in the far-off island of Taiwan. These latter were of course unintended consequences, but such a Mertonian phrase was never designed to further marginalise the already marginal. The destruction that came in the wake of the new technological demands for camphor was an isolated and localised version of the many incidents of the nineteenth and twentieth centuries that created the bifurcation of our world. It is also clear that at the core of the dynamism of the new globalism lay technology and its many connectivities, the densities of its networks and the character of its associations and institutions. No one in Birmingham knew anyone in eastern Taiwan. Yet they met in one way or another.

Conclusion

There are probably as many ways of doing global history as there are of doing history of technology. Thankfully no one has ever quantified an answer to such a claim. What does seem to almost indisputable is that global history in the years after the Great Exhibition of 1851 was very significantly underscored by an all-encompassing process of technological change monopolised by a relatively small group of fast-modernising Great Powers. And we might note that a term such as 'underscored' is in the end rather mealy-mouthed, liberally allowing the analytical power of other political or diplomatic or economic histories as identifying basic dynamics, whilst leaving the matter of some final explanation or interpretation entirely open-ended. At a time when no meta-thesis commands a position that could satisfy historians as the answer to this sort of question, most historians are anyway satisfied enough with the approaches we have. Technology might be wedded to both political and economic elements through clearly significant historical thematics that are increasingly well understood—colonialism and imperialism are obvious pathways, as is pure military power that, in the last analysis and despite the plethora of military manuals, came out of the mouth of a gun made in manufacturing industry. This volume uses case studies and connections to nuance the notion of 'underscored' in the absence of an agreeable meta-theory.

The character of these years was forged out of the linkages and interactions between early starters, late-developers, colonies and areas of recent settlement, and the respective fates of these groups may be described, measured and judged in terms of technologies, technological agents of a wide variety, technology transfers and failures, and so on. Most developmental dependency (e.g. Australia) and dependent underdevelopment (India) may be very well treated in accounts that have technological change at their basis. Likewise, in these chapters we see the reemergence to preeminence of civil engineering and public goods, especially as run by colonial authorities across the entire globe, as against the famous and individualistic mechanical engineering and private goods of the early-starter industrial revolutions. In this perspective, the First World War becomes a huge battle between early and late development, determined by technology, aimed at ownership of the wider world of colonies and areas of recent settlement.

The underdevelopment of most of the economies of the world and the fate of their peoples was significantly delineated in the failures of technologies to transfer in the years of aggressive global capitalism. Technology does not remove the historian to a safe technical haven. Indeed, the field is full of provocations. In these years, China is isolated and marginalised without official Western colonisation but using the tools of modernity—trade, locomotives, gunboats and armies, followed by the rise of industrial Japan on the wings of Western machines and institutions. The histories of the coloniser, the colonised and the indigenous periphery took place in an increasingly connected world in which individual experiences between such groups were gathered unconnected, in utterly separate worlds. As Jeremy Adelman has emphasised very recently and with brilliance, focus on historical global interaction, including that which was firmly technological, 'means reckoning with dimensions of networks and circuits that global historians—and possibly all narratives of cosmopolitan convergence—leave out of the story: lighting up corners of the earth leaves others in the dark. The story of the globalists illuminates some at the expense of others, the left behind, the ones who cannot move, and those who become immobilised because the light no longer shines on them.'[15] Technology is not light, and several of our authors have shown the darkness inherent in modernity and its connectivities: Inkster finds guns and violence in Taiwan as a result of high-tech camphor patents in Birmingham; Pretel finds amidst the engineering activity in patent systems the production of 'lower degrees of knowledge diffusion, limited competition and predatory strategies'. But, of course, this brings us back to the query: was technology underscoring such darker forces, or was technology itself underscored by the even more basic urges of capitalist modernity itself?

Notes

1. See especially for city/spatial emphases, Renn, Jürgen: 'Survey: The Place of Local Knowledge in the Global Community', in Renn, J. (ed.), *The Globalization of Knowledge in History*, Max Planck Research Library for the History and Development of Knowledge (online), 2012; McCann, P. and Acs, Z.: 'Globalization: Connectivities, Cities and Multinationals', *Regional Studies,* Vol. 45, No. 1 (2011), pp. 17–32.
2. Here we are pointing to the *compradore* as a special form of local agent whose power and income depend on command of linguistic, tacit and formal knowledge or information to extract profit or status from commercial, technical or cultural exchanges. In their key positions—which by nature are transitory—they may act almost in the same Janus-faced role as medieval guilds, in some cases being positive intermediaries between foreign and indigenous techniques, in other cases they may strongly inhibit passages of knowledge. LaBianca, O.S. and Scham, S.A. (eds.): *Connectivity in Antiquity – Globalization as a Long Term Historical Process,* London: Equinox, 2006. For more modern usage of the notion

of *compradore* in global history and connectivity which embrace brokerage of information and technique and apply mostly to East Asia from around the mid-nineteenth century—often in colonial settings—see Kai Yiu Chan, 'A Turning Point in China's Comprador System, 1912–1925', *Business History*, Vol. 43, No. 2 (2001), pp. 54–72; and for application to one case in some detail Goh Chŏr Boon, *Technology and Entrepot Colonialism in Singapore 1819–1940,* Singapore: Institute of Southeast Asian Studies, 2013, and Inkster, I.: 'The Trouble with Technology: Comments on the Experience of Singapore under Entrepot Capitalism', *Journal of the Malaysian Branch of the Royal Asiatic Society*, 72, No.1 (2000), pp. 107–15.

3. *The Practical Magazine* II (1873), pp. 431–2.
4. For some argument concerning the patterns of engineering agency see Inkster, I.: 'Engineering Identity, Intellectual Property, and Information Systems in Industrialization circa 1830–1914' in Cardoso de Matos, A., et al. (eds.), *The Professional Identity of Engineers: Historical and Contemporary Issues,* Lisbon: Edições Colibri, 2009, pp. 357–380.
5. It also lies comfortably with the strengthening thesis that much earlier non-Western knowledge from pure mathematics to steel-making contributed to central aspects of identifiably Western scientific evolution, this embracing the huge contributions of Chinese culture and the disturbing and renovating impacts of Islam in very early years, as well as the even less measurable impacts of the great diasporas.
6. Inkster, I.: 'Technology Transfer and Industrial Transformation: an Interpretation of the Pattern of Economic Development circa 1870–1914', in Fox, Robert (ed.), *Technological Change. Methods and Themes in the History of Technology,* London: Harwood Academic Publishers, 1996, pp. 177–201.
7. Lemon, Michael and Medina, Eden: 'Technology in an Expanded Field: A Review of History of Technology Scholarship on Latin America in selected English-Language Journals' in Medina, E., et al. (eds.), *Beyond Imported Magic. Essays on Science, Technology, and Society in Latin America*, Cambridge, MA: The MIT Press, 2014, quote p. 112.
8. In Solvay's ammonia-soda process an ammoniacal brine is fed into a stream of carbon dioxide to absorption, forming sodium hydrogen carbonate. This separates into fine crystals leaving ammonium chloride to be passed on, to be heated with lime, producing ammonia to be used again. The point here being that technically Solvay was superior and

used byproducts; industrially, however, it involved large expenditure in absorption towers and so on,. and some of its advantages could be superseded with Leblanc by savings on raw materials and production of a greater diversity of chemical products. This was a contemporary parallel that had definite technical parameters but commercial implications that were less measurable.

9. In this regard, it is worth mentioning that Frederick Halsey's statement in *The Metric Fallacy* concerning the 'iridescent dream' must be a direct reference to the phantasmagorical *Alice in Blunderland: An Iridescent Dream,* the novel by John Kendrick Bangs published in 1907 by Doubleday in New York. In this parody of Lewis Carroll, Alice enters a political economy of over-taxation and ludicrous regulation, so Halsey was aiming at his target with great contemporary satirical aplomb.
10. From 'Exposição' (Acta da sessão de 6 de Maio de 1899), *Revista de Obras Públicas e Minas* (1899), pp. 353–354 and 382–383.
11. For which see Inkster in this volume and Inkster, I: 'Machinofacture and Technical Change: The Patent Evidence', in Inkster et al. (eds.): *The Golden Age. Essays in British Social and Economic History 1850–1870*, London: Ashgate, 2000, pp. 121–142.
12. For outlines of the debate on this see the chapters by Inkster and Derbyshire in MacLeod, Roy and Kumar, Deepak (eds.), *Technology and the Raj: Western Technology and Technical Transfers to India, 1700–1947*, London: Sage, 1995.
13. Quoted in Das, M.N.: *Studies in the Economic and Social Development of Modern India*, Calcutta, 1959, pp. 35–36.
14. By inserting a clause that railway development in India should be in the hands of private companies under British government supervision with a guarantee of a minimum rate of return on capital of 5%—above the prevailing rates in the London money market—Lord Dalhousie's Railway Minute of April 1853 effectively 'killed effort for economy, promoted recklessness, and involved the country in liabilities much beyond what the people could bear'; quoted from Sanyal, Nalinaksha: *Development of Indian Railways*, Calcutta: University of Calcutta, 1930, p. 17; and see also Weld, W.E.: *India's Demand for Transportation*, New York, 1920 especially p. 65. For an excellent short summary see Dubey, Vinod: 'Railways', in Singh, V.B. (ed.), *Economic History of India 1857–1956*, Bombay: Allied Publishers, Bombay, 1965, pp. 327–347.
15. Adelman, Jeremy: 'What is Global History now?' *AEON* (2 March 2017).

Corrections to: Technology and Globalisation

David Pretel and Lino Camprubí

Correction to:
David Pretel and Lino Camprubí, Technology and Globalisation, Palgrave Studies in Economic History, https://doi.org/10.1007/978-3-319-75450-5

The Chapter titles for Chapter 4 & Chapter 12 were inadvertently published and now they have been updated as:

Chapter 4 Global Engineers: Professional Trajectories of the Graduates of the École Centrale des Arts et Manufactures (1830s–1920s)
Chapter 12 Engineers and Scientists as Commercial Agents of the Spanish Nuclear Programme

The updated online versions of the chapters can be found at
https://doi.org/10.1007/978-3-319-75450-5_4
https://doi.org/10.1007/978-3-319-75450-5_12

© The Author(s) 2018
D. Pretel, L. Camprubí (eds.), *Technology and Globalisation*, Palgrave Studies in Economic History, https://doi.org/10.1007/978-3-319-75450-5_15

Index[1]

[1] Note: Page numbers followed by 'n' refer to notes.

© The Author(s) 2018
D. Pretel, L. Camprubí (eds.), *Technology and Globalisation*, Palgrave Studies in Economic History, https://doi.org/10.1007/978-3-319-75450-5

L

M

N

O

P

CPI Antony Rowe
Chippenham, UK
2018-10-14 19:13